Finite State Machines in Hardware

Theory and Design (with VHDL and SystemVerilog)

Volnei A. Pedroni

The MIT Press
Cambridge, Massachusetts
London, England

MIT Press books may be purchased at special quantity discounts for business or sales promotional use. For information, please email special_sales@mitpress.mit.edu.

This book was set in Stone Sans and Stone Serif by Toppan Best-set Premedia Limited, Hong Kong. Printed and bound in the United States of America.

Library of Congress Cataloging-in-Publication Data

Pedroni, Volnei A.
Finite state machines in hardware : theory and design (with VHDL and SystemVerilog) / Volnei A. Pedroni.
pages cm
Includes bibliographical references and index.
ISBN 978-0-262-01966-8 (hardcover : alk. paper) 1. SystemVerilog (Computer hardware description language) 2. VHDL (Computer hardware description language) 3. Sequential machine theory—Data processing. 4. Computer systems—Mathematical models. I. Title.
TK7885.7.P443 2013
621.39'2—dc23

2013009431

10 9 8 7 6 5 4 3 2 1

Contents

Preface

This book deals with the crucial issue of implementing Finite State Machines (FSMs) in hardware, which has become increasingly important in the development of modern, complex digital systems.

Because FSM is a modeling technique for synchronous digital circuits, a detailed review of synchronous circuits in general is also presented, to enable in-depth and broad coverage of the topic.

A new classification for FSMs from a hardware perspective is introduced, which places any state machine under one of three categories: *regular machines, timed machines,* or *recursive machines*. The result is a clear, precise, and *systematic* approach to the construction of FSMs in hardware.

Many examples are presented in each category, from datapath controllers to password readers, from car alarms to multipliers and dividers, and from triggered circuits to serial data communications interfaces.

Several of the state machines, in all three categories, are subsequently implemented using VHDL and SystemVerilog. It starts with a review of these hardware description languages, accompanied by new, detailed templates. The subsequent designs are always complete and are accompanied by comments and simulation results, illustrating the design's main features.

Numerous exercises are also included in the chapters, providing an invaluable opportunity for students to play with state machines, VHDL and SystemVerilog languages, compilation and simulation tools, and FPGA development boards.

In summary, the book is a complete, modern, and interesting guide on the theory and physical implementation of synchronous digital circuits, particularly when such circuits are modeled as FSMs.

Acknowledgments

I want to express my gratitude to Bruno U. Pedroni for his invaluable help and suggestions during the initial phase of the book. I am also grateful to the personnel at MIT Press, especially Marc Lowenthal, acquisitions editor, for his assistance during the early phases of the book; and Marcy Ross, production editor, for her excellent work and endless patience during the editing and production phases.

1 The Finite State Machine Approach

1.1 Introduction

This chapter presents fundamental concepts and introduces new material on the finite state machine (FSM) approach for the modeling and design of sequential digital circuits.

A summary of the notation used in the book is presented in table 1.1.

1.2 Sequential Circuits and State Machines

Digital circuits can be classified as *combinational* or *sequential*. A combinational circuit is one whose output values depend solely on the present input values, whereas a sequential circuit has outputs that depend on previous system states. Consequently, the former is memoryless, whereas the latter requires some sort of memory (generally, D-type flip-flops [DFFs], reviewed in section 2.2).

An example of a combinational circuit is presented in figure 1.1a, which shows an *N*-bit adder; because the present sum is not affected by previous sums computed by the circuit, it is combinational. An example of sequential circuit is depicted in figure 1.1b, which shows a synchronous three-bit counter (it counts from 0 to 7); because its output depends on the system state (for example, if the current output is 5, then the next will be 6), it is a sequential circuit. Note the presence of a clock signal in the latter.

An often advantageous model for sequential circuits is presented in figure 1.2a, which consists of a combinational logic block in the forward path and a memory (DFFs) in the feedback loop. When this architecture is used, a *finite state machine* (FSM) results. Note that the state presently stored in the memory is called *pr_state*, and the state to be stored by the DFFs at the next (positive) clock transition is called *nx_state*.

An example of such a modeling technique is depicted in figure 1.2b, which shows the same circuit of figure 1.1b, now reorganized according to the architecture of

Table 1.1

Item	Representation	Examples
Signal names	In italic	*a*, *x*, *clk*, *rst*, *ena*, *WE*
Active-low signal names	In italic, followed by an *n*	*WEn*, *rstn*, *rst_n*
Single-bit values	Within a pair of single quotes	'0', '1', 'X', '–', 'Z'
Multi-bit values	Within a pair of double quotes	"00", "1000", "ZZZZ"
Integers	Without quotes	1000, 5, –256
Allowed bit values	'0' or 'L' for low logic level '1' or 'H' for high logic level 'X' or '–' for "don't care" 'Z' for high impedance	*y* = '0' or *y* = 'L' *y* = '1' or *y* = 'H' *y* = 'X' or *y* = '–' *y* = 'Z'
Bit indexing (outside VHDL or SystemVerilog codes)	Between parentheses, with a colon	*x*(7:0) means that *x* has 8 bits, *x*(7) is the most significant bit, *x*(0) is the least significant bit
Reset and clear signals	- Called *reset* (*rst*) when asynchronous (resets the circuit regardless of the clock) - Called *clear* (*clr*) when synchronous (effective only at the proper clock edge)	if *rst* = '1' then … if *clr* = '1' then …
Transition conditions in state diagrams	& means *and* | means *or* ! and ≠ mean *not* or *different* ¯ (bar) and ′ mean *not* or *inversion* 'X' and '–' mean "don't care" for a single-bit value "XX…" and "– –…" mean "don't care" for a multi-bit value – means "don't care" for an integer	if *a* = '1' & *b* = '0' then … if *a* = '1' | *b* = '0' then … if *x* != *a* then … or if *x* ≠ *a* then … *y* = *x*′ *x* = '1' & *y* = '–' If (*a* = "111" & *b* = "0–0") | *c* = "000" then … *m* = 5 & *n* = –

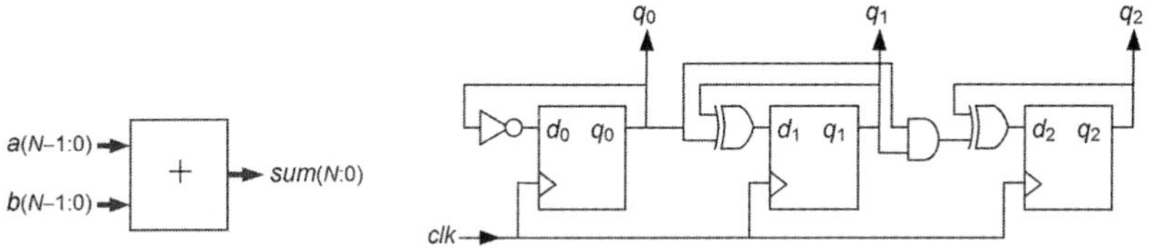
a(N–1:0)
b(N–1:0)
+
sum(N:0)
q0
q1
q2
d0 q0
d1 q1
d2 q2
clk

(a) *N*-bit adder (b) Synchronous 3-bit counter

Figure 1.1
Examples of (a) combinational and (b) sequential circuits.

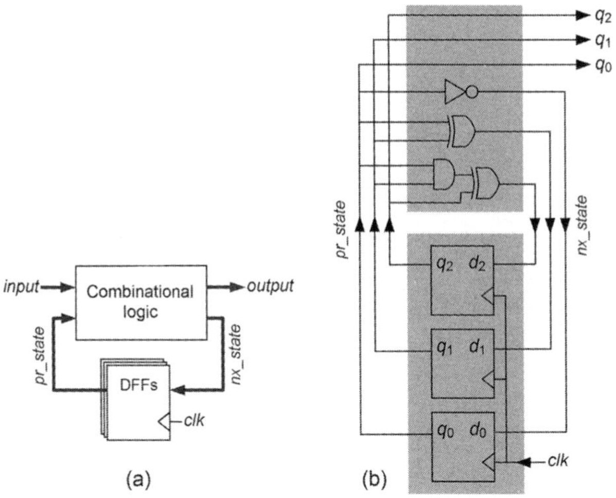
input
Combinational logic
output
pr_state
nx_state
DFFs
clk
(a)
q2
q1
q0
pr_state
nx_state
q2 d2
q1 d1
q0 d0
clk
(b)

Figure 1.2
(a) Sequential circuit with a registered feedback loop (a finite state machine). (b) Counter of figure 1.1b rearranged according to figure 1.2a.

figure 1.2a. Note that the lower section contains only flip-flops, whereas the upper section is purely combinational.

Concept

In short, a state machine is a *modeling/design* technique for sequential circuits. At any time, the machine sits in one of a finite number of possible states. For each state, both the output values and the transition conditions into other states are fully defined. The state is stored by the FSM, and the transition conditions are usually reevaluated at every (positive) clock edge, so the state-change procedure is always synchronous because the machine can only move to another state when the clock ticks. (Note: There has been some effort to develop asynchronous FSMs as well.)

Benefits

The FSM model provides a *systematic* approach (a *method*) for designing sequential circuits, which can lead to optimal or near-optimal implementations. Moreover, the method does not require any prior knowledge or specifics on how the general circuit (solution) for the problem at hand should look like.

When to Use the FSM Approach

This will be discussed in section 1.10.

Hardware- versus Software-Implemented State Machines

Designing and implementing correct state machines in hardware is generally (much) more complex than doing it in software. Some of the reasons for that are listed below.

1) It is physically impossible for the clock signal to arrive at all chip locations at exactly the same time (this is called clock skew), so some flip-flops will be activated before others, a concern that simply does not exist in software.
2) A naive design in hardware might lead to the inference of latches, which impair the time response. It might also lead to the other extreme, which consists of reregistering one or more signals, causing unwanted latency.
3) In hardware, signals might be subject to glitches, another concern that does not occur in software.
4) Contrary to software, hardware allows no abstraction. For example, if a state machine must produce in the next state the same output value produced in the current state, in software we can simply omit the corresponding expression; or if it requires an incrementer, we can simply write $x = x + 1$. In either case, an explicit expression would be required in hardware ($x = x$ or $x = x + 1$), which can only be evaluated if the value of x is available, so the machine itself must provide a means for storing (and properly retrieving) x.
5) In hardware, signals represent physical wires, so we cannot assign a signal source to an interconnection now and simply assign another later.
6) Many machines have asynchronous inputs, so the use of synchronizers (to avoid flip-flop metastability) must be considered, another concern that does not occur in software-implemented FSMs.
7) Some circuits need a special clock, obtained by "gating" the main clock. Depending on how it is done, the resulting clock might be subject to glitches, another issue that simply does not exist in software.
8) Several other concerns, such as "reset generation," "capturing the first bit," "keeping the final result stable," and "stretching the decision pulse", are also not a problem in software-implemented machines.

1.3 State Transition Diagrams

The *state transition diagram* (or simply *state diagram*) of a sequential circuit is a graphical representation of its functional specifications. Such a diagram must obey three fundamental principles:

1) It must include all possible system states.
2) All state transition conditions must be specified (unless a transition is unconditional) and must be truly complementary.
3) The list of output signals must be exactly the same in all states (for a standard architecture implementation).

To employ the FSM approach to design a sequential digital circuit, all three requisites listed above must be fulfilled, and also the list of states must not be too long. All sorts of controllers (including control units for datapath-based designs) are typical examples of circuits well suited for this design technique, as will become clear through the many examples presented in the book.

State machines can be of Moore or Mealy type. Both are described below.

Moore-Type State Machines

An FSM is said to be of Moore type when its output depends solely on the machine's present state. In other words, the output is not affected directly by the input (the input can only affect the machine's next state). The result is a fully synchronous circuit because the output can only change when the clock ticks.

An example of Moore FSM is presented in figure 1.3b. The circuit ports are shown in figure 1.3a, consisting of a data input, *x* (8-bit extended ASCII character), a data output, *y* (single bit), plus the conventional operational inputs of clock and reset. The circuit must produce *y* = '1' when the sequence "*abc*" occurs in *x*, that is, when *x* = "01100001" (ASCII code for *a*), followed by *x* = "01100010" (= *b*), then *x* = "01100011" (= *c*) occur.

The Moore-type state transition diagram of figure 1.3b contains four states, called *idle*, *char1*, *char2*, and *char3*. Each state tells the value that must be produced at the output (*y*) while the machine is in that state; note that only *char3* produces *y* = '1' because the machine only reaches that state if the correct sequence (*abc*) is detected. Finally, the transition conditions (on *x*, the input) are shown along the arrows.

The meaning of the state diagram of figure 1.3b is as follows. Say that the circuit is in the *idle* state; if *x* = *a* is received, it moves from *idle* to *char1*, otherwise it remains in *idle*; if it is in *char1* and *b* is received, it moves to state *char2*, otherwise it remains in *char1* if *a* was received or returns to *idle* if neither *a* nor *b* was received; and so on. Note in figure 1.3b that, because this is a Moore machine, the output depends only on the state in which the machine is, so the output values can be written inside the state circles.

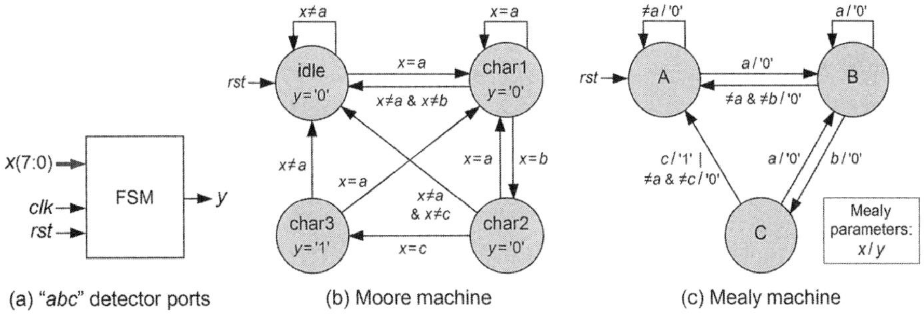
x(7:0)
clk
rst
FSM
y
(a) "abc" detector ports
x≠a
x=a
rst
idle
y='0'
x=a
char1
y='0'
x≠a & x≠b
x≠a
x=a
x=a
x=b
x≠a
& x≠c
char3
y='1'
x=c
char2
y='0'
(b) Moore machine
≠a / '0'
a / '0'
rst
A
a / '0'
B
≠a & ≠b / '0'
c / '1' |
≠a & ≠c / '0'
a / '0'
b / '0'
C
Mealy
parameters:
x / y
(c) Mealy machine

Figure 1.3
A finite state machine that detects the ASCII sequence "*abc*". (a) Circuit ports. Corresponding (b) Moore and (c) Mealy state transition diagrams.

It is important to mention that the state transitions are always synchronous (governed by a clock signal). For example, if the machine is in the *idle* state and the condition $x = a$ is true *at the moment when a (positive) clock edge occurs*, then the circuit moves to state *char1*. A machine can operate either at the positive or negative clock edge, or even at both clock edges if dual-edge flip-flops are employed. Unless specified otherwise, it will be assumed (default) that it is a positive-edge machine.

Mealy-Type State Machines

An FSM is said to be of Mealy type when its input can affect the output directly. In other words, the output now does not depend solely on the machine's state but also depends on the input value. The resulting circuit is no longer truly synchronous because the output might now change independently of the clock.

A Mealy-type solution for the same problem of figure 1.3a is depicted in figure 1.3c. Because the output can now exhibit more than one value for the same state (because the output also depends on the input value), the output values can no longer be written inside the state circles. Note that they are indeed marked on the arrows, along with the input (transition condition) values. Additionally, to simplify the notation, in the Mealy machine the signal names are generally omitted (they are indicated separately, as in the small rectangle of figure 1.3c). In this example the Mealy parameters are x/y, meaning "if x = value, then y = value"; for example, a/'0' means "if $x = a$, then y = '0'."

The meaning of the state diagram of figure 1.3c is as follows. If the circuit is in state A and the input is $x = a$, the output is y = '0', and the next state (at the next positive clock edge) will be B; otherwise, the output is still y = '0', but the next state will be A. Likewise, if the machine is in state C and the input is $x = a$, then the output is y = '0', and the next state will be B; otherwise, if the input is $x = c$, the output is y = '1', and the next state will be A; else, the output is y = '0', but the next state will still be A. A similar reasoning can easily be applied to state B. The direct dependence of the output on the input can easily be observed in the state diagram; for example, note that in state C the value of y varies with x, resulting in y = '1' when $x = c$ or y = '0' otherwise.

Because modern designs are generally synchronous, the Moore option tends to be preferred whenever the application permits.

Further details on Moore and Mealy constructions are seen in sections 1.7 and 1.8, in which the conversion from one to the other is described.

1.4 Equivalent State Transition Diagram Representations

Unconditional and equivalent representations for the state transition diagram are shown in figure 1.4, where a 1-to-5 counter is used as an example. Two cases are considered. The case in figure 1.4a has only clock and reset as inputs and as output has the 3-bit signal *outp* that encodes the counting. The case in figure 1.4c has an additional

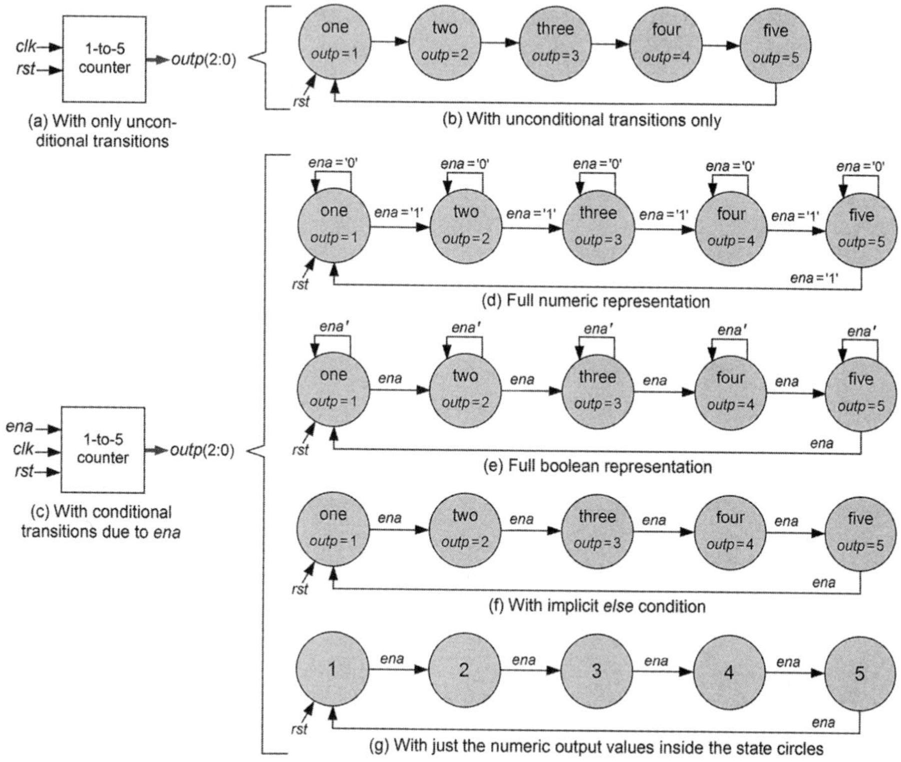
clk
rst
1-to-5 counter
outp(2:0)
(a) With only unconditional transitions
one outp=1
two outp=2
three outp=3
four outp=4
five outp=5
rst
(b) With unconditional transitions only
ena='0'
ena='1'
(d) Full numeric representation
ena'
ena
(e) Full boolean representation
ena
clk
rst
1-to-5 counter
outp(2:0)
(c) With conditional transitions due to ena
(f) With implicit else condition
1
2
3
4
5
(g) With just the numeric output values inside the state circles

Figure 1.4
(a, b) A 1-to-5 counter with only clock and reset as inputs (the transitions are unconditional). (c–g) The counter has an additional input (*ena*), which either enables the counter or causes it to stop. The representations in d–g are equivalent.

input, called *ena*, which enables the counter when asserted (*ena* = '1') or causes it to stop otherwise.

Figure 1.4b shows the FSM corresponding to the counter in figure 1.4a. Because there are no inputs in this circuit (except for the operational inputs, clock and reset), it can only be a Moore machine. Note that all possible states are included and that the value that must be produced at the output in each state is specified. However, there are no specifications for the transition conditions, which means that the transitions are *unconditional*, that is, they must occur at every (positive) clock edge.

Observe that a special (simplified) representation is reserved for the reset signal (not only in this example, but in all state transition diagrams). The reset signal is represented by a single arrow pointing to the state to which the machine is forced when *rst* = '1' occurs.

Figures 1.4d–g show equivalent representations for the FSM corresponding to the counter in figure 1.4c. Because now an external nonoperational input is present (*ena*, which lets the counter run when high or stops it when low), it can be modeled as either a Moore or a Mealy machine. However, because counters are inherently synchronous, the Moore approach is the natural choice.

The diagram in figure 1.4d is the most detailed, expressing, both by name and numerically, all transition conditions and output values. The representation in figure 1.4e expresses the transition conditions in Boolean form instead of numeric form. The representation in figure 1.4f assumes that *else* is implicit. Finally, the extreme simplification of figure 1.4g includes just the numeric output values inside the state circles, assuming again that *else* is implicit. The advantage of the first representation (figure 1.4d) is that it forces the designer to go over all possibilities more closely, whereas the advantage of the other representations is a simpler, neater diagram. To help the reader visualize small details, the first representation is used here more often than the others, but these representations are all equivalent and can be used interchangeably.

1.5 Under- and Overspecified State Transition Diagrams

This section describes a relatively frequent mistake that occurs while one is preparing the state transition diagram for a given problem, which consists of either under- or overspecifying it. An underspecification occurs when not all combinations of the transition control signals are covered, whereas an overspecification occurs when one or more combinations are included more than once.

Figure 1.5a shows an example of underspecification. Because the transition control signals are *a* and *b*, which are single-bit signals, the possible transition conditions are *ab* = {"00", "01", "10", "11"}. In state A, the AA transition is governed by the condition *a* = '0'; because this is independent of *b*, it is the same as writing *a* = '0' & *b* = '–', thus covering the cases *ab* = "0–" = {"00", "01"}. The AB transition is governed by the condition *a* = '1' & *b* = '1', thus covering the case *ab* = "11". Since there is no

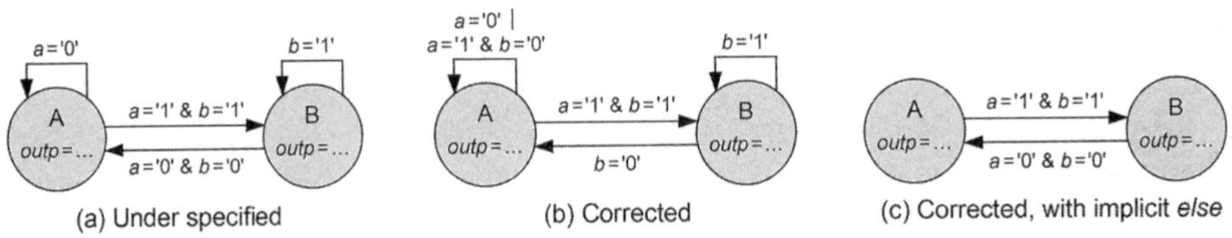
a='0'
b='1'
A
outp=...
a='1' & b='1'
a='0' & b='0'
B
outp=...
(a) Under specified
a='0' |
a='1' & b='0'
b='1'
A
outp=...
a='1' & b='1'
b='0'
B
outp=...
(b) Corrected
A
outp=...
a='1' & b='1'
a='0' & b='0'
B
outp=...
(c) Corrected, with implicit else

Figure 1.5
(a) Example of underspecified state transition diagram and (b, c) examples of possible solutions. In c, the *else* condition is implicit.

other outward transition in state A, we conclude that the condition *ab* = "10" was not covered in the state diagram. A similar analysis for state B shows that the transition condition *ab* = "10" was again not covered. Therefore, in this example, both states are underspecified. If the machine faces one of the unspecified combinations, it will either get stuck there or will proceed as defined (probably unconsciously) in the corresponding VHDL or SystemVerilog code.

Figure 1.5b shows a corrected version for the underspecified machine of figure 1.5a. It was considered that the missing condition for state A (*ab* = "10") should be associated to the AA transition, and the missing condition for state B (*ab* = "10") should be associated to the BA transition. Note that the latter caused the BA transition to become independent from *a*.

Another corrected version for the underspecified machine of figure 1.5a is presented in figure 1.5c. In this case the missing conditions for states A and B were associated to the AA and BB transitions, respectively. The representation with implicit *else* was used.

Figure 1.6a shows an example of overspecification. Again, *a* and *b* are the transition control signals. The AB transition is governed by the condition *a* = '1', thus covering the cases *ab* = "1–" = {"10", "11"}. The AC transition is governed by the condition *b* = '1', thus covering the cases *ab* = "–1" = {"01", "11"}. Note that *ab* = "11" appears in both AB and AC transitions, thus causing a conflict. To solve the problem, we must establish *priorities*.

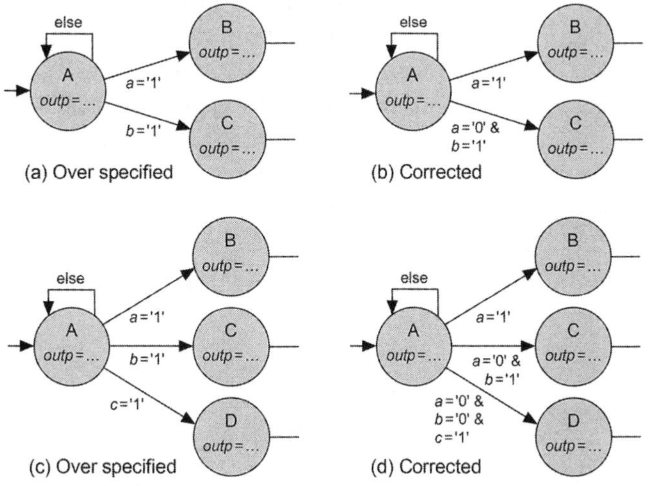

else
A
outp=...
a='1'
b='1'
B
outp=...
C
outp=...
(a) Over specified
else
A
outp=...
a='1'
a='0' &
b='1'
B
outp=...
C
outp=...
(b) Corrected
else
A
outp=...
a='1'
b='1'
c='1'
B
outp=...
C
outp=...
D
outp=...
(c) Over specified
else
A
outp=...
a='1'
a='0' &
b='1'
a='0' &
b='0' &
c='1'
B
outp=...
C
outp=...
D
outp=...
(d) Corrected

Figure 1.6
(a) Example of overspecified state transition diagram and (b) a possible solution. (c) Another overspecified machine and (d) a possible solution.

A corrected version for the overspecified machine of figure 1.6a is presented in figure 1.6b. In this example the following priority list was adopted (from highest to lowest): AB, AC, AA.

Figure 1.6c shows another example of overspecification. The transition control signals now are *a*, *b*, and *c*. The AB transition is governed by the condition *a* = '1' and thus covers the cases *abc* = "1– –" = {"100", "101", "110", "111"}. The AC transition is governed by the condition *b* = '1' and thus covers the cases *abc* = "–1–" = {"010", "011", "110", "111"}. Similarly, the AD transition is governed by the condition *c* = '1' and thus covers the cases *abc* = "– –1" = {"001", "011", "101", "111"}. Note that several conditions are repeated, causing conflicts. To solve the problem, we must again establish priorities.

A corrected version for the overspecified machine of figure 1.6c is presented in figure 1.6d. In this example the following priority list was adopted (from highest to lowest): AB, AC, AD, AA.

In summary, the outward transition conditions must be *exactly fully complementary*. In other words, they must include all possible combinations of the transition control signals, but without any repetitions.

In regard to underspecification, another example is shown in figure 1.7, in which an integer *t*, produced by a counter to represent time, is the transition control signal. The machine must stay in state A during *T* clock periods, moving then to state B. Because the timer's initial value is zero, it must count from 0 to $T - 1$ in order to span *T* clock cycles, which is the reason why $t = T - 1$ (instead of $t = T$) appears in the transition control conditions (specific details on timed transitions are given in chapter 8).

Note in figure 1.7a that the outward transition conditions from state A are not truly complementary because the $t > T - 1$ condition is not covered. It was fixed in figure 1.7b with the $t > T - 1$ condition associated with the AB transition. Another corrected option is shown in figure 1.7c, this time with the $t > T - 1$ condition associated with the AA transition.

There are two main reasons for not using non–truly complementary conditions. First, the machine can go into an undesirable state, even get deadlocked (for example, if the initial state is A and the timer is not properly reset, starting with $t > T - 1$,

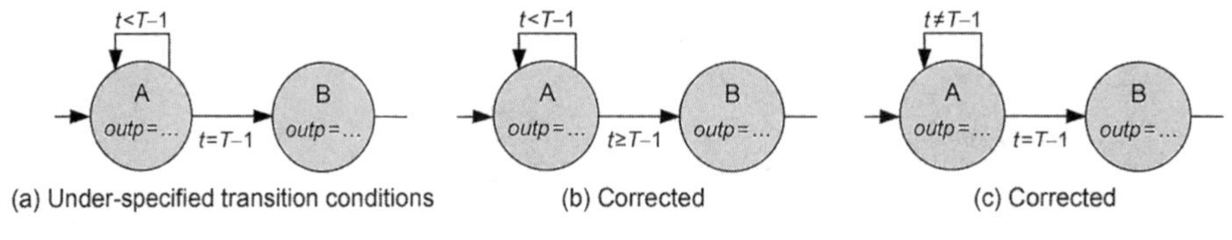
t<T–1
A
outp=...
t=T–1
B
outp=...
(a) Under-specified transition conditions
t<T–1
A
outp=...
t≥T–1
B
outp=...
(b) Corrected
t≠T–1
A
outp=...
t=T–1
B
outp=...
(c) Corrected

Figure 1.7
Noncomplementary transition conditions. (a) Condition $t > T - 1$ not covered. (b, c) Corrected versions with $t > T - 1$ associated with the AB and AA transitions, respectively.

depending on how this timer is controlled by the FSM, the machine can get stuck in state A forever). Second, it is more costly (in terms of hardware) to compute non–fully complementary conditions than otherwise. For example, if VHDL is used, the following sections of code could be employed for the three cases in figure 1.7:

```
For figure 1.7a:
if t=T-1 then
   nx_state <= B
elsif t<T-1 then
   nx_state <= A;
end if;
```

```
For figure 1.7b:
if t>=T-1 then
   nx_state <= B
else
   nx_state <= A;
end if;
```

```
For figure 1.7c:
if t=T-1 then
   nx_state <= B
else
   nx_state <= A;
end if;
```

Note that **else** was used to close the **if** statement in the last two codes, which means that all conditions are covered and only one comparison is needed. On the other hand, in the first code **elsif** was used instead, so an additional comparison is required; moreover, it does not cover all input combinations, so latches might be inferred by the compiler. In summary, the option in figure 1.7a produces an inferior circuit, and we still have to pay more for it.

Other common mistakes and problems that can occur while one is designing FSMs in hardware are described in chapter 4.

1.6 Transition Types

A very important classification for the transitions, from a hardware perspective, is introduced in this section. In section 3.6 this classification is used to separate any state machine into one of three categories, immensely easing its hardware-based design.

The state machine of figure 1.8a is used to describe the transition types, where x is the actual input, t is an auxiliary input generated by a timer, and y is the actual output. This machine contains all four possible types of transitions.

Transition AB *(conditional transition)* depends only on the actual input, x. If the machine is in state A, it must move to state B at the first (positive) clock edge that finds $x = x_1$.

Transition BC *(timed transition)* depends only on the timer, t. The machine must stay in state B during exactly T_1 clock cycles, moving then to state C. An auxiliary circuit (a timer, which is simply a counter, operating from 0 to $T_1 - 1$ in this transition) must be included in the design. By default, the timer is zeroed every time the FSM changes state; moreover, the timer is kept stopped at zero in states where it is not needed (states A and D of figure 1.8a, for example).

Transition CD *(conditional-timed transition)* is more complex because it depends on the actual input, x, and also on the timer, t. The machine must move to state D at the first (positive) clock edge that finds $x = x_2$ after staying in state C during T_2 clock

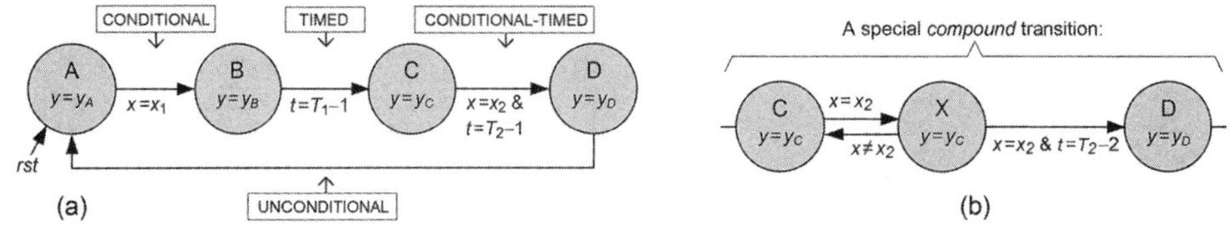
CONDITIONAL
TIMED
CONDITIONAL-TIMED
A
y=yA
x=x1
B
y=yB
t=T1−1
C
y=yC
x=x2 &
t=T2−1
D
y=yD
rst
UNCONDITIONAL
(a)
A special compound transition:
C
y=yC
x=x2
x≠x2
X
y=yC
x=x2 & t=T2−2
D
y=yD
(b)

Figure 1.8
(a) State machine containing all four types of transitions (from a hardware perspective): *conditional, timed, conditional-timed,* and *unconditional.* (b) A special *compound* transition, which checks whether a condition has been true *during the whole time.*

cycles. This implies that it will remain in state C during *at least* T_2 clock cycles, not necessarily during exactly T_2 clock cycles.

Transition DA *(unconditional transition)* is the simplest type of transition. The machine must move from state D to state A at the next (positive) clock edge, regardless of x and t, thus staying in D during exactly one clock period.

Note that even though t denotes time in the description above, it is not expressed in seconds but rather in "number of clock cycles." For example, if we want the machine to stay in a certain state during t_{state} = 2 ms, and the clock frequency is f_{clk} = 50 MHz, we simply adopt $T_{state} = t_{state} \times f_{clk} = 2{\cdot}10^{-3} \times 50{\cdot}10^{6} = 100{,}000$ clock cycles.

A special time-dependent transition is shown in figure 1.8b. Note that the conditional-timed transition CD in figure 1.8a only checks if $x = x_2$ after T_2 clock cycles. Say, however, that we want the machine to move from C to D only if $x = x_2$ has occurred *during the whole time* (i.e., during all T_2 clock cycles). To cover this case, a *compound* transition is needed that results from the combination of three pure transitions, as shown in figure 1.8b. This arrangement works well because the timer is zeroed every time the machine changes its state. Note that $T_2 - 1$ clock cycles are needed in the XD transition (so the timer must count from 0 to $T_2 - 2$) because one clock cycle is spent in the CX transition. Even though in many applications this "–2" factor in $t = T_2 - 2$ is not relevant, it is maintained here for the sake of accuracy. Much more on time-dependent transitions is presented in chapter 8.

In section 3.6, the transition types described above are used to classify any hardware-implemented FSM into one of the following three categories: *regular machines, timed machines,* or *recursive machines.* Two fundamental decisions must then be made when developing an actual design in hardware: the machine category (just listed) and the machine type (Moore or Mealy).

1.7 Moore-to-Mealy Conversion

Moore machines can be converted into corresponding Mealy machines. The latter will have the same number of states as the former if state merging is not possible, or fewer states otherwise.

When merging is not possible, the conversion is trivial, consisting simply of a change of notation (from Moore to Mealy style). To do so, just bring outside the output values marked inside the state circles and associate them with the corresponding *preceding* (inward) transitions. An example is presented in figure 1.9.

The merging of two states is possible when they fulfill the following two requisites:

1) Their sets of *outward* transitions are exactly equal.
2) The pairs of equal outward transitions (one from each state) go to the same states.

An example is presented in figures 1.10a–c. The original Moore FSM, with four states, is presented in figure 1.10a. Note that states A and B have the same set of *outward* transitions ($x = x_1$, $x = x_2$) and that the equal transitions go to the same states (from both A and B, the transitions governed by $x = x_1$ go to state C, while those governed by $x = x_2$ go to state D). Therefore, A and B can be merged. To do so, first

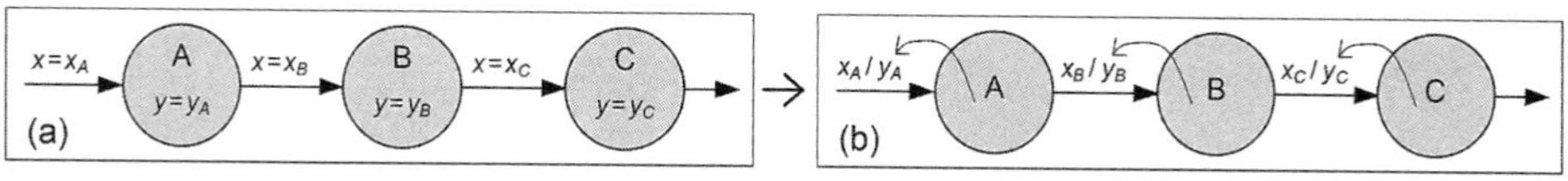

x=xA
A
y=yA
x=xB
B
y=yB
x=xC
C
y=yC
(a)
xA / yA
xB / yB
xC / yC
(b)

Figure 1.9
Moore-to-Mealy conversion when state merging is not possible (just a change of notation).

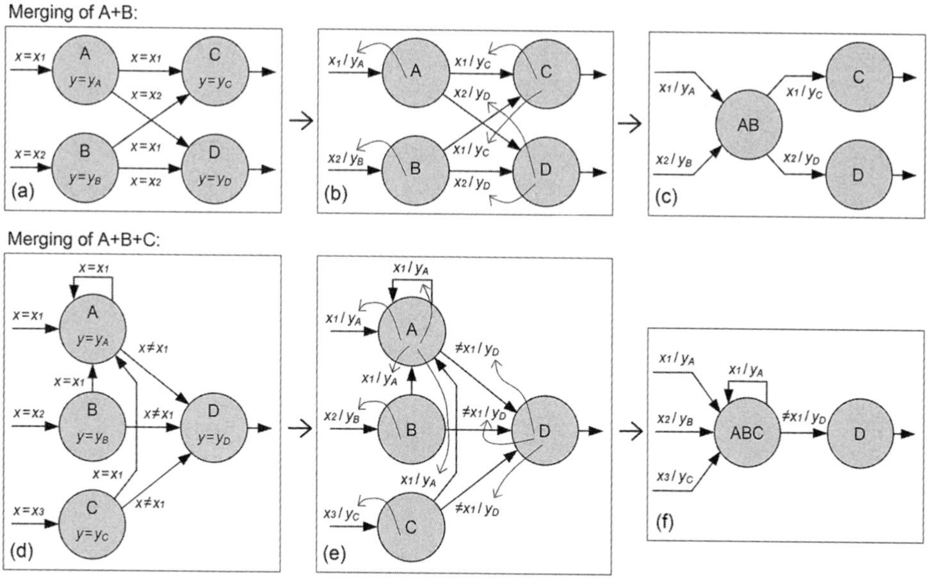

Merging of A+B:
(a)
(b)
(c)
AB
x1 / yC
x2 / yD
x1 / yA
x2 / yB
Merging of A+B+C:
(d)
(e)
(f)
x=x1
x=x2
x=x3
x≠x1
ABC
≠x1 / yD
x3 / yC

Figure 1.10
Moore-to-Mealy conversion principle. (a) Original Moore machine. (b) Moore-to-Mealy notation change. (c) Merging of states A and B. (d–f) Another example, following the same procedure.

the notation is changed from Moore to Mealy style, as shown in figure 1.10b, and then the merging is done in figure 1.10c.

Another example is presented in figures 1.10d–f following the same procedure. The analysis of this example is left to the reader. The reader is also invited to apply this procedure to the Moore machine of figure 1.3b and see if the Mealy machine of figure 1.3c results.

As expected, because of the highly restricting requirements for state merging (described above), in practical engineering problems the number of additional states in a Moore machine compared to its Mealy counterpart is generally very small.

1.8 Mealy-to-Moore Conversion

Mealy machines, too, can be converted into corresponding Moore machines. As seen above, the former can be smaller than the latter, although the difference (in number of states) in useful engineering applications is generally negligible.

The conversion principle consists again of two steps, illustrated in figures 1.11a–c. A Mealy FSM with three states is presented in figure 1.11a. The first step, shown in figure 1.11b, consists of changing the notation from Mealy to Moore style. Because in a Mealy machine the same state can exhibit more than one output value, the resulting Moore diagram might have states with *conditional* outputs, such as state A in the figure. The next step is to split each state into as many states as the possible output values.

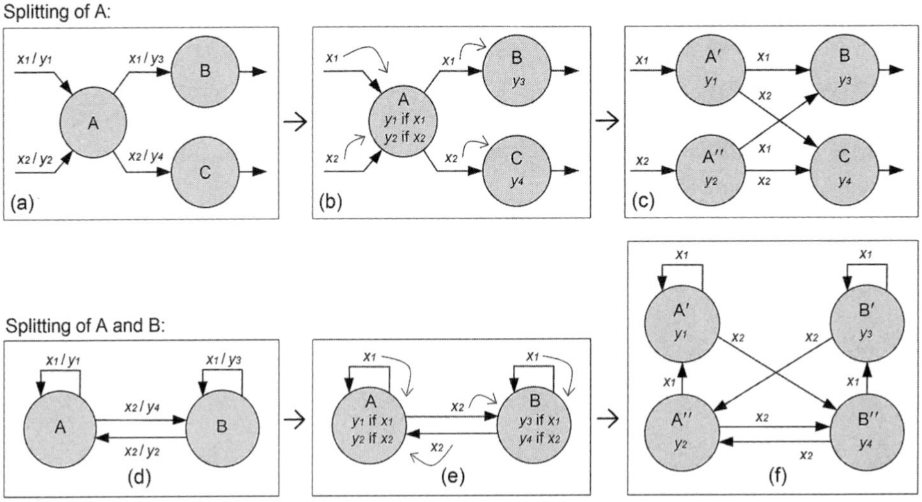
Splitting of A:
x1 / y1
x1 / y3
x2 / y2
x2 / y4
A
B
C
(a)
y1 if x1
y2 if x2
y3
y4
(b)
A′
A″
(c)
Splitting of A and B:
(d)
(e)
(f)
B′
B″

Figure 1.11
Mealy-to-Moore conversion principle. (a) Original Mealy machine. (b) Mealy-to-Moore notation change. (c) Splitting of state A into A′ and A″. (d–f) Another example, following the same procedure.

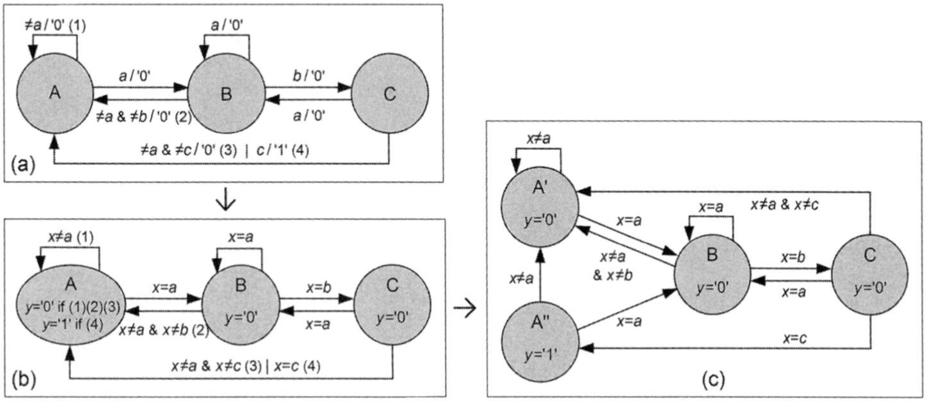
≠a / '0' (1)
a / '0'
A
a / '0'
B
b / '0'
C
≠a & ≠b / '0' (2)
a / '0'
≠a & ≠c / '0' (3) | c / '1' (4)
(a)
x≠a (1)
x=a
A
y='0' if (1)(2)(3)
y='1' if (4)
x=a
B
y='0'
x=b
C
y='0'
x≠a & x≠b (2)
x=a
x≠a & x≠c (3) | x=c (4)
(b)
x≠a
A'
y='0'
x=a
x≠a & x≠c
x=a
x≠a
& x≠b
B
y='0'
x=b
C
y='0'
x≠a
x=a
x=a
A"
y='1'
x=c
(c)

Figure 1.12
Example of Mealy-to-Moore conversion (“*abc*” detector of figure 1.3c).

This is shown in figure 1.11c, where only state A was split because it is the only state with multiple output values. Another example is presented in figures 1.11d–f, following the same procedure. The analysis of this example is left to the reader.

Just as a check, note in figure 1.11c that the outward transitions of states A′ and A″ are alike and that the equal transitions go to the same state, so A′ and A″ can be merged. Observe in figure 1.11f that the pairs A′-A″ and B′-B″ also fulfill the merging requirements, so they too can be merged.

A final example is presented in figure 1.12. In figure 1.12a, the same Mealy machine of figure 1.3c is shown (just reorganized horizontally). Note that there are four possible transitions into state A, of which the first three must produce y = ‘0’ while the last one must produce y = ‘1’ (hence with two possible values for the output). On the other hand, note that the transitions into states B and C must all produce a single output value (y = ‘0’). The resulting intermediate diagram, with Moore notation, is shown in figure 1.12b. Because only state A has more than one output value (two values), only A needs to be decomposed (into two states), resulting in the Moore machine shown in figure 1.12c. The reader is invited to compare it against that presented earlier, in figure 1.3b.

1.9 Algorithmic State Machine Chart

An algorithmic state machine (ASM) chart is another way of representing a state machine instead of using a state transition diagram. An ASM chart is a flowchart-like diagram containing information equivalent to that of the state diagram but generally in a more textual, algorithm-like form.

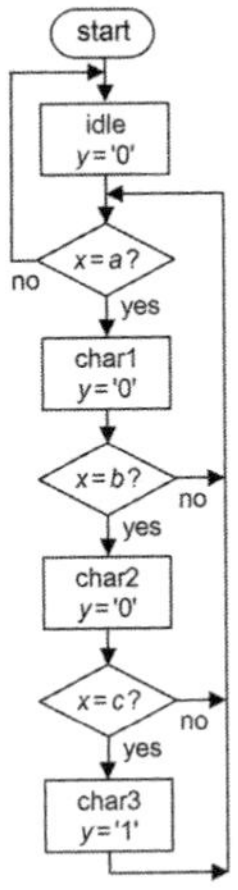
start
idle
y='0'
x=a?
no
yes
char1
y='0'
x=b?
no
yes
char2
y='0'
x=c?
no
yes
char3
y='1'

Figure 1.13
ASM chart for the "*abc*" detector of figure 1.3b.

As in flowcharts, the main elements of ASM charts are rectangles (representing the machine's states) and diamonds (representing condition checks). An example is presented in figure 1.13, which is equivalent to the "*abc*" detector of figure 1.3b.

In large and/or complex designs, ASM charts tend to be cumbersome. Moreover—and more importantly—they do not convey the hardware aspects as clearly as state transition diagrams. For these reasons, they are generally of limited use to hardware-based designs.

1.10 When to Use the FSM Approach

Even though any sequential circuit can be modeled/designed using the FSM approach, it is not always advantageous or necessary to do so. For example, if the circuit has too many states (say, over 100), it might be not viable to represent it as a state machine. Also, if it has very few states (say two or three), it might happen that a direct (experience-based) solution is straightforward. The number of control signals and the number of transitions are also determinant factors in the decision on whether or not to use the FSM approach.

Four candidates for the FSM approach are depicted in figure 1.14. The first candidate, in figure 1.14a, has only one (big) loop, with perhaps one control input. This is the case, for example, of regular counters (possibly with an enable input), which, as already mentioned, constitute a classical example of circuits for which the FSM technique is not needed.

The second candidate, in figure 1.14b, has few states, but proportionally more connections than the previous case due to more control inputs, which might even include

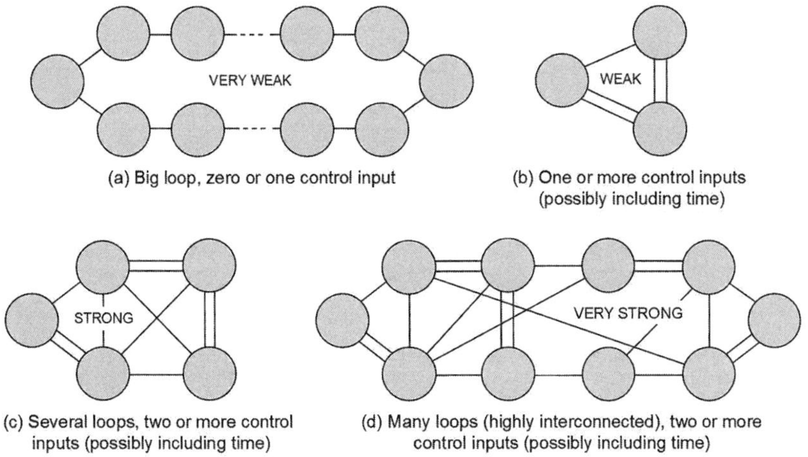
VERY WEAK
(a) Big loop, zero or one control input
WEAK
(b) One or more control inputs (possibly including time)
STRONG
(c) Several loops, two or more control inputs (possibly including time)
VERY STRONG
(d) Many loops (highly interconnected), two or more control inputs (possibly including time)

Figure 1.14
Weak and strong candidates for the FSM approach.

time (as will be shown in chapter 8, dealing with time-dependent transitions is more complex than dealing with regular transitions), so this candidate is not as weak as the previous one.

The candidate in figure 1.14c has more states and more control inputs than the previous one, resulting in a relatively strongly interconnected diagram. Consequently, this is a strong candidate for the FSM approach.

The final candidate, in figure 1.14d, has many states and several control inputs, resulting in a highly interconnected diagram. For this kind of candidate, the FSM approach is indispensable.

1.11 List of Main Machines Included in the Book

—Arbiter (bus access)
—Blinking light (with special features)
—Car alarms (basic and with chirps)
—Counters
—Datapath controller for a greatest common divisor
—Datapath controller for a largest-value detector
—Datapath controller for a sequential divider
—Datapath controller for a sequential multiplier
—Datapath controller for a square root calculator
—Datapath controller for an accumulator
—Debouncers (single and multiple, without and with one-shot conversion and memory)

—Divider
—Factorial calculator
—Flag monitor
—Garage door controller
—Greatest common divisor
—Hamming-weight calculator
—I^2C (inter-integrated circuits) interface
—Keypad encoder
—LCD (liquid crystal display) driver
—Leading-ones counter
—Light rotator
—Manchester encoders (regular and differential)
—Memory interfaces (SRAM and EEPROM)
—Multiplier
—One-shot circuits
—Parity detectors
—Password detector
—Pulse shifters
—Pulse stretchers
—Reference-value definers
—RTC (real-time clock) interface
—Serial data receivers
—Serial data transmitters
—SPI (serial peripheral interface)
—String detectors
—Strings comparators (short and long, with and without overlap)
—Temperature controller
—Traffic light controller
—Triggered circuits (bistable and monostable)
—Vending machine controller

1.12 Exercises

Exercise 1.1: FSM Architecture

Two sequential circuits are given in figure 1.15. Rearrange each of them according to the FSM architecture of figure 1.2.

Exercise 1.2: "*aabb*" Detector

Draw the state transition diagram for an FSM capable of detecting the sequence "*aabb*" (see example in figure 1.3) for the following cases:

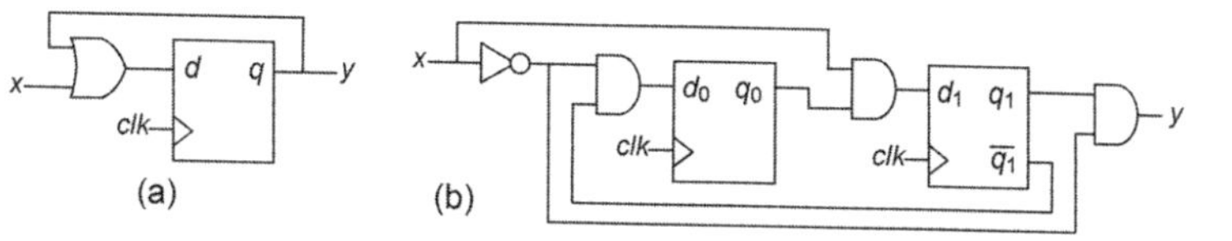
x
d
q
y
clk
(a)
x
d0 q0
clk
d1 q1
clk
y
(b)

Figure 1.15

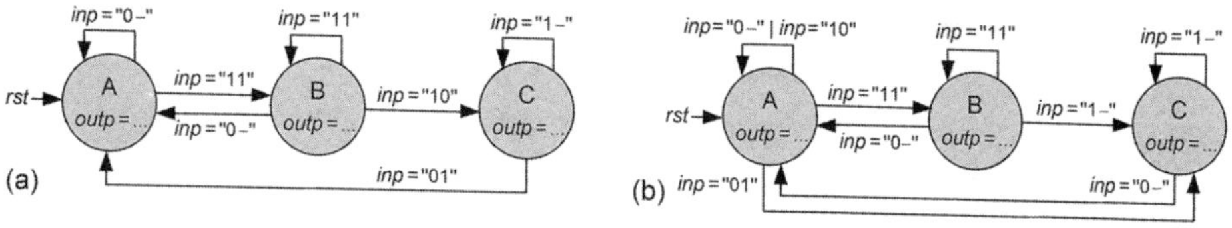
inp="0–"
inp="11"
inp="1–"
rst
A
outp=...
inp="11"
inp="0–"
B
outp=...
inp="10"
C
outp=...
inp="01"
(a)
inp="0–" | inp="10"
inp="11"
inp="1–"
rst
A
outp=...
inp="11"
inp="0–"
B
outp=...
inp="1–"
C
outp=...
(b) inp="01"
inp="0–"

Figure 1.16

a) Using a Moore machine.
b) Using a Mealy machine.

Exercise 1.3: Equivalent State Transition Diagrams

a) Present a simplified version (see figure 1.4) for the detailed state transition diagram of figure 5.4b.

b) Present fully detailed versions for the semidetailed state transition diagrams of figures 5.7c and 8.16c.

Exercise 1.4: Under- and Overspecified State Diagrams

a) Why is the state transition diagram of figure 1.16a said to be underspecified? Fix it.
b) Why is that of figure 1.16b said to be overspecified? Fix it.

Exercise 1.5: Transition Types

List the types (conditional, timed, etc.) of all transitions in the following FSMs:

a) Figure 8.12c.
b) Figure 8.14b.

Exercise 1.6: Moore-to-Mealy Conversion #1

Consider the Moore machine of figure 3.4a.

a) Are there states that can be merged in the Moore-to-Mealy conversion? Explain.
b) Do the conversion. After finishing it, compare your result to figure 3.6a.

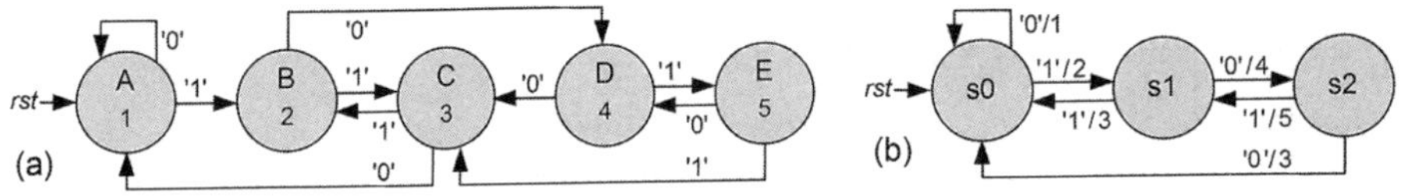
'0'
'0'
rst
A
1
'1'
B
2
'1'
'1'
C
3
'0'
'0'
D
4
'1'
'0'
'1'
E
5
(a)
'0'/1
rst
s0
'1'/2
'1'/3
s1
'0'/4
'1'/5
'0'/3
s2
(b)

Figure 1.17

Exercise 1.7: Moore-to-Mealy Conversion #2
Consider the Moore machine of figure 1.17a.

a) Are there states that can be merged in the Moore-to-Mealy conversion? Explain.
b) Do the conversion. Does your result have any relationship with figure 1.17b?

Exercise 1.8: Mealy-to-Moore Conversion #1
Consider the Mealy machine of figure 3.6a.

a) Are there any states that must be split in the Mealy-to-Moore conversion? Explain.
b) Do the conversion. After finishing it, compare your result to figure 3.4a.

Exercise 1.9: Mealy-to-Moore Conversion #2
Consider the Mealy machine of figure 1.17b.

a) Are there any states that must be split in the Mealy-to-Moore conversion? Explain.
b) Do the conversion. Does your result have any relationship with figure 1.17a?

2 Hardware Fundamentals—Part I

2.1 Introduction

This chapter and the one that follows discuss fundamental hardware-related aspects and introduce new material essential to fully understand and correctly design finite state machines in hardware. This chapter deals mainly with registers, and the next deals with the complete state machine structure.

The topics seen in these two chapters are used, reinforced, and expanded as the subsequent chapters unfold, particularly in chapters 5 (theory for category 1 machines), 8 (theory for category 2 machines), and 11 (theory for category 3 machines).

2.2 Flip-Flops

Flip-flops are available in four versions: SR (set-reset), D (data), T (toggle), and JK. The D-type flip-flop (DFF) is a general-purpose flip-flop and therefore the most commonly used. However, because counters are among the most common digital circuits, and counters are implemented with T-type flip-flops (TFFs), the TFF is also very popular. Nevertheless, because a TFF can be obtained from a DFF by simply connecting an inverted version of its output back to its input, the DFF is essentially the only flip-flop needed in most designs, no matter how big or how complex. For instance, the DFF is the only flip-flop fabricated in field programmable gate array (FPGA) devices.

The DFF is the flip-flop used to build the *state register* (that is, the memory that stores the machine's state) in hardware-implemented finite state machines (see figure 1.2). It is also the flip-flop used to build any other additional (optional or compulsory) register that the machine might require. Consequently, it is important to review its operation well.

Figure 2.1a shows the symbol and truth table for a basic positive-edge-triggered DFF. The inputs are *d* (data in) and *clk* (clock), while the output is *q* (data out). As can be seen in the truth table, *q*+ (which represents the next value of *q*) receives the value of *d* when a positive clock transition occurs (gray table line) but remains unchanged

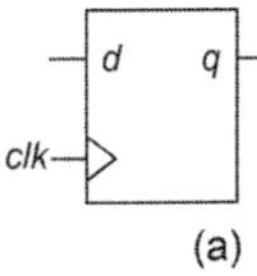
d
q
clk
(a)

clk	q+
0	q
1	q
↑	d
↓	q

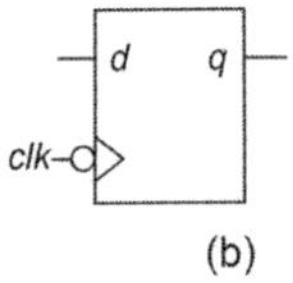
d
q
clk
(b)

clk	q+
0	q
1	q
↑	q
↓	d

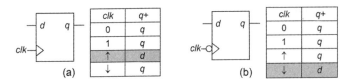

Figure 2.1
Symbol and truth table for basic (a) positive-edge and (b) negative-edge DFFs.

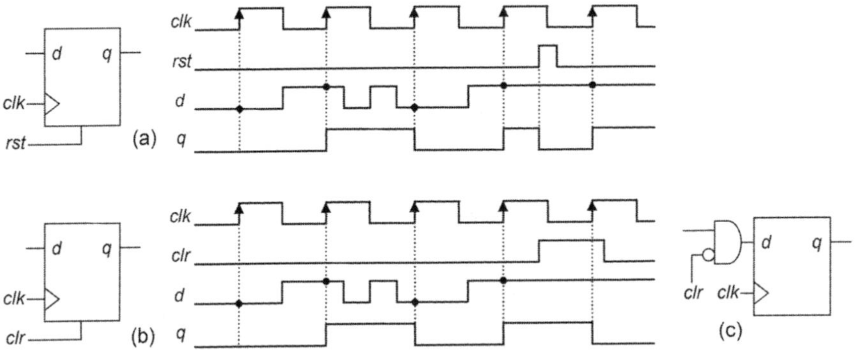
d
q
clk
rst
(a)
clk
rst
d
q
d
q
clk
clr
(b)
clk
clr
d
q
d
q
clr
clk
(c)

Figure 2.2
DFF symbol with (a) reset or (b) clear, followed by examples of *functional* response. (c) Diagram showing how clear can be implemented.

under any other condition. For this reason, it is said that a positive-edge-triggered (or simply positive-edge) DFF is "transparent" during positive clock transitions and "opaque" elsewhere.

Figure 2.1b shows the symbol and truth table for a basic negative-edge-triggered (or simply negative-edge) DFF (note the little circle at the clock input). This DFF is "transparent" during negative clock transitions (see last table line) and "opaque" elsewhere.

The behavior of any digital circuit can be expressed by means of its *functional* and *timing* responses. The former takes into account only the circuit's logical functions, thus conveying only its functional behavior, whereas the latter also takes into account the propagation delays as the signals travel through the circuit, thus expressing the circuit's actual behavior. Both types of responses (functional and timing) are illustrated next for flip-flops.

Figure 2.2 shows (on the left) two DFFs, the first having a reset (*rst*) input, and the second a clear (*clr*) input. In the context of this book the difference between reset and clear is that the former is asynchronous (it forces the output to zero regardless the clock value), whereas the latter is synchronous (the output is forced to zero when the proper clock transition occurs). It is important to mention, however, that these designations for reset and clear are not universal; for instance, FPGA companies usually call both "clear." The diagram in figure 2.2c shows how clear can be implemented; note that when *clr* = '1', *d* is forced to '0', so at the next positive clock edge this '0' will be copied to *q*, clearing the output.

Examples of functional response for both cases are included in figure 2.2. Arrows were placed on the clock waveforms to highlight the only moments at which the DFFs are transparent. As can be seen, the value of *d* is copied to *q* at each of these clock transitions. Note, however, that when *rst* is asserted (figure 2.2a), the output is forced to zero immediately, whereas when *clr* is asserted (figure 2.2b), the output is forced to zero at the next positive clock transition.

The *timing* response (with propagation delays taken into account) of a DFF is illustrated in figure 2.3. A DFF with reset, similar to that of figure 2.2a, was considered. The propagation delays are defined in figure 2.3a, where $t_{pCQ(HL)}$ and $t_{pCQ(LH)}$ represent the propagation delays from *clk* to *q* (time interval between the clock edge and the settling of *q* in the high-to-low and low-to-high transitions, respectively) and t_{pRQ} is the propagation delay between *rst* and *q*. In the example of figure 2.3b, the following values were assumed: $t_{pCQ(HL)} = t_{pCQ(LH)} = 2$ ns and $t_{pRQ} = 1$ ns (note the gray shades in the *q* waveform; the distance between the vertical lines is 1 ns).

A final pair of time-related parameters define the DFF's *forbidden* region (also called *aperture* or *transparency window*). As shown in figure 2.4, such parameters are called t_{setup} and t_{hold}, which specify, respectively, how long before and after the clock edge the

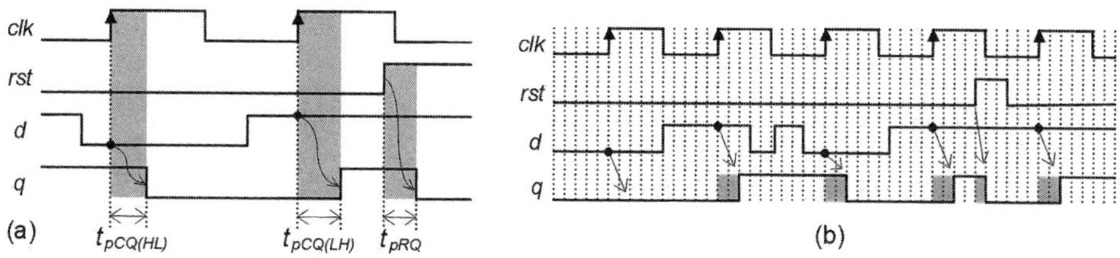
clk
rst
d
q
(a)
$t_{pCQ(HL)}$
$t_{pCQ(LH)}$
t_{pRQ}
clk
rst
d
q
(b)

Figure 2.3
(a) Time-related parameters of a DFF. (b) Example of *timing* response with $t_{pCQ(HL)} = t_{pCQ(LH)} = 2$ ns and $t_{pRQ} = 1$ ns.

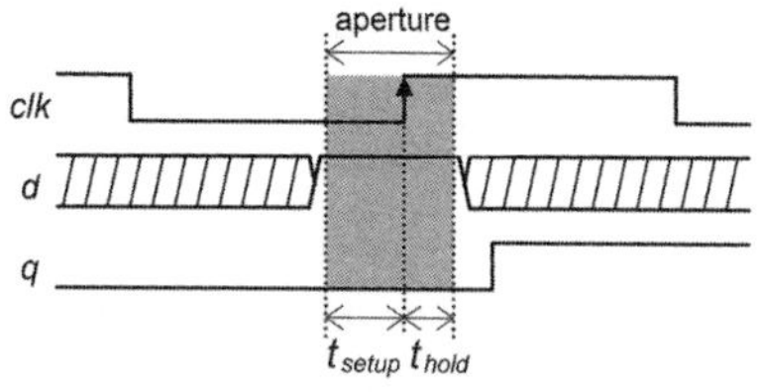
aperture
clk
d
q
t_{setup} t_{hold}

Figure 2.4
DFF's forbidden region (*d* must remain stable within the aperture window).

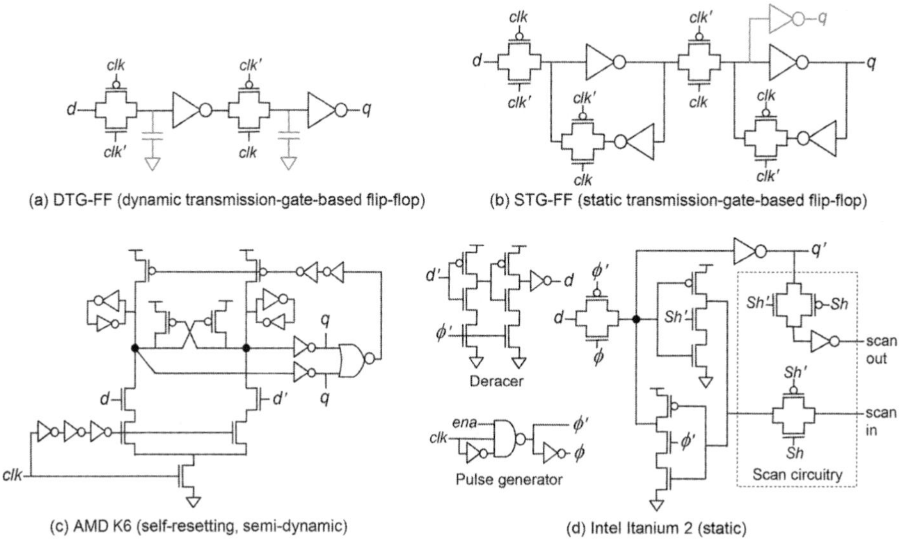
clk
clk'
d
q
(a) DTG-FF (dynamic transmission-gate-based flip-flop)
(b) STG-FF (static transmission-gate-based flip-flop)
d
q
clk'
Deracer
Pulse generator
ena
Sh'
Sh
scan out
scan in
Scan circuitry
(c) AMD K6 (self-resetting, semi-dynamic)
(d) Intel Itanium 2 (static)

Figure 2.5
Examples of actual DFF constructions. (a, b) Very popular implementations (dynamic and static versions, master-slave approach). (c, d) Two other commercial implementations (pulsed-latch approach).

input signal must remain stable. If *d* changes within the transparency window, the output value might be undetermined (further details on this are seen in the next section).

We conclude this section by presenting some examples of DFF constructions. The cases in figures 2.5a,b are among the most commonly used, consisting of dynamic and static versions for the same transmission-gate-based master-slave implementation. Two other commercial cases are shown in figures 2.5c,d, both based on the short-clock (pulsed latch) principle rather than on the master-slave approach.

2.3 Metastability and Synchronizers

Because many FSMs have control inputs that are asynchronous (that is, not related to the FSM's clock), such inputs can change during the state machine's DFFs' forbidden (aperture) window. This section describes what can happen in such cases and how its effect can be reduced.

This fact is illustrated in figure 2.6a, in which *d* changes precisely within the forbidden time interval (gray area). When this occurs, the output can go into an undetermined (metastable) state that lasts a relatively long time before finally resolving for '1' (path 1) or '0' (path 2). If the metastable state resolves within one clock period (as in the figure), at the next (positive) clock edge a valid value will be available (even

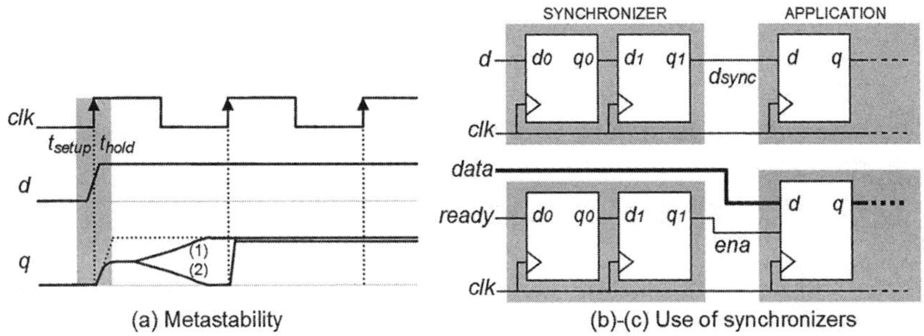

SYNCHRONIZER
APPLICATION
clk
t_{setup} t_{hold}
d
q
(1)
(2)
d
do qo
d1 q1
dsync
d q
clk
data
ready
do qo
d1 q1
ena
d q
clk
(a) Metastability
(b)-(c) Use of synchronizers

Figure 2.6
Illustration of flip-flop metastability and the use of synchronizers.

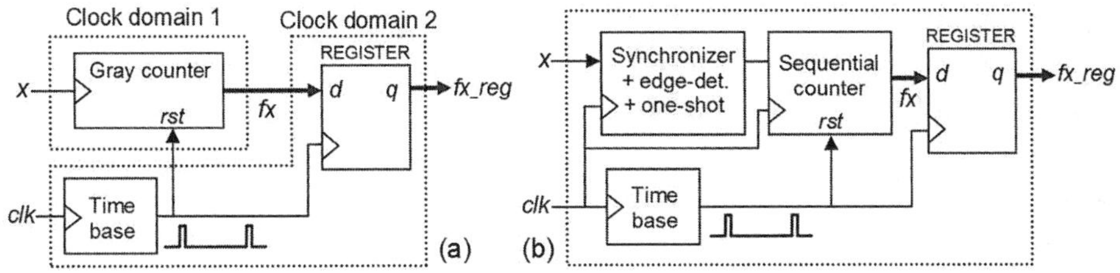

Clock domain 1
Clock domain 2
REGISTER
Gray counter
x
rst
fx
d q
fx_reg
clk
Time base
(a)
(b)
x
Synchronizer + edge-det. + one-shot
Sequential counter
rst
REGISTER
d q
fx
fx_reg
clk
Time base

Figure 2.7
Partial diagram for a frequency meter (two clock domains). (a) Solution with a Gray counter. (b) Solution with a synchronizer (so a regular counter can be employed).

though it might be different from the expected value); otherwise, an undetermined value will be read.

Synchronizers are circuits used to cope with metastability. The most common alternative is shown in figure 2.6b, consisting simply of a 2-stage shift register. In well-designed DFFs the probability of metastability is very small so the probability of having such a rare event going through both DFFs is extremely small. The obvious drawback is the two-clock-period latency imposed by this circuit (exercise 2.2). When the (multi-bit) data is accompanied by a control signal (data ready, figure 2.6c), only the control signal should be synchronized.

Another strategy to reduce the impact of having the input of a DFF change during its forbidden time window, applicable to counters, consists in using Gray counters instead of regular sequential counters (as reviewed in section 3.7, in a Gray counter only one bit changes from one codeword to the next—this applies also to Johnson counters). An example is depicted in figure 2.7a, which shows a partial diagram for a frequency meter. Because the system must measure the frequency *fx* of *x*, *x* acts as the clock to the corresponding counter; however, *x* and *clk* are uncorrelated, so a two-clock-domain situation results. The value of *fx* must be stored into the output register periodically (every 1 s, for example, resulting in a reading in Hz), and at the same time the counter must be reset in order to start a new counting. Because *x* and *clk* are uncorrelated, the storage of *fx* into the register might occur while *fx* is changing its value. If a sequential counter is used, several bits (or even all) can change from one codeword to the next, but because in a Gray counter only one bit changes, the value actually stored into the register cannot be off by more than one unit.

For comparison, a solution with a synchronizer (so a regular counter can be used) is included in figure 2.7b. Note that the synchronizer's output must be a short pulse (lasting only one clock period, T_{clk}); otherwise the counter could be incremented multiple times for the same pulse of *x*. Additionally, if a pulse in *x* might last less than T_{clk}, then an edge detector must also be included. (One-shot and edge-detecting circuits are described in the next section.)

A last class of circuits still involving the synchronous-asynchronous issue is presented in figure 2.8. They are *clock gaters*, needed in applications where the clock signal must be stopped (gated) during one or more clock periods (the I^2C and SPI serial data communications interfaces, studied in chapter 14, are examples where clock gating is necessary). The purpose here should not be confused with clock gating for power-saving reasons.

Figures 2.8a–f relate to *positive*-edge-triggered FSMs. Figure 2.8a highlights the facts that the machine operates at the positive clock edge and also that the clock-enabling signal *ena* is just one of its outputs. This signal (*ena*) must stop the clock when low, replacing the clock signal with a static-low value (analogous solutions can be easily derived for a static-high value).

The first solution, shown in figure 2.8b, is asynchronous and requires just an AND gate, where *clk* represents the main clock and *gclk* represents its gated version. The advantage of this solution is that the clock is interrupted at the same time that *ena* = '0' occurs (see gray shades in figures 2.8c–d). However, as depicted in the timing diagrams of figures 2.8c–d, the output is fine (glitch-free) only if the edge of *ena* reaches the gater *before* the edge of *clk* does, a situation that, though possible (due to long, unbalanced routings inside the chip), is very unlikely. Moreover, glitches in *ena* can propagate to the output. For these reasons, the clock gater of figure 2.8b is not recommended when the clock must be replaced with a static-low value.

The second solution, shown in figure 2.8e, is fully synchronous, so occasional glitches in *ena* are automatically filtered out. Also, note in figure 2.8f that the output is fine regardless of the delay (positive or negative) between *clk* and *ena*. For these reasons, the clock gater of figure 2.8e is recommended when the clock must be interrupted and replaced with a zero. Its drawback is that now *ena* = '0' must be produced in the *previous* clock cycle (previous FSM state—see gray shades in figure 2.8f), being therefore more error prone (requiring greater attention when preparing the corresponding state transition diagram).

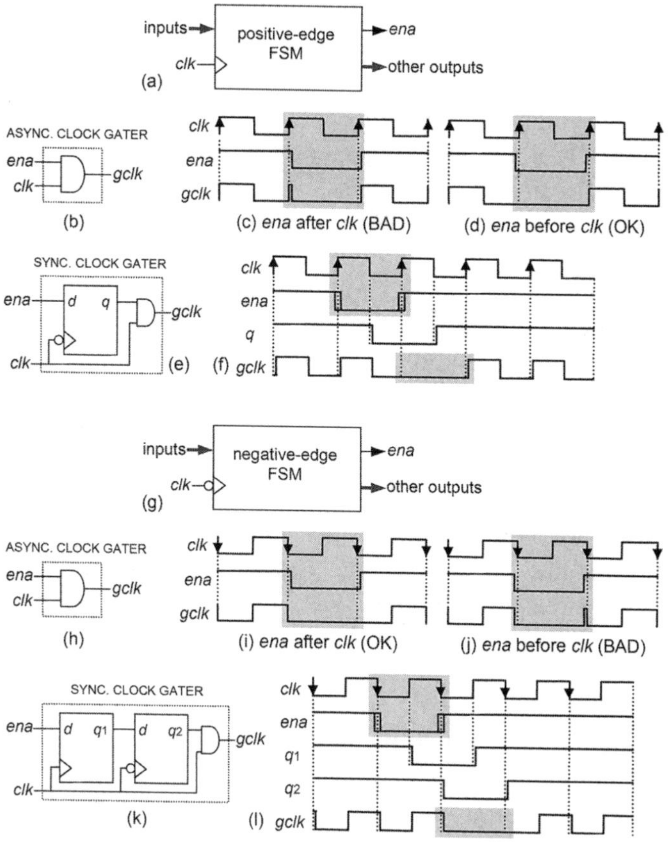
inputs
positive-edge FSM
clk
ena
other outputs
(a)
ASYNC. CLOCK GATER
ena
clk
gclk
(b)
clk
ena
gclk
(c) ena after clk (BAD)
(d) ena before clk (OK)
SYNC. CLOCK GATER
d
q
(e)
(f) gclk
inputs
negative-edge FSM
(g)
(h)
(i) ena after clk (OK)
(j) ena before clk (BAD)
q1
q2
(k)
(l) gclk

Figure 2.8

Clock gating circuits. (a–f) For positive-edge-triggered FSMs. (g–l) For negative-edge-triggered FSMs. Asynchronous (good and bad) solutions are shown in (b) and (h), and synchronous solutions (usually recommended) are presented in (e) and (k), all accompanied by illustrative timing diagrams.

Figures 2.8g–l relate to *negative*-edge triggered FSMs. Figure 2.8g highlights the facts that the machine operates at the negative clock edge and that *ena* is just one of its outputs. Again, *ena* must stop the clock when low, replacing it with a zero.

The first solution, shown in figure 2.8h, is the same as that in figure 2.8b. However, as depicted in the timing diagrams of figures 2.8i–j, the output is now fine (glitch-free) when the edge of *ena* reaches the gater *after* the edge of *clk* does, which is what normally occurs, so this solution is generally fine. Recall, however, that *ena* must be glitch-free.

The second solution, shown in figure 2.8k, is fully synchronous, so occasional glitches in *ena* are automatically filtered out. Also, note that in figure 2.8l the output is fine regardless of the delay (positive or negative) between *clk* and *ena* (two DFFs are needed here to guarantee that condition). This solution has the same drawback as that of figure 2.8e; that is, *ena* = '0' must be produced in the *previous* clock cycle (see gray shades in figure 2.8l), thus requiring a greater attention when developing the state transition diagram.

As a final comment, it is important to mention that in many applications the occurrence of metastability is not a problem, either because the metastable state cannot cause a malfunctioning or because the application itself is not critical.

2.4 Pulse Detection

Because many FSMs have asynchronous inputs, the duration of such inputs must be considered in the design. If an input pulse lasts at least one clock period ($T_{pulse} \geq T_{clk}$), its detection by the machine is guaranteed because at least one (positive) clock edge will occur while the pulse is present. On the other hand, if the pulse might last less than that, its detection is no longer guaranteed. Both cases are illustrated in figure 2.9a, where only the first of the two pulses is detected by the circuit. Note the small

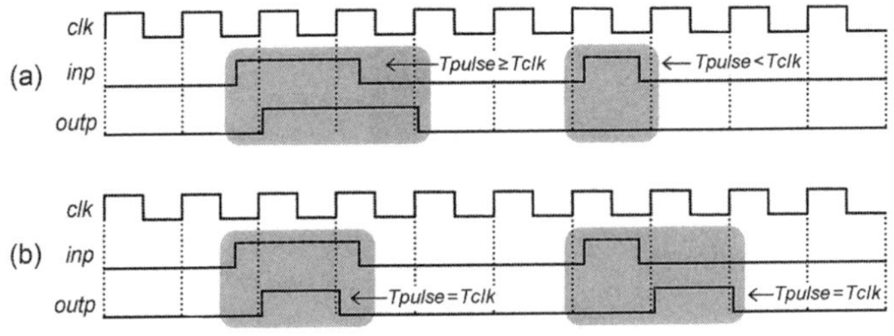
(a)
clk
inp
outp
Tpulse≥Tclk
Tpulse<Tclk
(b)
clk
inp
outp
Tpulse=Tclk
Tpulse=Tclk

Figure 2.9
(a) Only input pulses with duration $T_{pulse} \geq T_{clk}$ are guaranteed to be detected, and the output duration is proportional to the input duration. (b) Any pulse is detected, and the output duration is always T_{clk}.

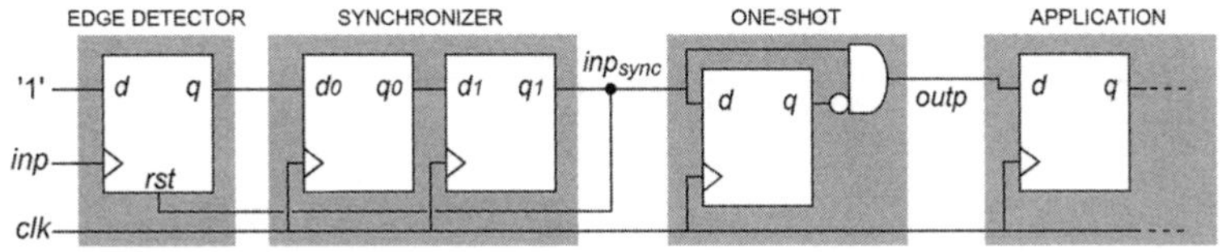
EDGE DETECTOR
SYNCHRONIZER
ONE-SHOT
APPLICATION
'1'
inp
clk
d
q
rst
d0
q0
d1
q1
inp_sync
outp

Figure 2.10
Circuit capable of detecting pulses of any width, producing a pulse with fixed length (one clock period) at the output (*outp*).

propagation delays left intentionally between the clock transitions and the corresponding responses in *outp* in order to portray a more realistic situation.

In some cases, the pulse (of any duration) must be detected and converted into a pulse whose duration is one clock period. This is illustrated in figure 2.9b, where both pulses are detected, and each produces an output pulse with duration T_{clk}.

A circuit that shortens the output pulse down to a predefined length (T_{clk} in the present case) is called a *one-shot* circuit, whereas one that detects short pulses is called an *edge detector*. Both are present in the example of figure 2.10, so pulses of any length can be detected and converted into pulses with one-clock-period duration.

The first circuit in figure 2.10 is the edge detector, which consists simply of a DFF plus a reset mechanism. Note that to be able to detect short pulses, *inp* is connected to the clock port instead of the data port. Because the data input is connected to V_{DD} ('1'), the output goes immediately to '1' when a positive edge occurs in *inp*. Some time later (see exercise 2.4), this '1' reaches inp_{sync}, resetting the input DFF, which will remain so until a new positive transition occurs in *inp*.

The second circuit in figure 2.10 is the synchronizer, already seen in the previous section.

The third and final circuit before the application is the one-shot circuit. Because of the AND gate, as soon as inp_{sync} goes to '1', *outp* goes to '1'. However, at the next (positive) clock edge, this value of inp_{sync} crosses the DFF, bringing *outp* back to '0', which it will remain until another pulse occurs at the input and the whole procedure is repeated. This one-shot circuit, however, works well only when the input is synchronous (see exercise 2.3), which is the case here.

Another circuit with the same purposes as that in figure 2.10 is discussed in exercise 2.5. In the chapters ahead, examples employing this kind of circuit are seen.

2.5 Glitches

Glitches are short voltage (or current) pulses produced involuntarily by combinational circuits. It is said that a hazard exists when the possibility of glitches in the circuit exists.

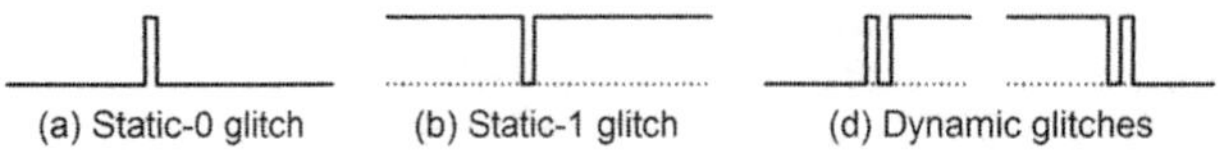
(a) Static-0 glitch
(b) Static-1 glitch
(d) Dynamic glitches

Figure 2.11
Glitch types.

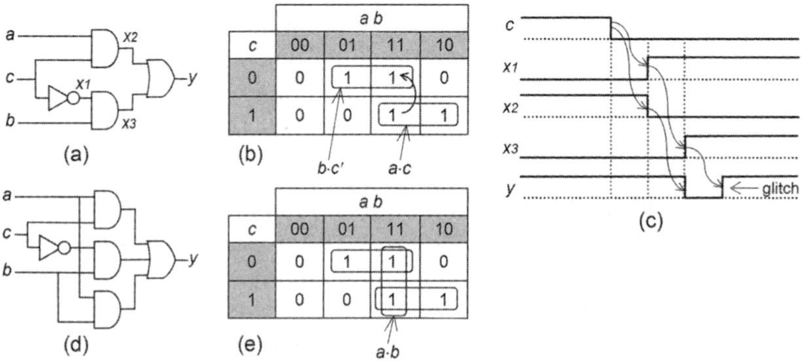
a
c
b
X1
X2
X3
y
(a)
ab
c
00
01
11
10
0
1
b·c'
a·c
(b)
x1
x2
x3
glitch
(c)
(d)
a·b
(e)

Figure 2.12
(a) Combinational circuit implementing the function $y = a{\cdot}c + b{\cdot}c'$. (b) Corresponding Karnaugh map. (c) Glitch generation when moving from *abc* = "111" to *abc* = "110". (d, e) Glitch eliminated by the addition of a redundant implicant.

Even though glitches are not a problem in many designs, it is important to be aware of their existence and understand how they can be eliminated when that is necessary. An example in which glitches can be disastrous is when a signal is used as a clock because then the associated flip-flops can be (improperly) triggered by the glitches.

Figure 2.11 shows the glitch types, which can be static (single pulse) or dynamic (multiple transitions). A static glitch is said to be of type static-0 when the signal should remain stable at '0' but a pulse toward '1' occurs. The meaning of static-1 glitches is analogous.

An example of a circuit subject to glitches is presented in figure 2.12a, which implements the function $y = a{\cdot}c + b{\cdot}c'$. The corresponding Karnaugh map is shown in figure 2.12b, where two prime implicants can be observed. Although the value of *y* is '1' for both *abc* = "111" and *abc* = "110", when the input transitions from the former to the latter, the involved propagation delays can produce a glitch at the output. This is illustrated in figure 2.12c, with *a* and *b* fixed at '1' and *c* changing from '1' to '0' (for simplicity, it was considered that the propagation delays of all gates are equal). Figure 2.12e shows a solution for this problem, which consists of including a redundant implicant covering the transition mentioned above, thus resulting in the circuit of figure 2.12d.

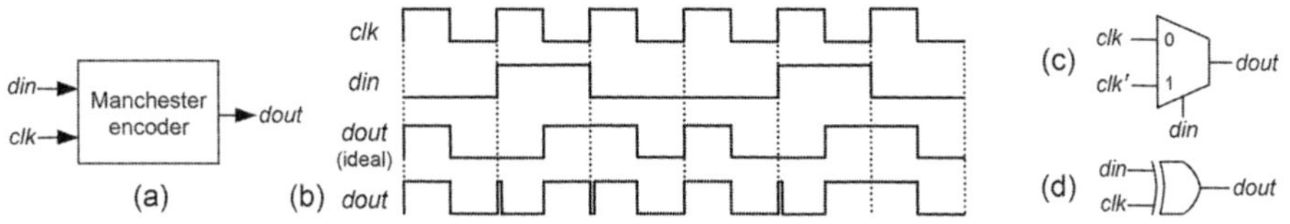
din
clk
Manchester encoder
dout
(a)
clk
din
dout (ideal)
(b) dout
(c)
clk
clk'
0
1
dout
din
(d)
din
clk
dout

Figure 2.13
A circuit (Manchester encoder) in which multiple inputs can change at the same time, subjecting the output to glitches that cannot be prevented with a combinational circuit.

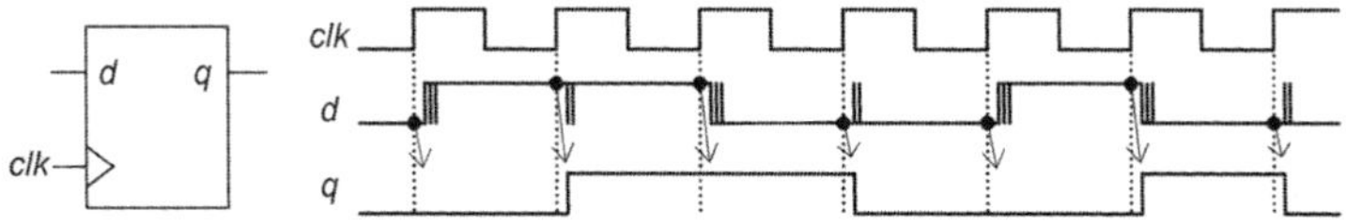
d
q
clk
clk
d
q

Figure 2.14
Glitch elimination with a flip-flop.

The problem with the solution above is that it covers only transitions in which just one input value changes. Because in actual designs multiple inputs can change (approximately) at the same time, this approach is of little practical interest.

An example in which more than one input can change is depicted in figure 2.13, which consists of a Manchester encoder. Figure 2.13a shows the circuit ports, and figure 2.13b shows the waveform that must be produced at the output (*dout* must be a '1'-to-'0' pulse when *din* = '0' or a '0'-to-'1' pulse if *din* = '1'). Looking at the waveforms, we verify that *dout* = *clk* when *din* = '0' or *dout* = *clk'* when *din* = '1', so this encoder can be implemented with a simple multiplexer, as depicted in figure 2.13c. Observe, however, in figure 2.13b, that *clk* and *din* can change at the same time, so unfixable glitches are potentially expected. Indeed, the last plot for *dout* in figure 2.13b takes into account such a possibility (due to different propagation delays), resulting in a series of glitches. Just for completeness, note that the trivial multiplexer of figure 2.13c, having the equation $dout = clk \cdot din' + clk' \cdot din$, can be implemented using just an XOR gate for *din* and *clk*, as shown in figure 2.13d.

If the combinational circuit is part of a synchronous system (as in state machines), then there is a simple—and, more importantly, systematic and guaranteed—solution for glitch elimination, which consists of passing the noisy signal through a DFF. Because glitches in a synchronous signal can only appear right after a clock edge, when such a signal is passed through a DFF the resulting output will be automatically free from glitches. This procedure is illustrated in figure 2.14, where *d* (synchronous) has glitches but *q* has not. Note that there is a price to pay, however, which is one clock cycle (if the same clock edge that produces *d* is used in the DFF) or one-half of a clock cycle (if the opposite clock edge is employed) of delay with respect to the original

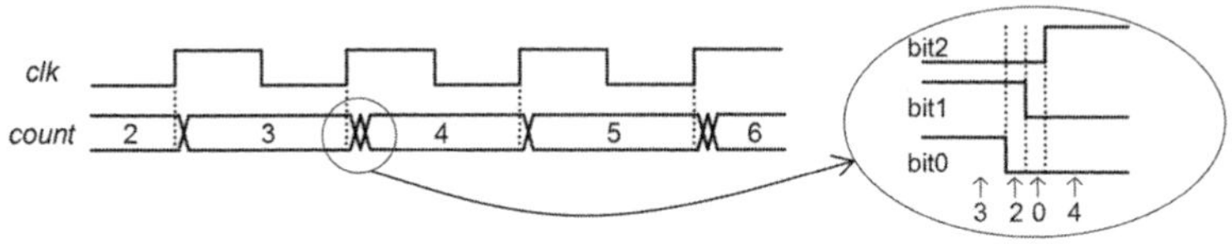
clk
count
2
3
4
5
6
bit2
bit1
bit0
3 2 0 4

Figure 2.15
Glitch interpretation (bits must be examined individually).

signal. As a final remark, observe that this technique is not OK when there are signal transitions at both clock edges, as in figure 2.13b, but that is not the case in state machines, so for FSMs this technique is fine.

We conclude this section by calling attention to a confusion that often occurs in inspecting simulation or measurement results. This is illustrated in figure 2.15, which shows a timing diagram for a three-bit counter. Because *count* is formed by more than one bit, it might exhibit glitch-like information. This, however, does not mean that actual glitches have occurred. Recall that two physical signals, due to different propagation delays, will never change exactly at the same time (and they are not perfect voltage steps anyway), so the value of *count* is expected to go through intermediate values before reaching the final value. As an example, in the inset of figure 2.15 *count* goes through $3 \rightarrow 2 \rightarrow 0 \rightarrow 4$ instead of moving straight from 3 to 4, even though glitches have not occurred. In conclusion, to inspect glitches, we must examine *only one bit* at a time.

2.6 Pipelined Implementations

Figure 2.16a shows a common architecture for high-speed synchronous systems. Each circuit—possibly designed by a different team or from an IP (intellectual property) cell—is constructed in RTL (register transfer logic) fashion, resulting in a pipelined implementation. In other words, combinational logic blocks (L_1, L_2, etc.) are followed/separated by registers (R_1, R_2, etc.) (registers are just DFF banks). The advantage of having a register as the final stage element is that the time behavior of DFFs is well known, so the overall timing response can be safely predicted, allowing the clock speed to be maximized.

To illustrate this, say that circuit 2 is constructed using only L_2-R_2-L_3. In this case, after a clock edge occurs, the total output propagation delay will be that through R_2 (t_{pCQ} of figure 2.3a) plus that through L_3. However, contrary to R_2, whose construction and parameters are known in advance, L_3 varies from one design to another and with the routing, making the time response more difficult to predict. Because the absence of R_2 increases the stage's propagation delay, the maximum clock speed gets reduced.

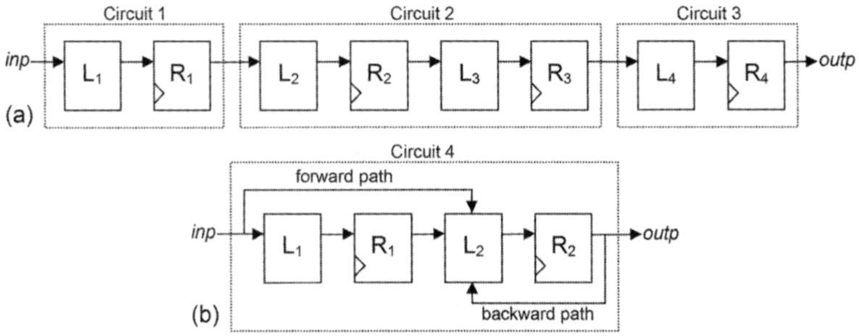
Circuit 1
Circuit 2
Circuit 3
inp
L1
R1
L2
R2
L3
R3
L4
R4
outp
(a)
Circuit 4
forward path
inp
L1
R1
L2
R2
outp
(b)
backward path

Figure 2.16
RTL pipeline (a) with loopless circuits only and (b) with a looped circuit.

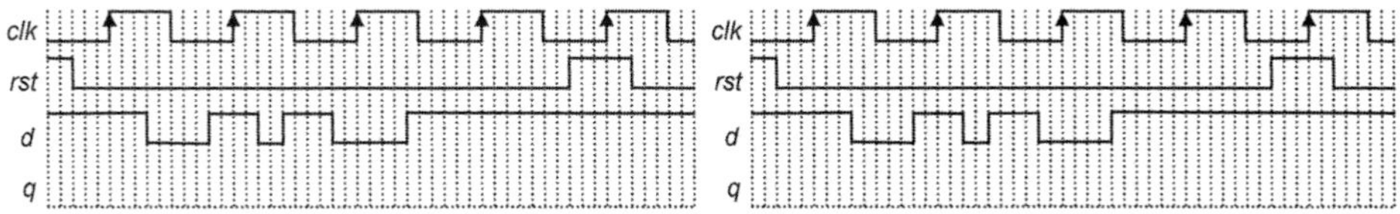
clk
rst
d
q
clk
rst
d
q

Figure 2.17

On the downside, pipelining increases the latency (number of clock cycles needed for a signal to travel though the system), which is not always acceptable. Consequently, both approaches (with and without the output register) are needed, and the choice of one or the other will be determined by the application.

The example in figure 2.16a contains only loopless stages, but looped circuits can also be found (generally, more difficult to design), as in circuit 4 in figure 2.16b. FSMs fall in the looped category.

To conclude this section, let us look at the order (input–output latency) of the synchronous circuits just described (figure 2.16). Circuits 1 and 3 are order-1 synchronous because the input–output transfer takes one clock cycle. Circuit 2 is order-2 synchronous because the transfer takes two clock cycles. Finally, circuit 4 is order-1 synchronous because its input affects L_2 directly, so its effect shows up at the output after just one clock cycle (L_2-R_2 pair).

2.7 Exercises

Exercise 2.1: DFF Response

Figure 2.17 shows waveforms for the clock, reset, and data inputs to the DFF of figure 2.2a.

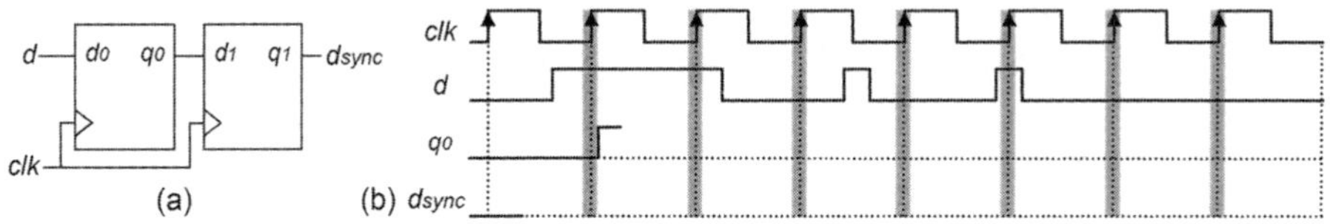
d
d0
q0
d1
q1
dsync
clk
(a)
clk
d
q0
(b) dsync

Figure 2.18

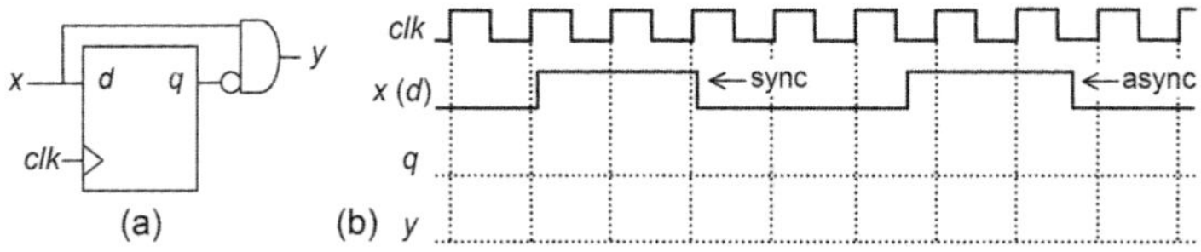
x
d
q
y
clk
(a)
clk
x (d)
←sync
←async
q
(b) y

Figure 2.19

a) Sketch, on the left, the DFF's functional response.

b) Sketch, on the right, the timing response. Assume $t_{pCQ(LH)}$ = 2 ns, $t_{pCQ(HL)}$ = 3 ns, and t_{pRQ} = 1 ns (the vertical lines are 1 ns apart).

Exercise 2.2: Metastability and Synchronizers

A popular synchronizer was presented in figure 2.6b and repeated in figure 2.18 along with an illustrative timing diagram.

a) What do the gray areas in figure 2.18b represent?

b) Which time parameters define the aperture window's width?

c) Is it desirable that the aperture window be as narrow as possible or as wide as possible?

d) Why must *d* remain stable during that time interval? What is metastability?

e) Why can synchronizers reduce the effect of metastability?

f) Given the asynchronous input *d* shown in the figure, draw the waveforms for q_0 and d_{sync}. (The initial part of q_0 was already drawn; the delay included between the clock edge and the signal edge is $t_{pCQ(LH)}$.)

g) At which positive clock edge (first, second, etc.) after *d* goes up does the signal actually delivered to the FSM (d_{sync}) go up?

h) Two short pulses (lasting less than one clock period) are included in the *d* waveform. Are they always detected? Explain.

Exercise 2.3: Basic One-Shot Circuit

Figure 2.19a shows the same elementary one-shot circuit seen in figure 2.10, which must produce at the output a pulse with a fixed (one clock period) duration every time the input goes up.

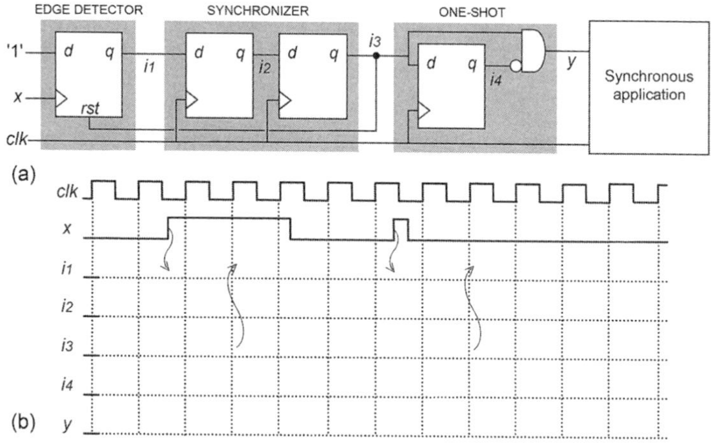
EDGE DETECTOR
SYNCHRONIZER
ONE-SHOT
'1'
x
clk
d
q
rst
i1
i2
i3
i4
y
Synchronous application
(a)
(b)
clk
x
i1
i2
i3
i4
y

Figure 2.20

a) It is said in section 2.4 that this circuit is fine only if the input is synchronous. In figure 2.19b, two illustrative pulses are shown for x, the first assumed to be synchronous (produced by the same clock that commands this DFF; the delays between the clock edges and the pulse edges are just t_{pCQ}) but the second not. Draw the waveforms for q and y, and confirm what was said about the output pulse duration.

b) It is also said that the input must last at least one clock period. Why?

c) Why is this circuit called a "one-shot" circuit? Does x need to return to '0' for it to produce the intended pulse?

Exercise 2.4: Fast Synchronized One-Shot Circuit #1

Figure 2.20a shows the same arrangement of figure 2.10, implementing a fast one-shot circuit with asynchronous, and possibly short, input. This circuit is capable of detecting input pulses shorter (and also longer, of course) than the clock period, producing at the output a pulse whose duration is always one clock period. Figure 2.20b shows the main signals involved in this circuit, with the plots for the clock and for the input (x) already completed (some helping arrows are also included in the figure).

a) Draw the waveforms for the internal (i_1 to i_4) and output (y) signals. Do not forget to leave a little delay between a signal transition and the corresponding response.

b) What are the minimum and maximum durations of i_1 (in clock periods)? Why is the initial (edge detector) DFF also called a "stretcher"?

c) What are the durations of i_2, i_3, i_4, and y (in clock periods)?

d) At which clock edge (first, second, etc.) after x goes up does y go up?

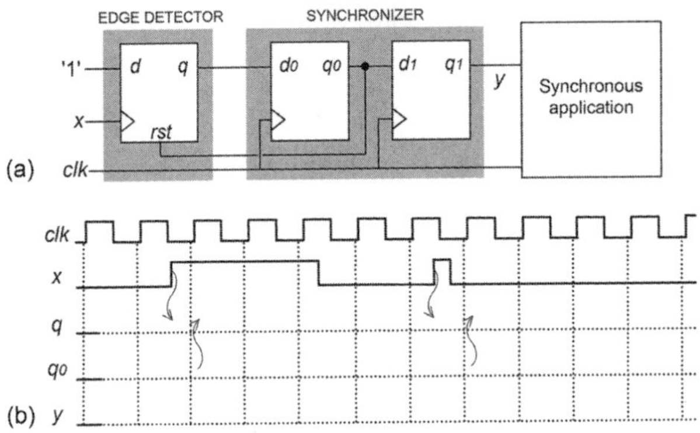
EDGE DETECTOR
SYNCHRONIZER
'1'
d
q
x
rst
(a)
clk
do
qo
d1
q1
y
Synchronous application
clk
x
q
qo
(b)
y

Figure 2.21

e) Note that the second pulse of x is shorter than one clock period and does not coincide with any positive clock transition. Is the overall circuit able to capture this pulse?

f) Is y subject to glitches? Explain.

Exercise 2.5: Fast Synchronized One-Shot Circuit #2

This exercise is an extension to the previous one. Figure 2.21a shows another synchronized one-shot circuit capable of detecting short pulses at the input while still producing a one-clock-period-long pulse at the output. All involved signals are included in figure 2.21b, with the plots for the clock and for the input (x) already completed (some helping arrows were also included in the figure).

a) Draw the waveforms for q, q_0, and y. Do not forget to leave a little delay between a signal transition and the corresponding response.

b) What are the minimum and maximum durations (in clock periods) of q?

c) What are the durations (in clock periods) of q_0 and y?

d) At which clock edge (first, second, etc.) after x goes up does y go up?

e) Note that the second pulse of x is shorter than one clock period and does not coincide with any positive clock transition. Is the overall circuit able to capture this pulse?

f) Is y subject to glitches? Explain.

g) Compare this circuit to that in figure 2.20 and comment on the respective advantages and disadvantages.

Exercise 2.6: Pipelined Construction

Figure 2.22 shows a complete two-stage pipeline, with L_i and R_i representing the logical blocks and the registers (DFFs), respectively. The propagation delays are also included in the figure, with low-to-high and high-to-low delays considered to be equal.

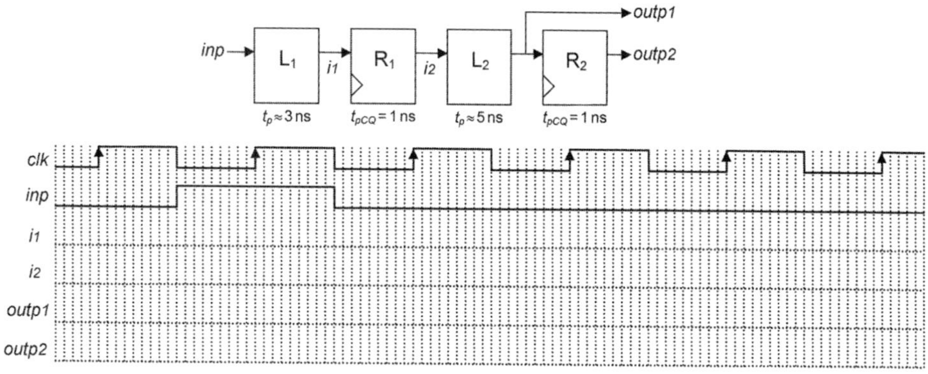
inp
L1
i1
R1
i2
L2
R2
outp1
outp2
tp≈3 ns
tpCQ=1 ns
tp≈5 ns
tpCQ=1 ns
clk
inp
i1
i2
outp1
outp2

Figure 2.22

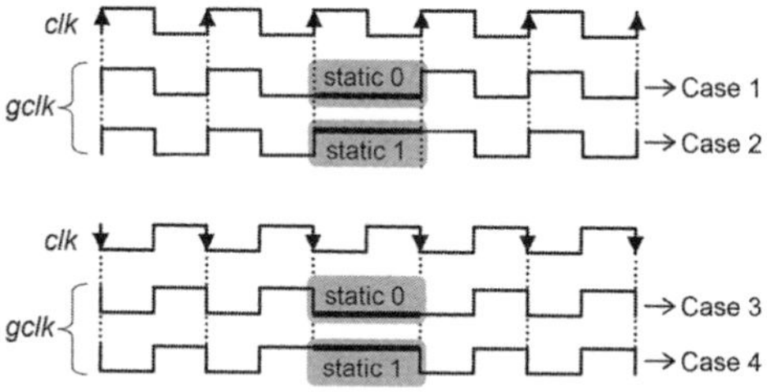
clk
gclk
static 0
static 1
→Case 1
→Case 2
clk
gclk
static 0
static 1
→Case 3
→Case 4

Figure 2.23

a) Draw the corresponding timing response in the lower part of the figure, where i_1 and i_2 represent internal signals. Consider that the vertical lines are 1 ns apart and, for sketching purposes, that L_1 and L_2 are just buffers.

b) After how many positive clock edges does the input pulse reach each output?

c) Which output (*outp1* or *outp2*) is fully pipelined? Which has superior time predictability? Why?

Exercise 2.7: Glitch-free Clock Gater

All four possible clock-gating cases are depicted in figure 2.23. Cases 1 and 2 relate to FSMs operating at the positive clock edge, whereas cases 3 and 4 are related to FSMs operating at the negative clock edge. In cases 1 and 3, the clock, when interrupted, is replaced with a zero, whereas in cases 2 and 4, it is replaced with a one.

a) Asynchronous and synchronous solutions were discussed/developed for cases 1 and 3 (see figure 2.8). Do the same for cases 2 and 4.

b) Can you devise other solutions (different from those presented in the book) for cases 1 and 3?

3 Hardware Fundamentals—Part II

3.1 Introduction

This chapter is a continuation of the previous chapter. It completes the study of fundamental hardware-related aspects that are essential to fully understand and correctly design finite state machines in hardware. Whereas chapter 2 dealt mainly with registers, chapter 3 deals with the complete state machine structure.

The topics seen in these two chapters are used, reinforced, and expanded as the subsequent chapters unfold, particularly in chapters 5 (theory for category 1 machines), 8 (theory for category 2 machines), and 11 (theory for category 3 machines).

3.2 Hardware Architectures for State Machines

State machines are looped circuits, as already illustrated in figure 1.2a. They can be of Moore or Mealy type, depending on how the input is connected to the combinational logic blocks. We want to verify here how FSMs are related to the pipeline models described in section 2.6.

Figure 3.1a shows a Moore machine, characterized by the fact that the input is connected only to block L_1 (recall that L_1 and L_2 represent logic blocks and that R_1 and R_2 are registers). Note the feedback loop from R_1 to L_1, which is the most fundamental characteristic of any FSM. Observe also that the machine itself contains only the L_1-R_1-L_2 stages, so if a full pipeline is desired, the optional register R_2 must be added. An equivalent representation is shown in figure 3.1b; the purpose of this arrangement is to emphasize the feedback loop.

Figure 3.1c shows a Mealy machine, characterized by the fact that the input is now also connected to block L_2. Note that again the machine itself contains only the L_1-R_1-L_2 stages. An equivalent representation is shown in figure 3.1d, again emphasizing the feedback loop.

The optional output register can be used to obtain a fully pipelined implementation with better time predictability and higher clock speed or for glitch removal, as seen

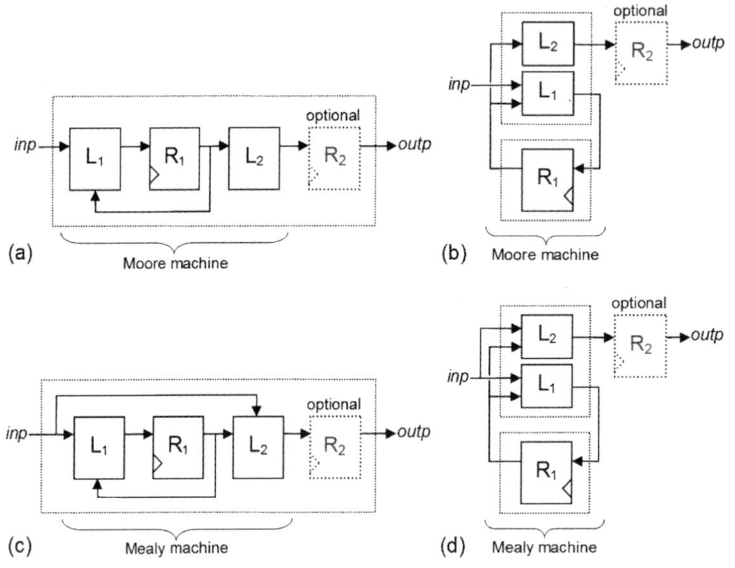
inp
L1
R1
L2
optional
R2
outp
(a)
Moore machine
(b)
Moore machine
(c)
Mealy machine
(d)
Mealy machine

Figure 3.1
Pipelined hardware representations for (a) Moore and (c) Mealy machines. (b, d) Equivalent representations highlighting the feedback loop.

in section 2.5. The consequence of this extra register is that the new output will be delayed with respect to the original output by either one clock cycle (if all registers operate at the same clock edge) or by one-half of a clock cycle (if they operate at opposite clock edges). If it is a Moore machine, and this extra delay is a problem in the application, the Mealy option should be considered. Obviously, if the extra register is being added to obtain a pipelined construction, such a latency increase is probably not a problem.

The state machine diagram of figure 3.1d is shown with additional details in figure 3.2. The original machine is depicted in figure 3.2a, while figure 3.2b shows the machine plus the optional output register (identified as *out-registered FSM* or *pipelined FSM*). Note the following in figure 3.2a:

1) The lower section is purely sequential (contains only DFFs) and comprises the state register (it stores the state of the FSM). Therefore, clock and reset signals are connected only to that section.
2) The signal stored in the DFFs is called *pr_state* (present state), whereas the signal to be stored in the DFFs at the next clock tick is called *nx_state* (next state).
3) The upper section is purely combinational (hence memoryless).
4) The upper section is divided into two parts, one producing the actual (outward) output, the other producing the circuit's next state.

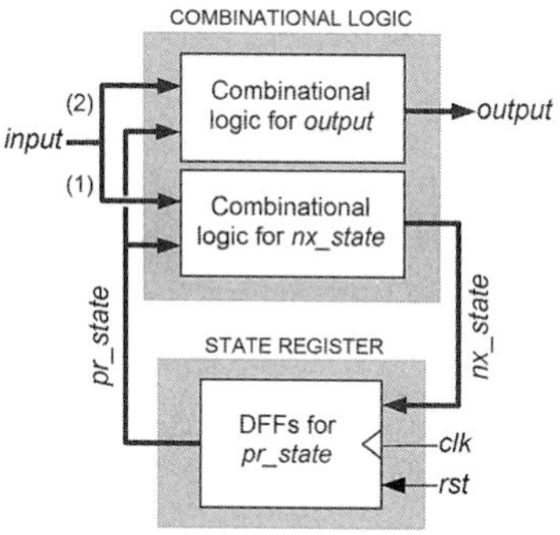
COMBINATIONAL LOGIC
Combinational logic for output
Combinational logic for nx_state
output
input
(2)
(1)
pr_state
nx_state
STATE REGISTER
DFFs for pr_state
clk
rst

(a) Pure Moore or Mealy machine

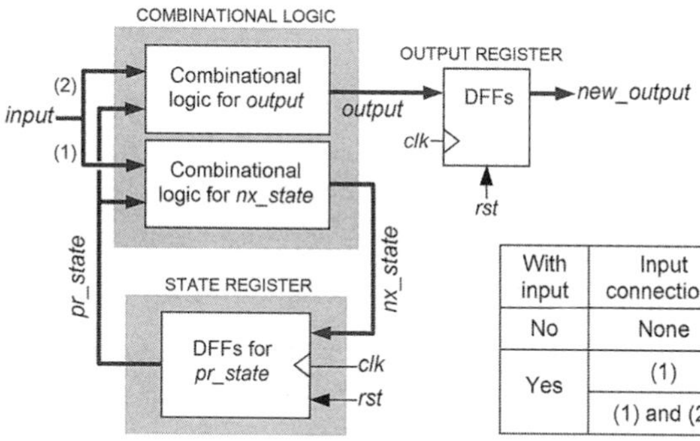
COMBINATIONAL LOGIC
Combinational logic for output
Combinational logic for nx_state
OUTPUT REGISTER
DFFs
new_output
output
clk
rst
input
(2)
(1)
pr_state
nx_state
STATE REGISTER
DFFs for pr_state
clk
rst

(b) Out-registered Moore or Mealy machine

With input	Input connections	Machine type
No	None	Moore
Yes	(1)	Moore
	(1) and (2)	Mealy

(c) Input connections x machine types

Figure 3.2
FSM architectures and input connection options.

5) The upper section has two inputs. That called *input* is external and, depending on the application, might not exist (as in the example of figure 1.2b). That called *pr_state* is internal and mandatory.
6) There are three input options (figure 3.2c): with no external input (Moore); with the input connected only to the *nx_state* logic block (connection 1; Moore); and with the input connected to both logic blocks (connections 1 and 2; Mealy).
7) Finally, note the extra register at the output (figure 3.2b) for glitch removal or pipelined implementation.

3.3 Fundamental Design Technique for Moore Machines

This section describes a fundamental design technique for Moore machines. It is a "by hand" design; in the succeeding chapters, the designs are developed with VHDL and SystemVerilog.

As seen above, from a hardware perspective a Moore machine can be represented as in figures 3.2a,b, but having only connection 1 or no external input at all (except, of course, for clock and reset). The corresponding design procedure, consisting of five steps, is summarized below. The first four steps relate to the FSM proper, and the last step regards the optional output register.

Step 1: Draw the state transition diagram.
Step 2: Based on the state diagram, write two truth tables, one for the next state and the other for the output. Then rearrange the truth tables, replacing the state names with signal names (q for flip-flop outputs, d for flip-flop inputs) and using corresponding binary values. To do this, choose first the encoding style (described in section 3.7).

Step 3: Extract, from the rearranged truth tables, the optimal Boolean expressions for *nx_state* and for the output.
Step 4: Draw the corresponding circuit, placing all flip-flops (DFFs only) in the lower section and the combinational logic for the expressions derived above in the upper section (as in figure 3.2a).
Step 5 (optional): Analyze the application and include the extra register (for glitch removal or pipelining) if you conclude that it is necessary.

To illustrate this design technique, let us consider the circuit of figure 3.3a, which must detect the sequence "010" in the single-bit data stream x, producing y = '1' at the output when such a sequence occurs. As depicted in figure 3.3b, overlaps are not allowed (if they were allowed, the trivial solution of figure 3.3c could be used).

Step 1: The corresponding Moore diagram is presented in figure 3.4a.
Step 2: The truth tables for *nx_state* and y are in figures 3.4b,c. To make the procedure clearer, the tables were written first using the state names, based directly on the diagram of figure 3.4a; then they were rearranged using signal names (q for flip-flop outputs, d for flip-flop inputs) and binary values. Reset is connected directly to the flip-flops, so it does not appear here.
Step 3: From the truth tables, the Karnaugh maps of figure 3.4d are drawn, from which the expressions for d_1, d_0, and y are obtained. Note that because this is a Moore machine, the input should not affect the output directly, which can be confirmed in the expression for y ($= q_1 \cdot q_0$), which does not contain x.
Step 4: Based on the expressions obtained above, the circuit is drawn (figure 3.4e).
Step 5: If glitches are not acceptable in this application, we must analyze the circuit to check whether the implementation is subject to glitches. In the present example, we saw that $y = q_1 \cdot q_0$. Moreover, we know that q_1 and q_0 can both change at the same time (see in figure 3.4c that the transition between states *one* and *two* requires both to change, but the output must remain low in both cases; depending on the routing inside the chip, q_1 might go to '1' before q_0 comes down to '0', so the AND gate will produce a momentary '1'—a glitch). The output DFF of figure 3.2b can then be added

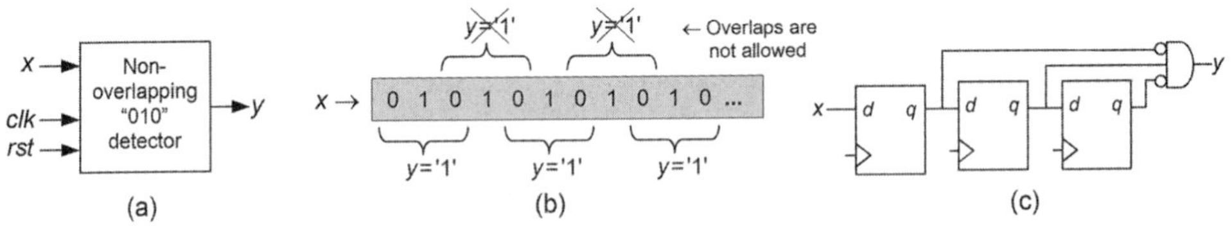
x
clk
rst
Non-overlapping "010" detector
y
(a)
y≠'1'
y≠'1'
← Overlaps are not allowed
x →
0 1 0 1 0 1 0 1 0 1 0 ...
y='1'
y='1'
y='1'
(b)
x
d q
d q
d q
y
(c)

Figure 3.3
FSM that detects the sequence "010". (a) Circuit ports. (b) Operation example (overlaps not allowed). (c) Trivial solution if overlaps were allowed.

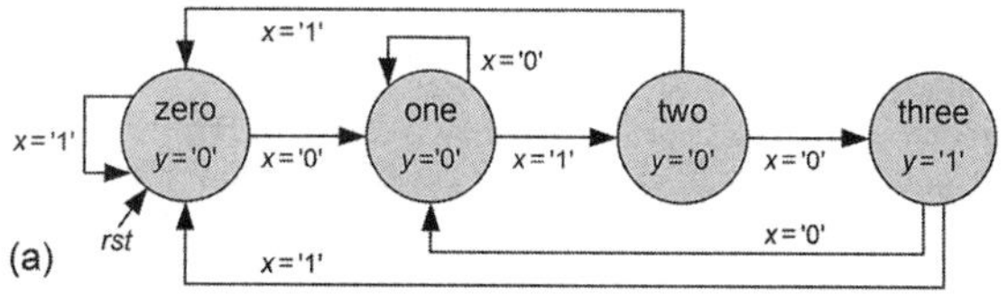
x='1'
x='0'
zero
y='0'
one
y='0'
two
y='0'
three
y='1'
x='1'
x='0'
x='1'
x='0'
rst
x='0'
x='1'
(a)

Truth table for next state

pr_state	x	nx_state
zero	0	one
	1	zero
one	0	one
	1	two
two	0	three
	1	zero
three	0	one
	1	zero

→

Truth table for next state

pr_state q_1 q_0	x	nx_state d_1 d_0
0 0	0	0 1
	1	0 0
0 1	0	0 1
	1	1 0
1 0	0	1 1
	1	0 0
1 1	0	0 1
	1	0 0

(b)

Truth table for output

pr_state	y
zero	0
one	0
two	0
three	1

→

Truth table for output

pr_state q_1 q_0	y
0 0	0
0 1	0
1 0	0
1 1	1

(c)

Karnaugh for d_1:

x \ q_1 q_0	00	01	11	10
0	0	0	0	1
1	0	1	0	0

$d_1 = q_1' \cdot q_0 \cdot x + q_1 \cdot q_0' \cdot x'$

Karnaugh for d_0:

x \ q_1 q_0	00	01	11	10
0	1	1	1	1
1	0	0	0	0

$d_0 = x'$

Karnaugh for y:

q_0 \ q_1	0	1
0	0	0
1	0	1

$y = q_1 \cdot q_0$

(d)

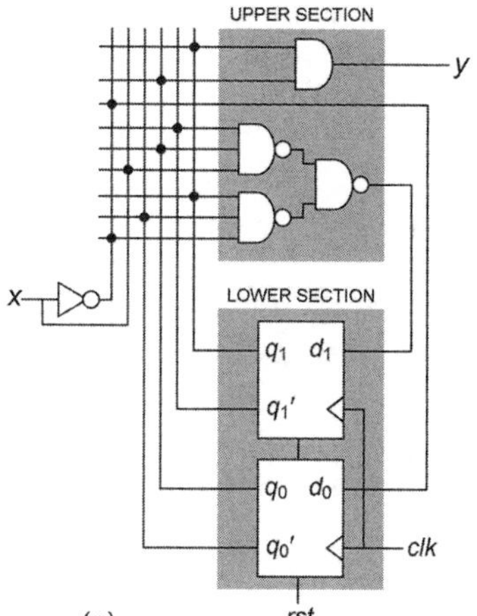
UPPER SECTION
y
LOWER SECTION
x
q1 d1
q1'
q0 d0
q0'
clk
rst

(e)

Figure 3.4

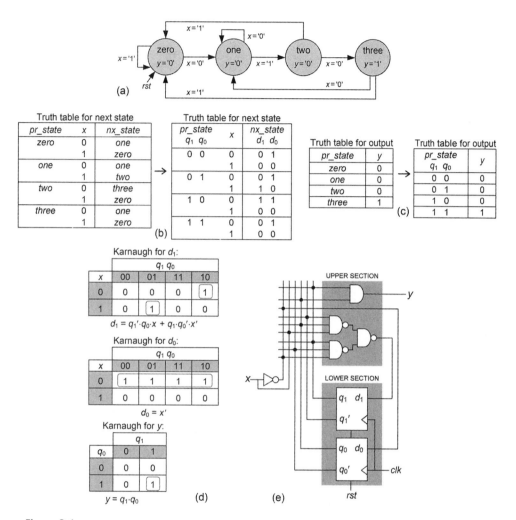

Moore machine for a nonoverlapping "010" detector. (a) State transition diagram. (b, c) Truth tables for next state and output. (d) Corresponding Karnaugh maps and minimal Boolean expressions. (e) Resulting circuit.

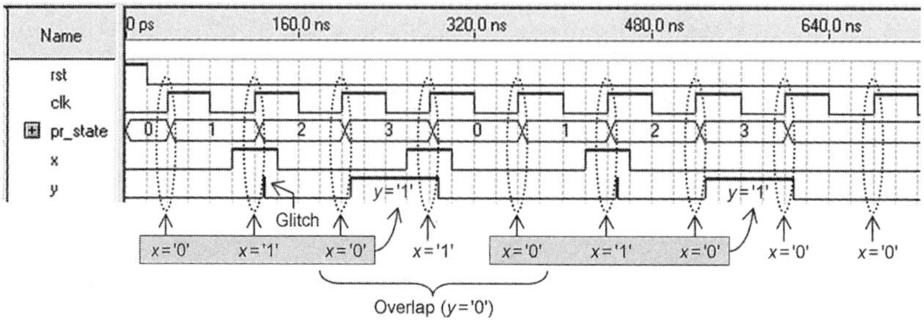
Name
0 ps
160,0 ns
320,0 ns
480,0 ns
640,0 ns
rst
clk
pr_state
x
y
0
1
2
3
0
1
2
3
y='1'
y='1'
Glitch
x='0'
x='1'
x='0'
x='1'
x='0'
x='1'
x='0'
x='0'
x='0'
Overlap (y='0')

Figure 3.5
Simulation results from the nonoverlapping "010" detector of figure 3.4e (Moore type).

to render a glitch-free *y*. (Recall the comments made earlier on increased latency and possible use of a Mealy machine in this kind of situation.)

Simulation results from the circuit of figure 3.4e, without the extra DFF, synthesized using VHDL, are shown in figure 3.5. Note the following (expected) results:

1) The output changes only at (positive) clock edges (Moore machines are synchronous).
2) The output goes to '1' at the first (positive) clock edge after the sequence "010" occurs.
3) Overlaps do not cause the output to go to '1'.
4) Without the optional output DFF, glitches do occur at the output.

3.4 Fundamental Design Technique for Mealy Machines

This section describes a fundamental design technique for Mealy machines. It is a "by hand" design; in the succeeding chapters, the designs are developed with VHDL and SystemVerilog.

As seen above, from a hardware perspective a Mealy machine can be represented as in figures 3.2a,b with both input connections (1 and 2). The corresponding design procedure is the same seen for Moore machines, with just one difference, in step 2, as follows.

Step 2: Based on the state diagram, *write a single truth table,* including both the next state and the output. Then rearrange the truth table . . . etc.

If the Moore-to-Mealy conversion technique introduced in section 1.7 is applied to the Moore diagram of figure 3.4a, the Mealy diagram of figure 3.6a results. Its truth table (including both *nx_state* and *y*) is shown in figure 3.6b. From this table, the

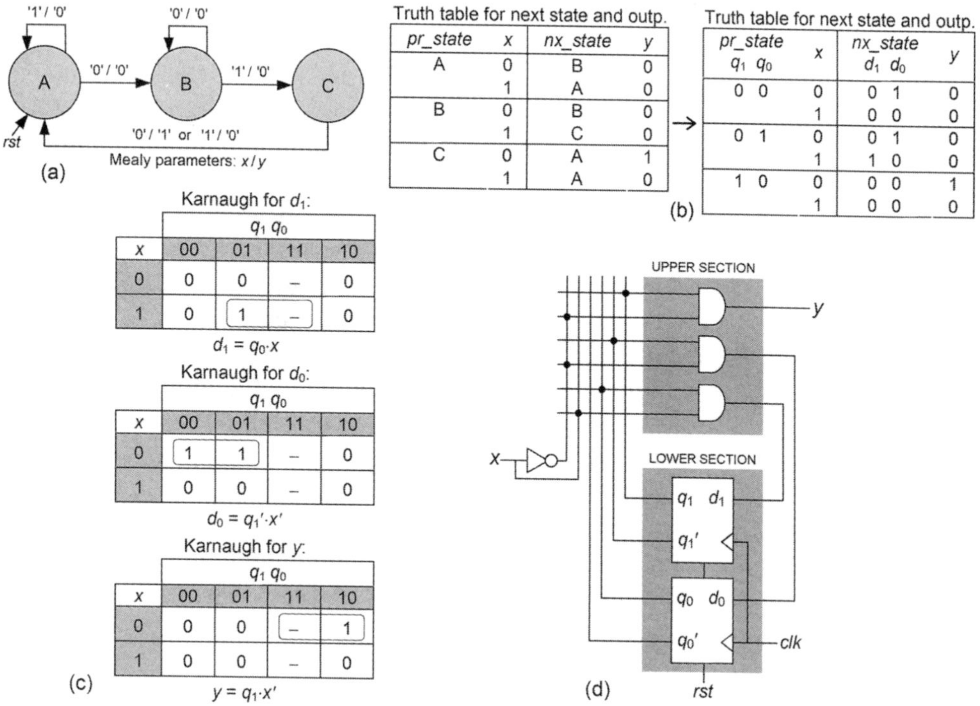
'1' / '0'
'0' / '0'
A
'0' / '0'
B
'1' / '0'
C
rst
'0' / '1' or '1' / '0'
Mealy parameters: x / y
(a)
Truth table for next state and outp.
pr_state x nx_state y
A 0 B 0
1 A 0
B 0 B 0
1 C 0
C 0 A 1
1 A 0
Truth table for next state and outp.
pr_state q1 q0 x nx_state d1 d0 y
0 0 0 0 1 0
1 0 0 0
0 1 0 0 1 0
1 1 0 0
1 0 0 0 0 1
1 0 0 0
(b)
Karnaugh for d1:
d1 = q0·x
Karnaugh for d0:
d0 = q1'·x'
Karnaugh for y:
y = q1·x'
(c)
UPPER SECTION
y
x
LOWER SECTION
q1 d1
q1'
q0 d0
q0'
clk
rst
(d)

Figure 3.6
Mealy machine for a nonovelapping "010" detector. (a) State transition diagram. (b) Truth table for both next state and output. (c) Corresponding Karnaugh maps and minimal Boolean expressions. (d) Resulting circuit.

Karnaugh maps and optimal Boolean expressions of figure 3.6c result. Finally, the corresponding Mealy circuit is presented in figure 3.6d.

Note that, contrary to the Moore case (previous section), here the expression for the output (y) does include the input (x); that is, it is now $y = q_1 \cdot x'$, whereas in the Moore case it was $y = q_1 \cdot q_0$. This means that y can change asynchronously (that is, as soon as x changes, independently from the clock), which occurs when q_1 = '1', because then $y = x'$.

The decision on using or not using the extra register (step 5) is similar to that for Moore machines. However, because Mealy machines are asynchronous, if a project accepts this type of circuit, glitches are generally of no relevance. An interesting use for out-registered (pipelined) Mealy machines is to implement glitch-free Moore-like circuits (details are shown in the next section).

3.5 Moore versus Mealy Time Behavior

It is very important to understand the differences between Moore and Mealy solutions well. To do that, let us compare the timing responses of the Moore and Mealy circuits designed above for the nonoverlapping "010" detector (figures 3.4e and 3.6d, respectively).

An example of timing response for the Moore circuit is presented in figure 3.7a. The input sequence is *x* = {'0', '1', '0', '1', '0', '1', '0', '0'}. Following the state transition diagram of figure 3.4a, the sequence obtained for the present state is *pr_state* = {*one, two, three, zero, one, two, three, one*}, shown in the corresponding plot of figure 3.7a. Because *y* is '1' when the machine is in state *three* and '0' elsewhere, the resulting output sequence (after a small propagation delay) is *y* = {'0', '0', '1', '0', '0', '0', '1', '0'}. Note that *y* is indeed synchronous because it can change only when the clock ticks (at positive clock transitions in this example), and then remains fixed during the whole clock period. Because *y* is synchronous, glitches in *y* can happen only right after (positive) clock transitions (as already illustrated in figure 3.5). If the optional output register is used (to remove glitches, for example), then a registered output *y_reg* results, which is simply a (clean) shifted version of *y*. In other words, *y* goes to '1' at the next clock edge after the condition "010" occurs, while *y_reg* goes to '1' at the second clock edge after that sequence happens.

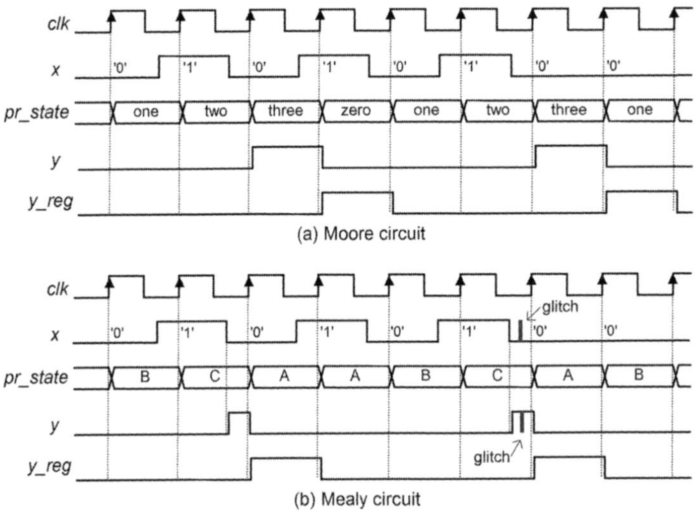
clk
x
'0'
'1'
'0'
'1'
'0'
'1'
'0'
'0'
pr_state
one
two
three
zero
one
two
three
one
y
y_reg
(a) Moore circuit
clk
x
'0'
'1'
'0'
'1'
'0'
'1'
'0'
'0'
glitch
pr_state
B
C
A
A
B
C
A
B
y
y_reg
glitch
(b) Mealy circuit

Figure 3.7
Time behavior of the nonoverlapping "010" detector. (a) For the Moore circuit of figure 3.4e. (b) For the Mealy circuit of figure 3.6d.

A similar analysis is presented for the Mealy circuit in figure 3.7b. The input sequence is again x = {'0', '1', '0', '1', '0', '1', '0', '0'}. Following the state transition diagram of figure 3.6a, the sequence obtained for the present state is *pr_state* = {B, C, A, A, B, C, A, B}. As shown in the state diagram, the output is '0' in states A and B, and it is x' in state C, thus resulting (after a small propagation delay) in the output sequence y = {'0', x', '0', '0', '0', x', '0', '0'}. Note that x appears in the list for y, which was expected because in a Mealy machine the input can affect the output directly. For the same reason, this machine is asynchronous (note that in state C, y changes independently of the clock). Observe also that, because it is asynchronous, the shape of y can be quite strange, with values lasting less than one clock period. On the other hand, observe the interesting shape of *y_reg*, which is exactly the same as the shape of y in the Moore case. This means that when we want to get rid of glitches and the consequent extra delay of an out-registered (pipelined) Moore machine is not acceptable in the application, an out-registered (pipelined) Mealy machine can be used.

Another drawback of the original Mealy machine is that input glitches can propagate to the output, as depicted in the plot for y in figure 3.7b.

3.6 State Machine Categories

We have already seen that state machines can be classified into two types, based on their *input connections*, as follows.

1) *Moore machines:* The input, if it exists, is connected only to the logic block that computes the next state.
2) *Mealy machines:* The input is connected to both logic blocks, that is, for the next state and for the actual output.

In this section we introduce a new classification, into three categories, based on the *transition types* and *nature of the outputs*.

As mentioned in section 1.2 (see Hardware- versus Software-Implemented State Machines), designing a hardware-implemented FSM is generally (much) more complex than designing a software-implemented FSM. To ease such a task, a very important new classification, from a hardware perspective and covering *any* state machine, is introduced here (this classification dictates the organization of the subsequent chapters). Figure 3.8 is used as an illustration.

Category 1: Regular (pure) state machines (studied in chapters 5–7): This category, illustrated in figure 3.8a, is the simplest. It consists of machines with only untimed transitions and outputs that do not depend on previous (past) output values.
Category 2: Timed state machines (studied in chapters 8–10): These consist of machines with one or more transitions that depend on time, but still having only outputs that

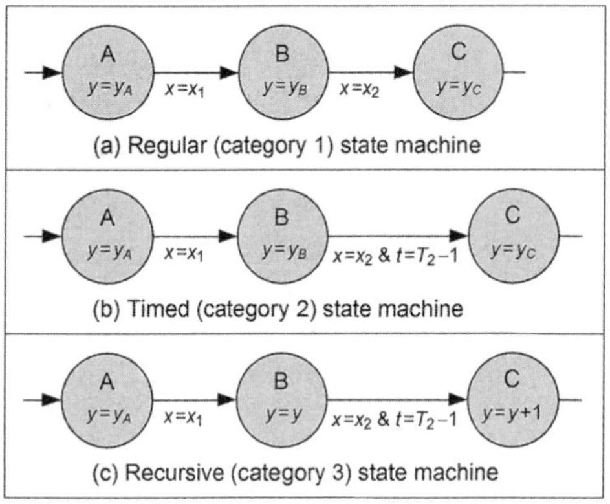
A
y=yA
x=x1
B
y=yB
x=x2
C
y=yC
(a) Regular (category 1) state machine
A
y=yA
x=x1
B
y=yB
x=x2 & t=T2−1
C
y=yC
(b) Timed (category 2) state machine
A
y=yA
x=x1
B
y=y
x=x2 & t=T2−1
C
y=y+1
(c) Recursive (category 3) state machine

Figure 3.8
State machine categories.

do not depend on previous (past) output values. In the example of figure 3.8b, the first transition is conditional, but the second is conditional-timed. Recall that, by default, the timer is zeroed every time the FSM changes state and is kept stopped at zero in the states where it is not needed (states A and C in figure 3.8b). Category 2 machines can have all four types of transitions (conditional, timed, conditional-timed, and unconditional).

Category 3: Recursive state machines (studied in chapters 11–13): In this case, illustrated in figure 3.8c, the transitions can be timed, as in category 2, but now one or more outputs depend on previous (past) output values. Such dependency is usually expressed by means of recursive equations (the output is a function of itself). In this example the output must keep in state B the same value that it had when the machine left state A, whereas in state C the output must present the incremented version of the value that it had when the machine left state B. This implies that some sort of extra memory is required because the expressions $y = y$ and $y = y + 1$ can only be evaluated if the value of y is available somewhere. Because this is a hardware implementation, such auxiliary memory must be provided along with the state machine.

As will be seen in coming chapters, the classification presented above will ease immensely the design of hardware-based state machines (and no other classification is needed). The two fundamental decisions before starting a design are then the following:

1) First, the state machine category (regular, timed, or recursive);
2) Second, the state machine type (Moore or Mealy).

3.7 State-Encoding Options

This section describes the main codes used for encoding the states of an FSM. The most common encoding alternatives are *sequential* (also called *binary*), *Gray*, *Johnson*, and *one-hot*, all illustrated in figure 3.9a for an eight-state FSM. Note that the first two require three bits, the third requires four bits, and the last one requires eight bits.

To illustrate the encoding options further, let us consider a machine with the following five states (using VHDL notation):

type state **is** (A, B, C, D, E);

3.7.1 Sequential Binary Encoding

The states are encoded using the conventional binary code (increasing order of corresponding decimal values; see figure 3.9a). For the type *state* above, three bits would be needed, resulting A = "000" (decimal value = 0), B = "001" (= 1), C = "010" (= 2), D = "011" (= 3), and E = "100" (= 4).

The advantage of this encoding is that it requires the smallest number of flip-flops; with *N* flip-flops (*N* bits), up to 2^N states can be encoded. The disadvantage is that it might require more combinational logic than other encoding options (illustrated in exercises 3.2 and 3.3), so the resulting circuit might be slightly slower.

(a)

State	Encoding			
	Sequential	Gray	Johnson	One-hot
0	000	000	0000	00000001
1	001	001	0001	00000010
2	010	011	0011	00000100
3	011	010	0111	00001000
4	100	110	1111	00010000
5	101	111	1110	00100000
6	110	101	1100	01000000
7	111	100	1000	10000000

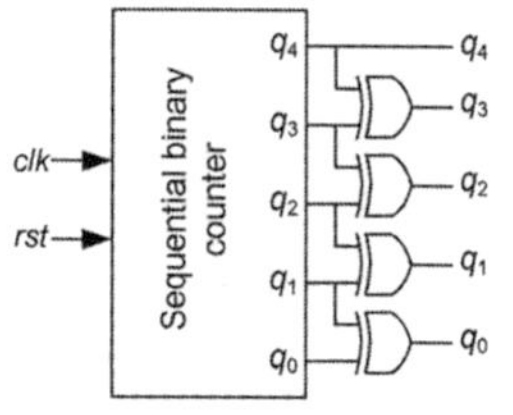
clk
rst
Sequential binary counter
q4
q3
q2
q1
q0

(b)

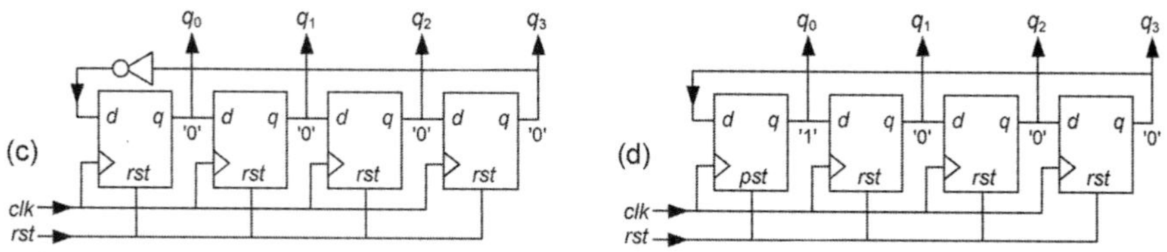
(c)
q0
q1
q2
q3
d
q
rst
'0'
clk
rst
(d)
pst
'1'

Figure 3.9

(a) Main encoding options for an eight-state machine. (b) Regular sequential binary counter with outputs converted to Gray code. (c) Johnson counter. (d) One-hot counter.

3.7.2 One-Hot Encoding

At the other extreme is the one-hot code, in which only one bit is high in each codeword, so with *N* flip-flops only *N* states can be encoded (see figure 3.9a). For the type *state* above, five bits would be needed, resulting in A = "00001", B = "00010", C = "00100", D = "01000", and E = "10000".

This code demands the largest number of flip-flops, but the amount of combinational logic tends to be smaller than that of other encodings (illustrated in exercise 3.2), often leading to a slightly faster implementation. For big machines (say, over 40 or 50 states), the hardware for this type of encoding tends to be prohibitively large.

Just to illustrate the one-hot code, figure 3.9d depicts a one-hot counter, which consists simply of a shift register with a direct feedback loop. Note that the initial state is $q_3q_2q_1q_0$ = "0001" because the reset signal is connected to the reset port of three DFFs and to the preset port of the other.

3.7.3 Johnson Encoding

This is an implementation in between the two above. With *N* flip-flops, 2*N* states can be encoded (see figure 3.9a). It does not require much more combinational logic than the one-hot alternative, but it can encode twice the number of states. Each codeword is obtained by circularly shifting the previous codeword to the left and inverting the incoming bit. For the type *state* above, three bits would be needed, resulting A = "000", B = "001", C = "011", D = "111", and E = "110".

Just to illustrate the Johnson code, a Johnson counter is depicted in figure 3.9c, which consists simply of a shift register with an inverter in the feedback loop.

An important property of this code is that the Hamming distance (number of bits that are different) between any two adjacent codewords is just 1 (see figure 3.9a), so it can be useful in the same applications as the Gray code, described below.

3.7.4 Gray Encoding

Gray code is similar to the sequential code in the sense that it too requires the least number of flip-flops (with *N* flip-flops, up to 2^N states can be encoded), but the amount of combinational logic can be slightly larger (illustrated in exercise 3.3) and the speed slightly lower.

This code too exhibits the property of unitary Hamming distance between any two adjacent codewords, useful in certain implementations involving multiple clock domains (recall comments of section 2.3). Because of this property, a Gray *counter* is free from glitches during state transitions (except when returning to the initial state if the counter's modulo is not a power of 2); consequently, if a Gray-encoded FSM has a long path without branching, transitions along that path are glitch-free.

In a Gray code, each codeword is obtained by modifying the value of the rightmost bit in the previous codeword such that a new codeword results (see figure 3.9a). For the type *state* above, three bits would be needed, resulting in A = "000", B = "001", C = "011", D = "010", and E = "110".

Just to illustrate the Gray code, a Gray counter is presented in figure 3.9b, which consists simply of a regular sequential counter whose output is converted into Gray code by means of the following expressions (see XOR gates in figure 3.9b): $q(N-1)_{Gray} = q(N-1)_{Seq}$; $q(i)_{Gray} = q(i+1)_{Seq} \oplus q(i)_{Seq}$ for $i = N-2$ to $i = 0$.

3.7.5 Modified One-Hot Encoding with All-Zero State

Figure 3.10a shows an example using the true one-hot code described above for a four-bit system (a four-state FSM). As expected, the bits of *pr_state* are {"0001", "0010", "0100", "1000"}. A modified version, with bit zero inverted, is depicted in figure 3.10b. The encoding is now {"0000", "0011", "0101", "1001"}, thus containing the all-zero codeword. This code has the same properties as the true one-hot code in the sense that it too has a Hamming distance of 2 between any two codewords, and all codewords can be identified based on a single bit.

The alternative of figure 3.10b is used, for example, by Altera's Quartus II compiler when synthesizing state machines using the one-hot option. The reason for doing so is that all DFFs in Altera's FPGAs (and Xilinx's as well for that matter) are initialized to a low output on power-up, so if an explicit reset port was not included in the design, the machine will still be able to start from a specific state, avoiding improper initialization and deadlock. More details on this are seen in sections 3.8 and 3.9, which discuss the importance of reset in FSMs and how to implement safe FSMs.

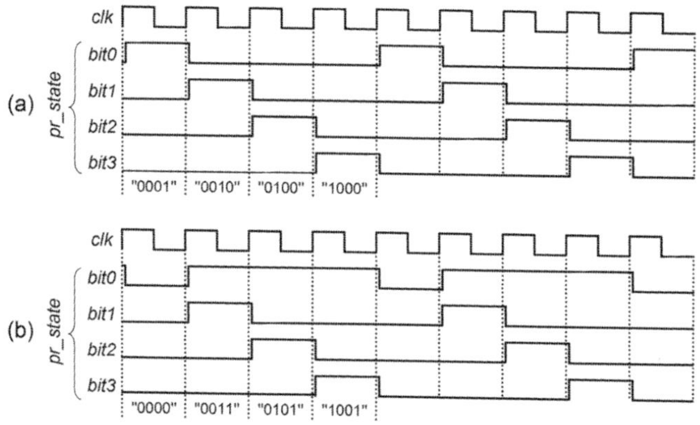
(a)
pr_state
clk
bit0
bit1
bit2
bit3
"0001"
"0010"
"0100"
"1000"
(b)
pr_state
clk
bit0
bit1
bit2
bit3
"0000"
"0011"
"0101"
"1001"

Figure 3.10

(a) True one-hot encoding. (b) Modified one-hot encoding (bit zero inverted), containing the all-zero codeword.

3.7.6 Other Encoding Schemes

Besides the encoding schemes described above, VHDL and SystemVerilog synthesis compilers have at least two other options, known as *user* and *auto*. The former is a user-defined encoding (the codeword for each state is specified by the user), whereas the latter is used to let the compiler choose the best encoding scheme based on the target device. Typically, *auto* employs sequential encoding for small machines (for example, up to four or five states), then one-hot for medium-sized machines (for example, up to 40 or 50 states), and finally sequential again (or an equivalent, such as Gray) for larger machines. In general, *auto* is the compiler's default option.

The one-hot style is common in applications where flip-flops are abundant, such as field programmable gate arrays (FPGAs), whereas minimal-bit encodings (such as sequential and Gray) are common in complex programmable logic devices (CPLDs) and in compact, low-cost application-specific integrated circuits (ASICs).

Chapters 6 and 7 show how to select the encoding scheme when using VHDL or SystemVerilog, respectively.

3.8 The Need for Reset

If no reset signal is provided and no intentional circuit asymmetry exists (such that a specific output state is favored), the initial state (output either low or high) of a flip-flop, on power-up, might be arbitrary. Because flip-flops are used to construct the state register, the machine's initial state would then also be arbitrary. In this case, one of two situations will result: either the initial state is *internal* (that is, belongs) to the machine or is *external* (does not belong) to it. Of course, if N bits are used to encode a machine that has 2^N states, then the initial state can only belong to the machine.

When the initial state is internal, deadlock can still happen, but only in rare cases, so the usual main consequence is a possibly undesirable sequence of events during the first few state transitions. If the initial state is external, however, deadlock is much more likely, obviously in addition to the possibly undesirable sequence of events during the first few state transitions after the system converges to one of the FSM states (assuming that deadlock has not occurred).

To further illustrate this discussion, let us consider the four-state counter of figure 3.11a, whose states are encoded using one-hot code (the corresponding $pr\_state = q_3q_2q_1q_0$ vectors are shown below the state circles). The equations for $nx\_state = d_3d_2d_1d_0$ can be easily obtained using the method described in section 3.3, resulting $d_3 = q_2$, $d_2 = q_1$, $d_1 = q_0$, and $d_0 = q_3$. Consequently, if the initial state is $q_3q_2q_1q_0$ = "0000", for example, $d_3d_2d_1d_0$ = "0000" results; because $nx\_state = pr\_state$, a deadlock then occurs. Indeed, based on the equations above, note that with one-hot encoding any time the initial state falls outside the FSM the machine gets deadlocked.

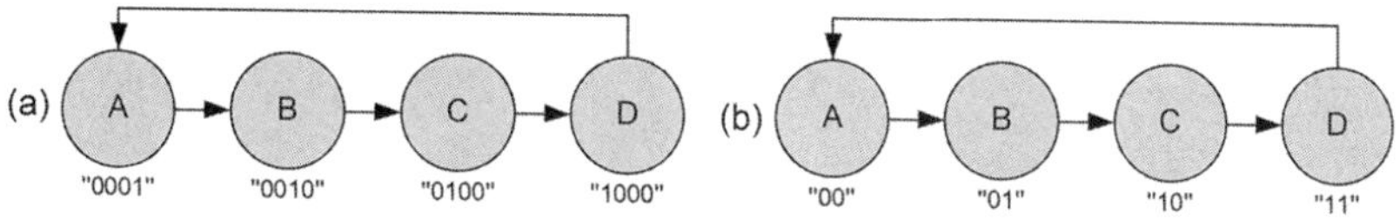
(a)
A
B
C
D
"0001"
"0010"
"0100"
"1000"
(b)
A
B
C
D
"00"
"01"
"10"
"11"

Figure 3.11
Four-state counter encoded with (a) one-hot and (b) sequential code.

Another example is shown in figure 3.11b, now using regular sequential encoding. Because the number of states is a power of two, all possible two-bit values belong to the machine; since all internal states are deadlock-free, the whole machine is deadlock-free. It is important to mention, however, that with sequential or Gray encoding full code usage is not needed for an outside initial state to eventually converge to one of the actual machine states, thus without deadlock.

Deadlock prevention is necessary in any FSM, but in many applications it is also required that the FSM start from a specific state and with proper transition control conditions. For example, one might want the FSM used to implement a traffic light controller to start from a state that keeps the lights red in all directions for a few seconds (on power-up, after an energy failure, for example) before it proceeds to its regular operation. This requires an explicit reset signal if the DFFs initial state is arbitrary.

It is important to mention that there are devices (such as Altera's and Xilinx's FPGAs) whose DFFs are automatically reset to '0' on power-up. In that case, if the all-zero codeword belongs to the code and is assigned to the intended initial state, the FSM will be reset automatically.

Still regarding the FPGAs mentioned above, whose DFFs' initial state is '0', note that that does not reset the machine automatically when one-hot encoding is used (see in figure 3.11a that "0000" is not part of the codewords list). However, contrary to Xilinx's XST synthesis compiler (of the ISE suite), which uses true one-hot code, Altera's Quartus II synthesis compiler uses the modified one-hot code seen in figure 3.10b, which does include the all-zero codeword, thus allowing automatic reset.

Figure 3.12 shows typical sources for the reset signal. In figure 3.12a it is generated by a resistor-capacitor (RC) circuit; when the power is turned on, the voltage on C is zero, so the full V_{DD} voltage is applied to the *rst* input, resetting the circuit; the voltage on C then grows (with a time constant R·C = 0.1 s), thus eventually reducing the voltage on R (*rst*) to zero and so freeing the circuit from the reset command. In figure 3.12b, a specialized integrated circuit is used to generate the reset signal; if V_{DD} falls below a predefined threshold voltage, *rst* = '1' is produced, resetting the target circuit. Finally, in figure 3.12c, the reset signal is produced by another circuit that might belong to the same system as the FSM (for example, another FSM to which this one

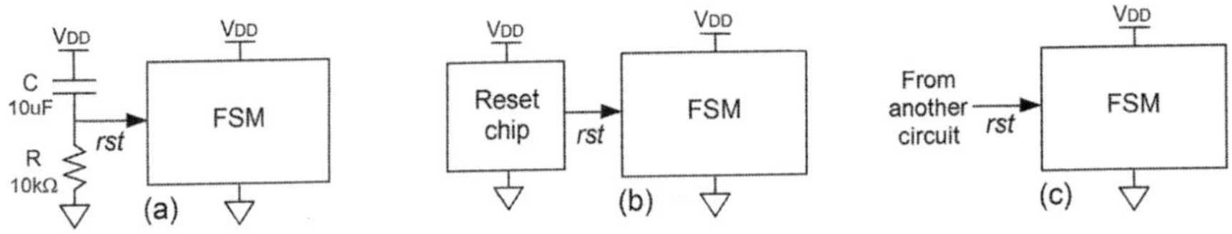
VDD
C
10uF
R
10kΩ
rst
FSM
(a)
Reset chip
rst
FSM
(b)
From another circuit
rst
FSM
(c)

Figure 3.12
Reset options (a) generated by an RC circuit, (b) generated by a specialized reset chip, and (c) generated by another circuit possibly belonging to the same system as the FSM.

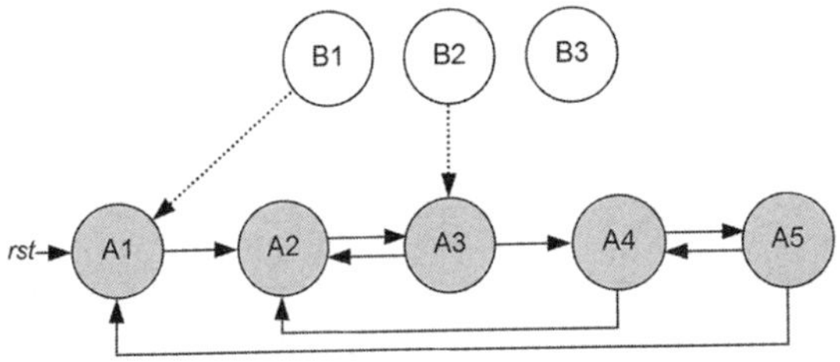
B1
B2
B3
rst
A1
A2
A3
A4
A5

Figure 3.13
FSM with a nonconvergent external state (B3 causes deadlock).

works as a secondary machine, or a watchdog circuit). The first two cases are related to the power supply, so they are power-related resets, whereas the last one can be produced at any time.

As a final comment, it is important to mention that there are applications in which reset is not necessary, either because the initial state is not important or because it is set automatically by another signal (assuming that the machine is not subject to deadlock). However, that is rarely the case, so the exclusion of reset should only be done after very careful design analysis.

Additional details on FSMs' initial state and deadlock are given in the next section.

3.9 Safe State Machines

The concept of safe state machines concerns deadlock-proof implementations.

Figure 3.13 shows an FSM with five states (A1–A5), assumed to be encoded with three-bit values, thus resulting in three states (B1–B3) *outside* the machine. The machine can start from or move to an external state when no proper reset is provided (when the DFFs' initial state is arbitrary) or because of noise during regular operation (which might flip one or more encoding bits). In this example the machine is able to automatically recover when state B1 or B2 occurs (converging to state A1 or A3, respectively), but it gets deadlocked if state B3 happens.

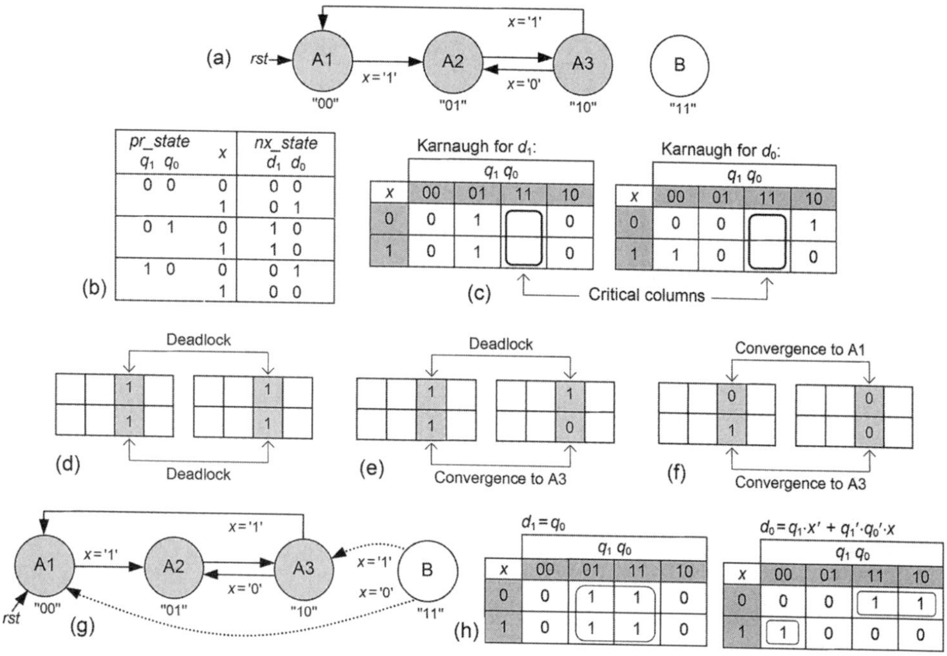
(a)
rst
A1
A2
A3
B
x='1'
x='1'
x='0'
"00"
"01"
"10"
"11"
(b)
pr_state
q1 q0
x
nx_state
d1 d0
Karnaugh for d1:
Karnaugh for d0:
(c)
Critical columns
Deadlock
Deadlock
(d)
Deadlock
(e)
Convergence to A3
Convergence to A1
(f)
Convergence to A3
(g)
(h)
d1=q0
d0=q1·x' + q1'·q0'·x

Figure 3.14
Safe-machine analysis. (a) Example with three states and sequential encoding. (b) Corresponding truth table for *nx_state*. (c) Resulting Karnaugh maps. (d–f) Three compositions for the critical (free-filling) columns. (g) Resulting system if the option in f is used. (h) Optimal expressions for *nx_state*, which coincide with the (bad) case of e.

Note that the use of an explicit reset signal only prevents the first possibility, not the effect of noise. Consequently, for a truly safe FSM, the external states must also be considered when the expressions for *nx_state* are developed. Other, less usual mechanisms also exist, including the use of a watchdog that resets the machine in case it remains in the same state longer than a predefined time limit, but this would obviously be applicable only if the machine were expected to change its state periodically.

A detailed example is presented in figure 3.14. The actual machine states are A1–A3. Because sequential encoding is used (note the codewords below the state circles in figure 3.14a), two bits are needed, resulting in one external state (B). We want to examine the conditions needed for this machine to be able to recover in case state B happens.

By the method of section 3.3, the truth table for *nx_state* shown in figure 3.14b is obtained, from which the Karnaugh maps of figure 3.14c result. Note in the latter that

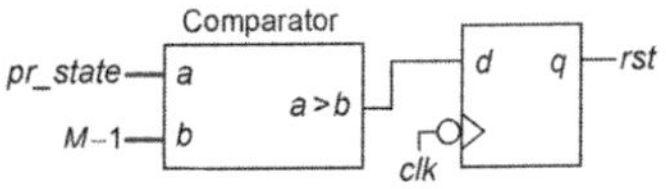
Comparator
pr_state
M−1
a
b
a>b
d
q
rst
clk

Figure 3.15
A simple solution for truly safe state machines when using sequential encoding.

the column q_1q_0 = "11" is the critical column because all elements in that column can be filled indifferently with '0' or '1', so a number of different equations can be used for d_1 and d_0. We have to make certain that the condition *nx_state* = *pr_state* (i.e., $d_1d_0 = q_1q_0$) never happens, either directly or through a loop in the external states, because then deadlock will occur.

The maps of figure 3.14c were repeated in figures 3.14d–f for three different compositions of the critical columns. In figure 3.14d, all positions were filled with '1's, so d_1d_0 = "11" results when q_1q_0 = "11" (i.e., *nx_state* = *pr_state*) for both x = '0' (upper row) and x = '1' (lower row), causing deadlock. In figure 3.14e, d_1d_0 = "11" for x = '0' (upper row) and d_1d_0 = "10" for x = '1' (lower row), so a deadlock would occur in the former, and a convergence to state A3 would occur in the latter. Finally, the case in figure 3.14f has d_1d_0 = "00" for x = '0' and d_1d_0 = "10" for x = '1', so both would converge (to A1 and A3, respectively).

Figure 3.14g shows the complete resulting state diagram (internal plus external states) in case the encoding of figure 3.14f is adopted. As seen above, B converges to A1 if x = '0' or to A3 if x = '1', so this is a truly safe implementation.

To conclude, the Karnaugh maps are repeated in figure 3.14h, in which the optimal (minimal) expressions were adopted for *nx_state* (i.e., d_1 and d_0). Note that the resulting situation is that of figure 3.14e, subject, therefore, to deadlock. In summary, in this example the optimal implementation from a hardware-saving perspective is not the best implementation from a safety point of view.

We conclude this section by showing, in figure 3.15, a simple solution that is applicable to FSMs that employ sequential encoding. M is the machine's number of states. If $M = 20$, for example, five bits are needed, and the state values will range from 0 to 19. If at any moment (either at initialization or during regular operation) *pr_state* happens to be above 19, it means that an *external* state has occurred, so a reset pulse is produced (this reset can be combined with the original reset signal, if it exists, by means of an OR gate). The flip-flop in figure 3.15 is needed because the comparator can have glitches at the output during state transitions (see, for example, figure 2.15).

3.10 Capturing the First Bit

This section discusses a common hardware need that occurs, for example, when data is processed serially. In such cases there is often an input bit (here called *dv*, for data valid) that is asserted during one clock cycle to inform that the data is ready (to be stored, processed, etc.). In many cases the first data bit (or vector) is made available at the same time that *dv* is asserted, so one must be very careful not to miss that first bit (or vector).

An example is shown in figure 3.16a, which consists of a serial data receiver. The FSM must store the input data *x* in a register *y*. Note that it is a timed machine, which must stay in state B during *T* clock cycles, where *T* is the number of bits of *x* to be stored in *y*. This first solution is of Moore type and employs the default clocking scheme (everybody clocked at the same clock edge—positive by default—as indicated in the rectangle in the upper part of the figure).

An illustrative timing diagram for the FSM of figure 3.16a is presented in figure 3.16b, for *x* = "1011" (see the shaded area on the *x* waveform). The problem with this solution is that it misses the first bit of *x* because the machine only moves to state B at the first (positive) clock edge after *dv* = '1' occurs (note in the figure that when the

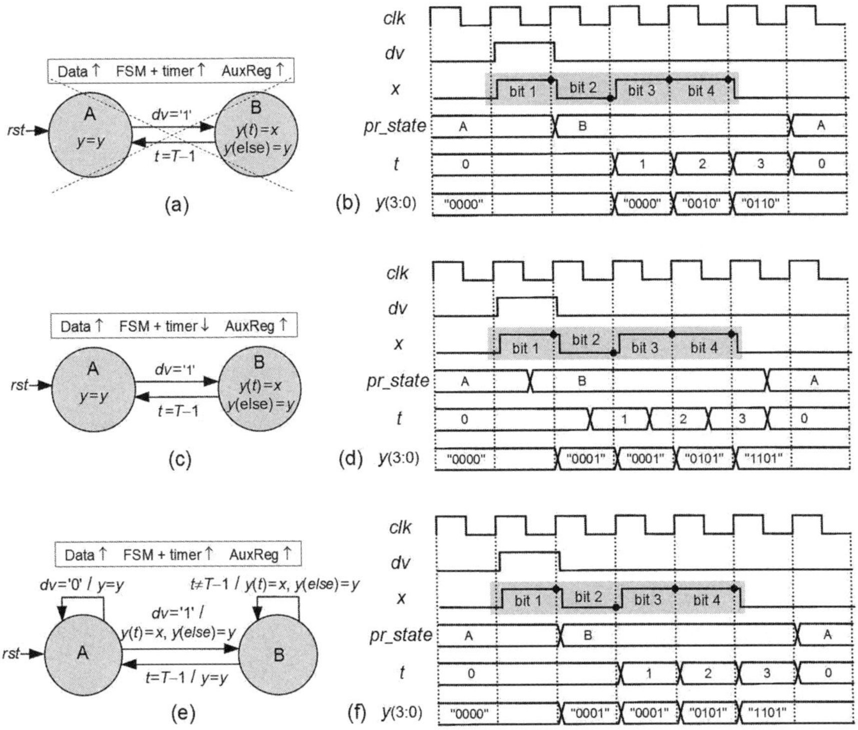

Data ↑ FSM + timer ↑ AuxReg ↑
rst
A
y=y
dv='1'
B
y(t)=x
y(else)=y
t=T–1
(a)
clk
dv
x
bit 1
bit 2
bit 3
bit 4
pr_state
A
B
t
0
1
2
3
(b) y(3:0)
"0000"
"0010"
"0110"
Data ↑ FSM + timer ↓ AuxReg ↑
(c)
"0001"
"0101"
"1101"
(d)
dv='0' / y=y
t≠T–1 / y(t)=x, y(else)=y
dv='1' /
y(t)=x, y(else)=y
t=T–1 / y=y
(e)
(f)

Figure 3.16

Techniques for capturing the first bit when it coincides with *dv*. (a, b) Bad Moore solution. (c, d) Fine Moore solution. (e, f) Fine Mealy solution.

first clock edge occurs *after* the machine is in state B, bit 2 is captured by *y*(0) instead of bit 1).

The same FSM is used in figure 3.16c. However, even though the data (*dv* and *x*) and the auxiliary register (for *y*) are still updated at the positive clock edge, the FSM and its timer operate now at the negative clock transition. Note in the accompanying timing diagram in figure 3.16d that again the proper values are produced for *t* (recall that, by default, the timer is zeroed every time the FSM changes state, and here it is kept stopped at zero in the states where it is not needed—state A in this example), but now all bits are captured properly, resulting, in the end, *y*(3:0) = "1101".

A final solution is shown in figure 3.16e, which consists of using a Mealy machine instead of a Moore machine. Note in the accompanying timing diagram, in figure 3.16f, that this machine too works well, and now the default clocking scheme (everybody operating at the same clock edge) is employed.

Both solutions above require an auxiliary register (for *y*). This kind of FSM will be studied in detail in chapter 11.

3.11 Storing the Final Result

This section discusses another need that sometimes arises in hardware implementations. It consists of wanting the final result from one run of a process to remain stable (constant, exhibited on a display, for example) until another run is completed, with the new result replacing the old one and also remaining unchanged until the next value is produced, and so on. It is also common in such cases to have a control signal (*dv*, data-valid) that indicates when the input data is ready, so data processing should commence. The *dv* signal, which generally lasts one clock period but can also last for the entire process, can help produce the desired feature.

Figures 3.17a,b illustrate the case of *dv* lasting for the entire process (as usual, a small propagation delay was included between a clock edge and its response in order to portray a realistic situation). Note that in figure 3.17a, *dv* is updated at the positive clock edge, which is the same edge at which the circuit operates (see the waveform for *output*), whereas in figure 3.17b, *dv* is updated at the negative clock edge. As shown in figure 3.17c, the negative edge of *dv* is used to store the final result (assumed to be *output* = *e*; see gray shade) into an auxiliary register. This result will obviously remain unchanged until a new result overwrites it.

The problem with the alternative in figure 3.17a is that the output value is unlikely to have had enough time to settle, so an incorrect value will probably be registered. This alternative would be unsafe even if *dv* lasted an extra clock period because of the time delay between the edges of *clk* and *dv*.

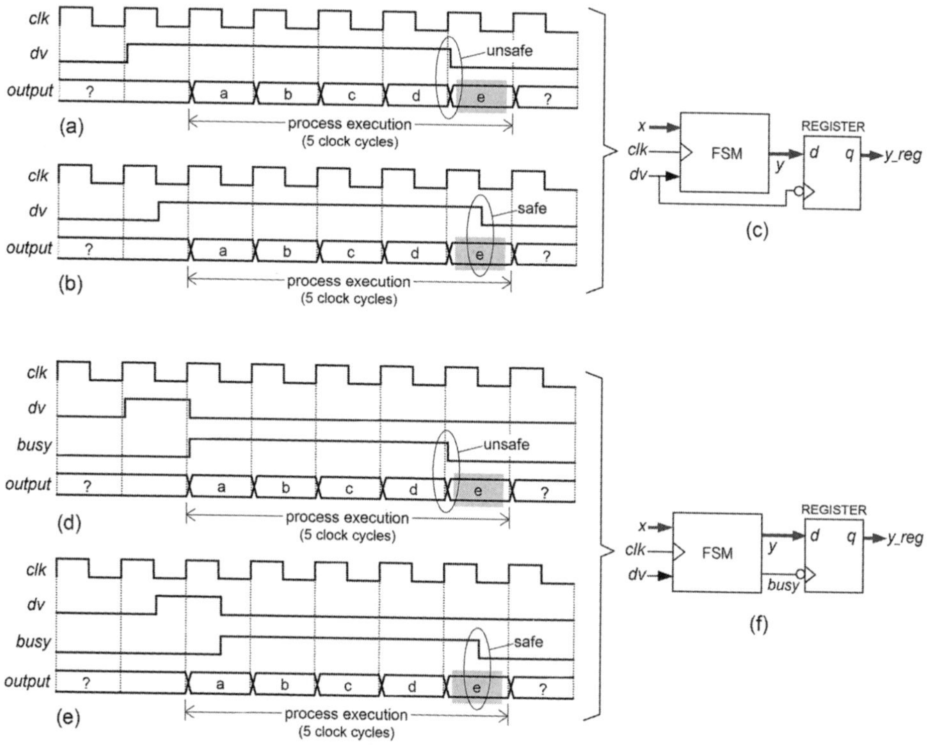
clk
dv
unsafe
output
?
a
b
c
d
e
?
(a)
process execution
(5 clock cycles)
clk
dv
safe
output
(b)
process execution
(5 clock cycles)
x
clk
dv
FSM
y
REGISTER
d
q
y_reg
(c)
clk
dv
busy
unsafe
output
(d)
process execution
(5 clock cycles)
clk
dv
busy
safe
output
(e)
process execution
(5 clock cycles)
x
clk
dv
FSM
y
busy
REGISTER
d
q
y_reg
(f)

Figure 3.17
Storing the final result. (a–c) *dv* lasts the entire process, so its falling edge is used to activate the output register. (d–f) *dv* lasts only one clock period, so the falling edge of *busy* is used to activate the register.

The alternative in figure 3.17b is obviously safe, but using the negative edge implies that the output value has to be ready within $T_{clk}/2$, which might reduce the circuit's maximum speed.

Figures 3.17d,e illustrate the more usual case in which *dv* lasts only one clock cycle, so an auxiliary signal (called *busy* in the figures) is needed to activate the register in figure 3.17f. To produce *busy* from *dv*, a pulse stretcher (studied in section 8.11.10) can be used. However, because *busy* behaves exactly like *dv* in figures 3.17a,b, the same limitations apply here.

A completely different approach is presented in figure 3.18. Note that, contrary to the alternatives in figure 3.17, which do not employ the actual clock to store the final result, *clk* itself is used to activate the output register in figure 3.18b, with *dv* used

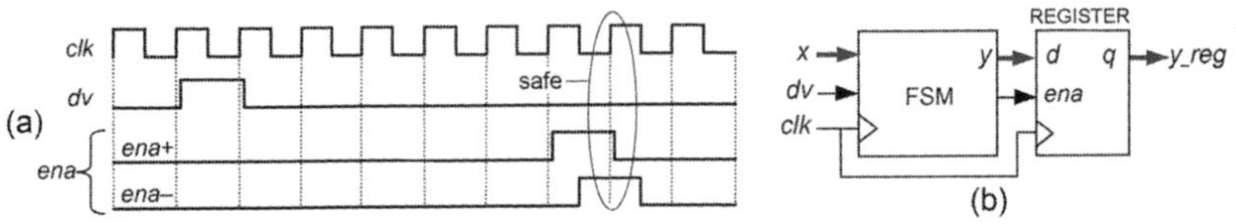
clk
dv
safe
(a)
ena
ena+
ena–
x
dv
clk
FSM
y
REGISTER
d
q
ena
y_reg
(b)

Figure 3.18
Truly synchronous alternative for storing the final result, which consists of producing an enable signal, so the actual clock is responsible for activating the register.

simply to produce an enable (*ena*) signal. The result is a truly synchronous, safe implementation. Two versions of *ena* are shown, updated either at the positive (*ena*+) or negative (*ena*–) clock edge, which can be selected according to the application. This kind of circuit (pulse shifter) is studied in section 8.11.9.

3.12 Multimachine Designs

State machine decomposition, also called state machine factoring, refers to FSMs that are split into two or more machines to ease the design or to take advantage of machines that have been designed previously. In more general terms, two or more smaller FSMs are associated in order to produce the same results as a larger, more complex machine.

Typical associations/decompositions are depicted in figures 3.19a–c. A series (cascade) association is illustrated in figure 3.19a; a parallel association is shown in figure 3.19b; finally, an internal association (one machine is called as part of the other) is depicted in figure 3.19c.

An actual example is depicted in figure 3.19d, which shows an association that falls in the case of figure 3.19c. The main FSM is a factorial ($f = n!$) calculator (details can be seen in exercise 11.9), so a multiplier is needed; because a multiplier can also be implemented using the FSM approach (see section 11.7.5), the latter is called as part of the former.

Another interesting example is presented in figure 3.20a, which shows a machine with a pair of states that need to be repeated a number of times. If T is large (say 64; therefore, 130 states), it is impractical to represent this circuit as a regular FSM. A solution for this problem is presented in figure 3.20b, with the original circuit decomposed into two machines, the first with three states, the other with two states, regardless of the value of T. Note that the main machine has a "superstate" (SS) that simply enables the secondary machine to run. When *ena* = '0', the secondary machine remains in the reset state, whereas *ena* = '1' causes it to flip back and forth between states B and C during $2T$ clock cycles (recall that the timer's initial value

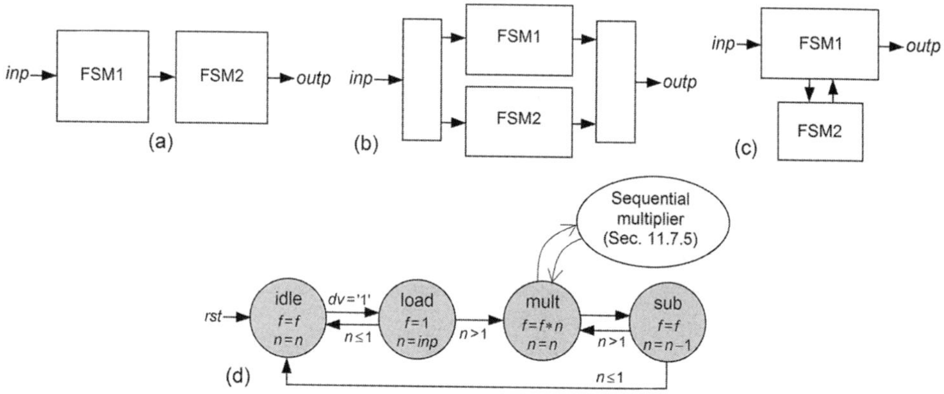
inp
FSM1
FSM2
outp
(a)
inp
FSM1
FSM2
outp
(b)
inp
FSM1
outp
FSM2
(c)
Sequential multiplier (Sec. 11.7.5)
rst
idle
f=f
n=n
dv='1'
load
f=1
n=inp
n≤1
n>1
mult
f=f*n
n=n
sub
f=f
n=n–1
n>1
n≤1
(d)

Figure 3.19
Multimachine implementations with the FSMs associated (a) in series, (b) in parallel, or (c) with one machine called as part of the other. (d) Example showing a factorial calculator, which needs a multiplier, so the latter is called as part of the former.

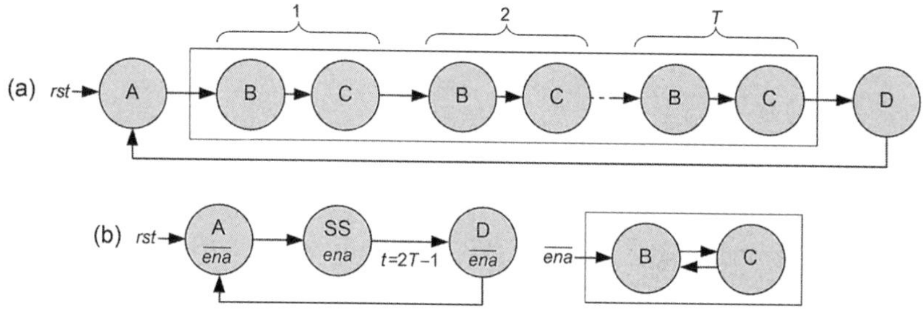
1
2
T
(a)
rst
A
B
C
B
C
B
C
D
(b)
rst
A
ena
SS
ena
t=2T–1
D
ena
ena
B
C

Figure 3.20
(a) FSM with a repetitive pair of states. (b) Solution with FSM decomposition where the secondary machine operates as a "superstate" to the main machine.

is zero, so it runs from 0 to $2T$–1 in this case), after which the machine moves to state D.

An example where this arrangement can be useful is in serial data communications, as in the I^{2}C interface (chapter 14), because a data vector must be transmitted (state B) by the master, then an acknowledgment bit must be received from the slave (state C), with these operations repeated until all data vectors have been sent out.

Another area in which the use of multiple machines is relatively common is in control units for CPUs, in which simpler instructions (e.g., *load* and *store*) are part of more elaborate CPU instructions. Several examples with multiple machines are seen in chapters 8 and 11.

3.13 State Machines for Datapath Control

The purposes of this section are to review datapath-related concepts and to describe how state machines can be used to build the *control unit* that controls a datapath. Because the control unit is generally the most complex circuit to design in this kind of application, and the FSM approach is employed to do it, a study of state machines for datapath control is indispensable.

A popular datapath is that of microprocessors and microcontrollers, needed to construct the CPU, write/read data to/from memory, communicate with peripherals, and so on. Fundamental components for datapath construction are depicted in figure 3.21. Note that they all have some type of control input (*sel, ALUop, wrR, rdM, wrM, wrPC, wrIR*).

Multiplexers are digital switches used to route data from one location to another; in the case of figure 3.21, when the selection (*sel*) input is '0', the upper input is passed to the output, whereas a '1' causes the lower input to be passed to the output. The arithmetic logic unit (ALU), as the name says, is responsible for executing arithmetic (+, −, *, /, ...) and logic (AND, OR, XOR, ...) operations; the operation is selected by the ALU's operational code, *ALUop*. Registers are simply DFF banks: for example, a 32-bit register is simply a set of 32 parallel DFFs; note the write-register (*wrR*) input, which must be asserted for the input data to be stored into the DFFs (at the next positive clock edge). The data memory is used to store data during datapath operations; it contains two control inputs, for reading from (*rdM*) or writing into (*wrM*) the

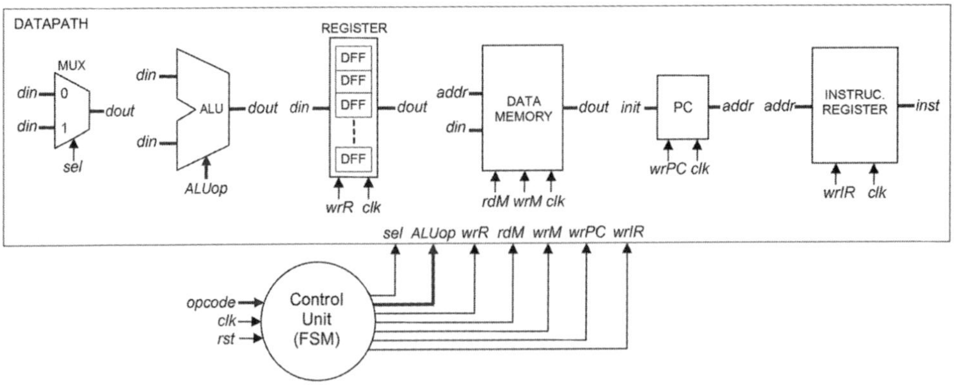
DATAPATH
MUX
din
0
1
sel
dout
din
ALU
ALUop
dout
REGISTER
DFF
DFF
DFF
DFF
din
dout
wrR clk
addr
din
DATA MEMORY
dout
rdM wrM clk
init
PC
addr
wrPC clk
addr
INSTRUC. REGISTER
inst
wrIR clk
sel ALUop wrR rdM wrM wrPC wrIR
opcode
clk
rst
Control Unit (FSM)

Figure 3.21
Main datapath components.

memory. The program counter (PC) is a counter that keeps the instruction address; observe the write-PC (*wrPC*) control input. Finally, the instruction register is responsible for storing and decoding the instructions; note the write-instruction-register (*wrIR*) control input.

The control unit, shown in the lower part of the figure, is responsible for producing all control signals. Its main input (besides clock and reset) is an opcode, based on which the whole sequence of events needed for the datapath to perform the desired computations is provided. An important aspect to observe is that the control unit, although responsible for sequencing all of the computations, does not access the data directly (except for some occasional trivial data monitoring). Instead, it just makes the proper path manipulations such that the datapath itself produces the intended results.

A simple datapath is depicted in figure 3.22a, containing an ALU, two registers (A, B), and a multiplexer. Four control signals are involved: *selA* (selects the data source for register A), *wrA* and *wrB* (enable writing into registers A and B, respectively—at the next positive clock edge), and *ALUop* (some of the ALU operations are listed in figure 3.22b). These four control signals must be generated by the control unit, based on *opcode*. For simplicity, clock and reset are generally omitted in datapath representations (as in figure 3.22a), but they are obviously needed.

As an example, say that the following computation must be performed by the datapath of figure 3.22a: When an external input, called *dv* (data-valid bit), is asserted

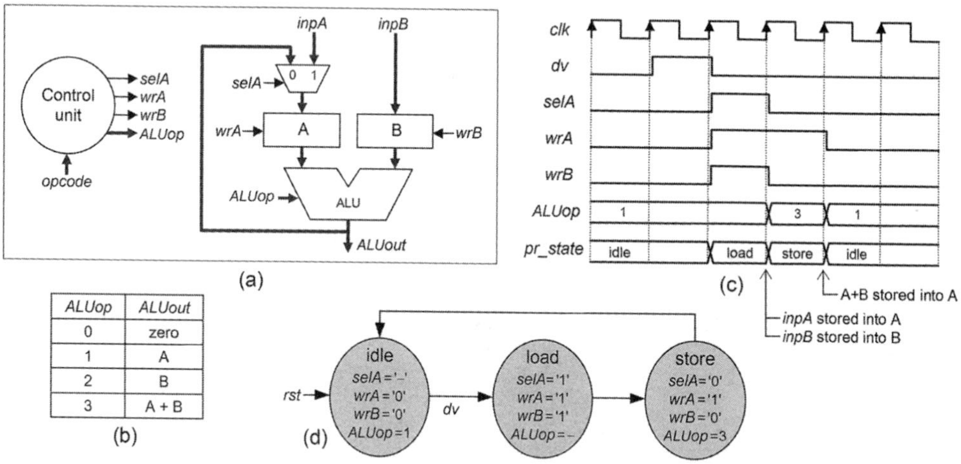

Control unit
selA
wrA
wrB
ALUop
opcode
inpA
inpB
0 1
A
B
ALU
ALUout
(a)
clk
dv
selA
wrA
wrB
ALUop
pr_state
idle
load
store
idle
(c)
A+B stored into A
inpA stored into A
inpB stored into B
(b)
rst
idle
selA='-'
wrA='0'
wrB='0'
ALUop=1
dv
load
selA='1'
wrA='1'
wrB='1'
ALUop=-
store
selA='0'
wrA='1'
wrB='0'
ALUop=3
(d)

ALUop	ALUout
0	zero
1	A
2	B
3	A + B

Figure 3.22
(a) Datapath example. (b) Partial ALU's opcode table. (c) Illustrative timing diagram for the following computation: inputs are stored into A and B, then added, with the result then stored into A. (d) Corresponding Moore machine.

(during one clock cycle), *inpA* and *inpB* must be added and the result stored in register A (so we can assume *opcode* = *dv*).

The overall datapath operation is illustrated in the timing diagram of figure 3.22c (as usual, small propagation delays were left between the clock transitions and the corresponding responses to portray a more realistic situation), which shows the system clock (*clk*), the data-valid bit (*dv*), the four signals to be produced by the FSM (*selA*, *wrA*, *wrB*, *ALUop*), and, finally, the machine's state (*pr_state*) after every positive clock transition.

A very important aspect to observe, which often causes confusion and leads to incorrect designs, is how data is stored in datapaths. Because the control unit does not access the data directly but just provides the proper path for it, the storage occurs *at the end* of any write-enabling state. For example, note in figure 3.22c that *wrA* = *wrB* = '1' in the *load* state, which means that at the *next* clock edge (thus, at the *end* of the *load* state) the inputs will be stored in A and B (see comments at the bottom of the figure). The same is true in the *store* state; because *wrA* = '1' in it, at the *next* clock edge (*end* of that state), the sum will be stored in A.

A corresponding Moore-type solution is presented in figure 3.22d, which is a direct translation of figure 3.22c (compare the values in the plots against those in the state transition diagram). Note also in figure 3.22d that the list of outputs is *exactly the same* in all three states, as required for hardware-implemented (as opposed to software-implemented) FSMs, otherwise latches would be inferred (unless the optional output register is included).

The next example (figure 3.23) shows a relatively complete CPU datapath (based on MIPS [Patterson & Hennessy, 2011]). It contains an ALU, four multiplexers (memory address source, register data source, ALU source A, and ALU source B), several registers (instruction register—IR, memory data register—MDR, general purpose register file,

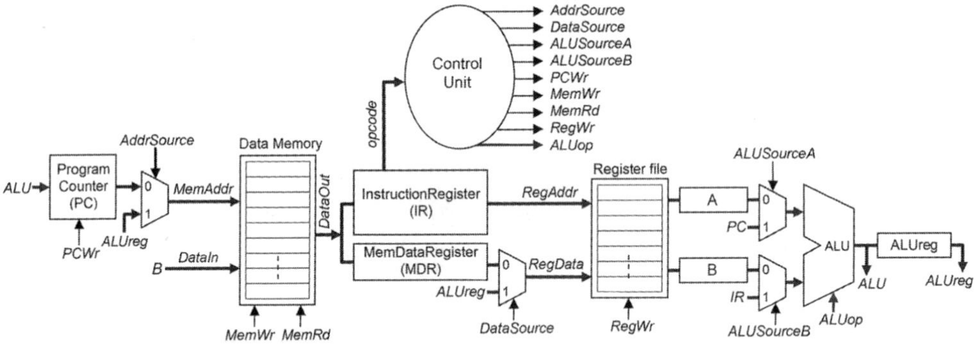
Control Unit
AddrSource
DataSource
ALUSourceA
ALUSourceB
PCWr
MemWr
MemRd
RegWr
ALUop
opcode
AddrSource
Data Memory
Program Counter (PC)
ALU
0
1
MemAddr
ALUreg
PCWr
DataIn
B
DataOut
MemWr
MemRd
InstructionRegister (IR)
MemDataRegister (MDR)
RegAddr
0
1
RegData
ALUreg
DataSource
Register file
RegWr
ALUSourceA
A
PC
B
IR
ALUSourceB
ALU
ALUop
ALUreg
ALU
ALUreg

Figure 3.23
A CPU datapath (based on the MIPS architecture).

general purpose A and B registers, and the ALU register—ALUreg), plus a general purpose data memory.

As usual, a control unit (FSM) is needed to control the datapath, which produces nine signals: *AddrSource* (to control the memory address source mux), *DataSource* (to control the register data source mux), *ALUSourceA* (to control the ALU input-A mux), *ALUSourceB* (to control the ALU input-B mux), *PCWr* (to enable writing into the program counter register), *MemWr* (to enable writing into the memory), *MemRd* (to enable reading the memory), *RegWr* (to enable writing into the register file), and *ALUop* (to define the ALU operation). The control unit produces these signals based on the opcode received from the instruction register/decoder.

Each CPU instruction is broken down into a series of clock cycles, with each cycle limited to one ALU operation plus storage or one memory/register access plus storage, such that at the end of each cycle the data needed in the next cycle will be available in one of the registers or in memory. For example, note in figure 3.23 that the data read from the memory is stored in MDR, the data from the register file is stored in A or B, and the ALU result is stored in ALUreg.

Two partial examples of instructions executed by the CPU of figure 3.23 are depicted in figure 3.24, concerning the store word (SW) and load word (LW) instructions. As shown in figure 3.24a, SW is composed of two main parts: SetAddr, in which the memory address to which data will be written is set, and WriteMem, which causes the data to be effectively stored (at the end of that state, as explained earlier). Note the following in the SetAddr state: *ALUSourceA* = '0', so port A of the ALU is fed by register A; *ALUSourceB* = '1', so port B of the ALU is fed by the IR; and *ALUop* = add, meaning that the actual opcode will cause the ALU to add its inputs, thereby producing the intended memory address stored in ALUreg. Note the following in the WriteMem state: *AddrSource* = '1', so the address comes from the ALU register (as expected); and *MemWr* = '1', so writing is enabled and will occur at the next (positive) clock edge. As a final remark, note that the list of outputs is not the same in both states (it should be for hardware-based implementations), which was done here just for simplicity.

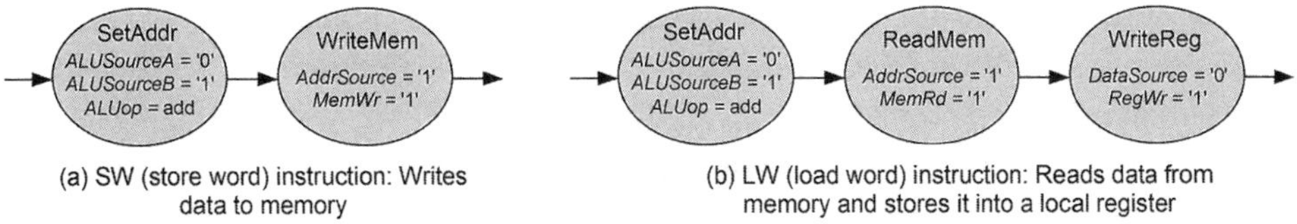
SetAddr
ALUSourceA = '0'
ALUSourceB = '1'
ALUop = add
WriteMem
AddrSource = '1'
MemWr = '1'
(a) SW (store word) instruction: Writes data to memory
SetAddr
ALUSourceA = '0'
ALUSourceB = '1'
ALUop = add
ReadMem
AddrSource = '1'
MemRd = '1'
WriteReg
DataSource = '0'
RegWr = '1'
(b) LW (load word) instruction: Reads data from memory and stores it into a local register

Figure 3.24

Partial state machines implementing memory access using the datapath of figure 3.23. Recall that in an actual design the list of outputs has to be exactly the same in all states of an FSM.

The LW instruction, depicted in figure 3.24b, is composed of three main parts: SetAddr, responsible for setting the memory address from which data must be read; ReadMem, which causes the data to be effectively read (at the end of this state); and WriteReg, which causes the read data to be stored in one of the IR registers (again, at the end of the state). The SetAddr state is similar to the previous case. Note the following in the ReadMem state: *AddrSourceA* = '1', so the address comes from the ALU register (as expected); and *MemRd* = '1', so reading is enabled and the data will be stored in MDR at the next (positive) clock edge. Finally, note the following in the WriteReg state: *DataSourceA* = '0', so the data must come from MDR (as expected); and *RegWr* = '1', so the data will be written into the register pointed to by the IR at the next (positive) clock edge.

The reasoning used in the instructions above can be extended to all instructions of a CPU, resulting in a generally large set of small state machines, collectively responsible for implementing all of a CPU's instructions.

Points to Remember when Designing a Control Unit

We close this section with some comments that can be helpful for the proper understanding and correct design of FSMs for datapath control.

1) *Sequential circuit and Moore machine:* The control unit is normally the only sequential circuit in a datapath-based design (except for the PC, but this is just a basic counter; registers are also clocked and so can be memories, but these act just as data storage elements), and its design is normally based on the FSM approach. Moreover, because control units are inherently synchronous, the Moore approach is generally preferred (over Mealy). In the comments that follow, it is assumed that the Moore model was adopted.

2) *No direct data access:* Even though the control unit is responsible for sequencing all datapath computations, it normally does not access the actual data directly (except for occasional trivial data monitoring).

3) *Late data storage:* Data storage in a datapath-based design is controlled by a write-enable signal produced by the control unit. Because such a signal will be ready only *after* the clock edge that causes the machine to enter the write-enabling state, the actual writing will only occur at the *next* clock edge, that is, *at the end* of that state (in other words, it will occur just prior the moment at which the machine *leaves* the write-enabling state).

4) *Dependency on input data:* In some applications the machine must read/write data and, based on the data value, make a decision on which state to go to next. Because the data will be available only when the machine *leaves* the read/write state, the decision can obviously not be made yet. In such cases, a wait state must be included before the decision can be made. This is illustrated in figure 3.25. Assume that we are using a datapath similar to that in figure 3.22a, where the inputs must be stored in registers

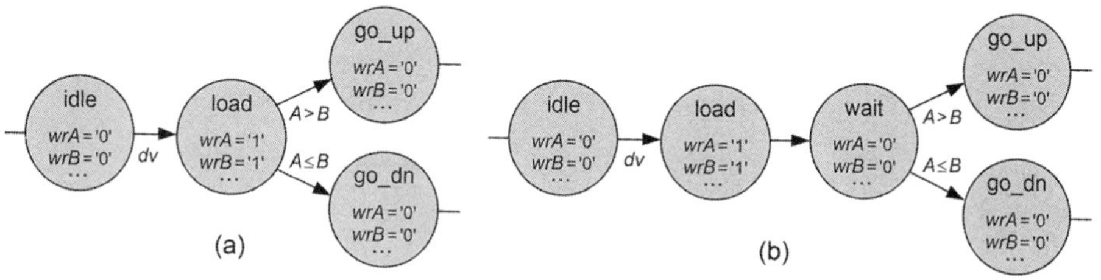
idle
wrA='0'
wrB='0'
...
dv
load
wrA='1'
wrB='1'
...
A>B
A≤B
go_up
wrA='0'
wrB='0'
...
go_dn
wrA='0'
wrB='0'
...
(a)
idle
wrA='0'
wrB='0'
...
dv
load
wrA='1'
wrB='1'
...
wait
wrA='0'
wrB='0'
...
A>B
A≤B
go_up
wrA='0'
wrB='0'
...
go_dn
wrA='0'
wrB='0'
...
(b)

Figure 3.25
Input-data-dependent decision in a Moore-type control unit. (a) Incorrect. (b) correct.

A and B and then compared to decide where the machine should go next. In figure 3.25a the data-valid bit causes the machine to move from the *idle* state to the *load* state; in the latter, *wrA* = *wrB* = '1', so when the machine *leaves* that state, the proper data will be available for comparison. Therefore, in this machine the comparison will actually be between the data values previously stored in A and B (a mistake). This was fixed in figure 3.25b with the inclusion of a wait state. (Recall that in a Mealy machine the wait state is not needed, but the Mealy approach is rarely adopted in datapath-related applications.)

3.14 Exercises

Exercise 3.1: Moore and Mealy Circuits

a) Just by looking at the circuit of figure 3.4e, how can you tell that it is a Moore machine?

b) How can you tell that the circuit of figure 3.6d is a Mealy machine?

Exercise 3.2: By-Hand Design of a Moore Machine #1

We saw in section 3.7 that the number of DFFs and the amount of combinational logic needed to build an FSM can vary substantially with the encoding style chosen. In the "by-hand" design of section 3.3, sequential binary encoding was employed (e.g., *pr_state* = "00" for state *zero*, "01" for state *one*, "10" for state *two*, and "11" for state *three*).

a) Redo that design, again "by hand," using Gray code (state *zero* → "00", *one* → "01", *two* → "11", *three* → "10").

b) Redo it again, now using true one-hot code (state *zero* → "0001", *one* → "0010", *two* → "0100", *three* → "1000").

c) Compare these three solutions (sequential, Gray, and one-hot). Which requires the fewest DFFs? Which requires the least combinational logic? Which has the best time predictability for the output?

Exercise 3.3: By-Hand Design of a Moore Machine #2

Consider the Moore machine of figure 5.7c, which implements a short-pulse generator.

a) Design it "by hand" using sequential encoding. Show that $y = q_0$.
b) Design it using Gray encoding. Show that $y = q_1' \cdot q_0$.
c) Design it using the following user-defined encoding: A = "–0", B = "01", C = "11". Show that $d_0 = x$, $d_1 = q_0$, and $y = q_1' \cdot q_0$.
d) Draw all three circuits and show that the last one is the simplest.
e) Which of these circuits is/are guaranteed to have a glitch-free output, with better time predictability? Explain.

Exercise 3.4: By-Hand Design of a Mealy Machine

a) Draw a Mealy-type state transition diagram for the parity detector of figure 5.5.
b) Design a circuit that implements this machine, with sequential encoding.

Exercise 3.5: Time Behavior of a Moore Machine

Say that the parity detector of figure 5.4b operates with the clock signal of figure 3.26, receiving at the input the signal x also included in the figure. Draw the other two waveforms (machine's present state and output; the initial part of *pr_state* was already filled). Does the output change only when the state changes?

Exercise 3.6: Time Behavior of a Mealy Machine

This exercise is a continuation of the one above.

a) Draw a Mealy-type solution for the parity detector of figure 5.4.
b) Say that this machine is operating with the clock of figure 3.27, receiving the signal x also included in the figure. Draw the other two waveforms (machine's present state and output). Does the output change only when the state changes?
c) Compare the time behavior of this Mealy solution against that of the Moore counterpart developed in the previous exercise. Which is different (*pr_state* or y or both) from one solution to the other?

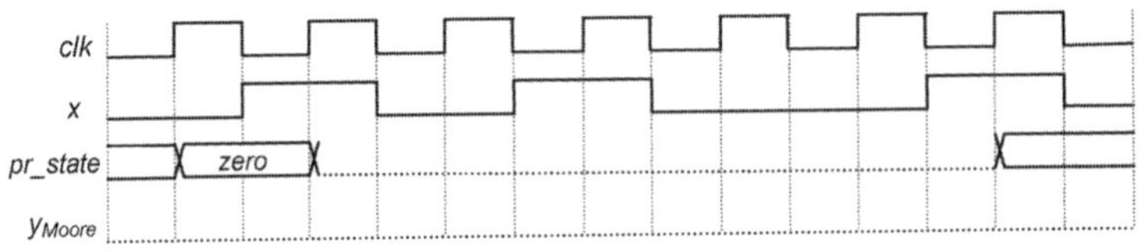
clk
x
pr_state
zero
yMoore

Figure 3.26

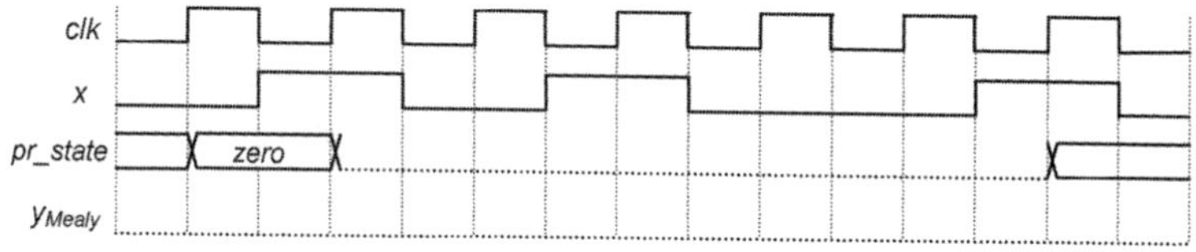
clk
x
pr_state
zero
yMealy

Figure 3.27

Exercise 3.7: State Machine Categories

What is the category of the machines in figures 1.3b, 1.3c, 3.4a, and 3.6a? Why are the machines of figures 8.12c and 8.14b said to be of category 2? Why are the machines of figures 11.5b and 11.7b said to be of category 3?

Exercise 3.8: State Encoding

List the codewords used to encode the states of the garage door controller of figure 5.9c in the following cases:

a) With sequential encoding.
b) With Gray encoding.
c) With Johnson encoding.
d) With one-hot encoding.

Exercise 3.9: Number of Flip-Flops

Calculate the number of DFFs needed to encode an FSM with $M = 8$ or $M = 33$ states, in the following cases:

a) With sequential encoding.
b) With Gray encoding.
c) With Johnson encoding.
d) With one-hot encoding.

Exercise 3.10: Need for Reset #1

Say that the FSM of figure 5.4b, which implements a basic parity detector, is encoded using regular sequential encoding (so *pr_state* = '0' in state *zero* and *pr_state* = '1' in state *one*).

a) Are there any states *outside* the machine (i.e., unused codewords)?
b) Is an explicit reset signal needed when this machine is implemented in an FPGA (flip-flops reset automatically on power-up)? Can deadlock occur in this case?
c) Answer the questions of part b when the flip-flops' initial state is arbitrary.

Exercise 3.11: Need for Reset #2

This exercise concerns the parity detector of figure 5.5c, which has a data-valid (*dv*) input. Assume that a reset input is not provided and that the circuit is implemented in a device whose flip-flops' initial state (on power-up) is arbitrary.

a) If the initial state falls *inside* the machine (in state *zero*, *one*, or *hold_one*), will the circuit operate properly? Does this answer depend on the encoding scheme?
b) Answer the same questions above for the case when the initial state falls *outside* the machine.
c) Prove that this FSM works well in both cases mentioned above if sequential encoding is used and optimal (minimal) expressions are used for *nx_state* (d_1, d_0).
d) Prove that it also works well in both cases mentioned above if one-hot encoding is used and optimal (minimal) expressions are used for *nx_state* (d_2, d_1, d_0).
e) Consider that sequential encoding is used. Show that if the "don't care" bits are all filled with '0's the machine is not subject to deadlock, but if they are all filled with '1's then deadlock can occur.
f) Consider now that one-hot encoding is used. Show that in both cases ("don't care" bits all filled with '0's or all filled with '1's) deadlock can occur.

Suggestion to solve parts c and d: First, review sections 3.8 and 3.9; next, use the method seen in section 3.3 to find the machine's optimal expressions for *nx_state*; then apply the values of *pr_state* (i.e., q_1 and q_0 in c or q_2, q_1, and q_0 in d) and of the transition conditions (*dv* and *x*) for the cases not used in the FSM encoding (states *outside* the machine) to the expressions derived to show that the results always converge to states *inside* the machine.

Exercise 3.12: Capturing the First Bit

Two options for processing data correctly when the first data bit is made available at the same time that the data-valid (*dv*) bit is asserted were presented in figure 3.16. Show that the option in figure 3.16c will no longer work if the auxiliary register that stores *x* operates at the falling clock edge.

Exercise 3.13: Storing the Final Result

Explain why the option in figure 3.18 is better than any of the options in figure 3.17 for registering a process's final result.

Exercise 3.14: Multimachine Design

In figure 3.20a, an FSM with a repetitive pair of states is shown, for which a solution using two FSMs was presented in figure 3.20b. Complete the timing diagrams of figure 3.28 for the machines of figure 3.20b, assuming that $T = 3$.

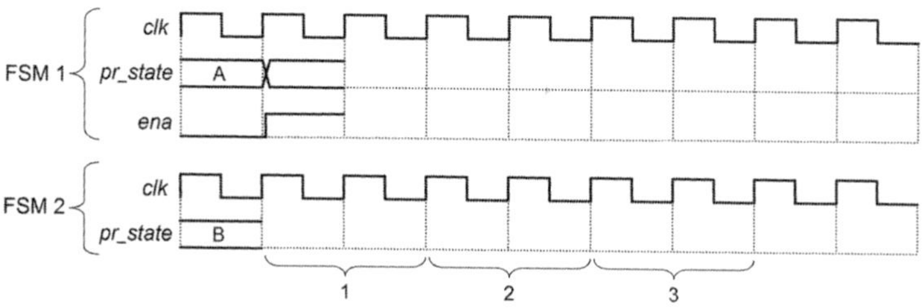
FSM 1
clk
pr_state
A
ena
FSM 2
clk
pr_state
B
1
2
3

Figure 3.28

Exercise 3.15: Datapath Control

Assume that the datapath of figure 3.22a must operate as an add-and-accumulate circuit (ACC), accumulating in A four consecutive values of *inpB*. The data-valid pulse (*dv*), lasting only one clock period, must again start the four-iteration procedure, after which the resulting value must remain displayed at *ALUout* until another *dv* pulse occurs. In summary, the operations are: 0+B→A, A+B→A, A+B→A, and A+B→A.

a) Draw an illustrative timing diagram (as in figure 3.22c) for an FSM that controls this datapath.
b) Draw a corresponding state transition diagram (as in figure 3.22d) for this machine.

After solving the problem, check section 5.4.7.

4 Design Steps and Classical Mistakes

4.1 Introduction

This chapter presents a list of classical problems and mistakes that might occur in the design of hardware-based finite state machines. Subsequently, a summary regarding the main design steps is also presented.

4.2 Classical Problems and Mistakes

4.2.1 Skipping the State Transition Diagram

Probably the most error-prone step in the design of a circuit based on the state machine approach is to think that it is fine to go straight from the specifications to the design without sketching the state transition diagram (students sometimes believe that they can "see" the state diagram in their minds). With this approach, states can be missed, or, more likely, output values and state transitions may be ill specified. This step is critical because any error in the state transition diagram will invalidate the whole effort, no matter how well the rest is done.

4.2.2 Wrong Architecture

Once one has been convinced that sketching the state transition diagram is indispensable, the next step is not to draw it but first to decide which type the machine architecture should be used (a major mistake is to think that all machines are "just the same"). A great effort has been made in this book to show that, when one is using hardware (as opposed to software) to implement an FSM, the circuit architecture can vary substantially from one problem to another. For that reason, a classification into three categories, which covers *any* state machine, was introduced in section 3.6 (see figure 3.8), as follows: category 1 for regular state machines; category 2 for timed state machines; and category 3 for recursive state machines.

The first decision is to select correctly in which of these categories the machine to be designed falls. That not only will lead to the right circuit (optimal resources usage) but will also immensely reduce the design effort. The second important decision is to choose between the corresponding Moore and Mealy architectures. The third and final architectural decision is whether to include or not in the FSM the optional output register (figure 3.2b).

4.2.3 Incorrect State Transition Diagram Composition

As seen in section 1.3, the state diagram must obey three fundamental principles:

1) It must include all possible system states.
2) All state transition conditions must be specified (unless a transition is unconditional), and such conditions must be truly complementary.
3) The list of output signals must be exactly the same in all states (standard architecture).

Failing to comply with requisite 1 above will lead inevitably to an incorrect circuit. Even though this seems an obvious step, there are situations in which subtle details are involved, such as the inclusion of wait states to hold until the data to be inspected is ready (as in figure 3.25, for example) or to suppress state bypass (as in figure 4.2, for example).

Condition 2 above requires that the complete set of transition conditions be neither under- nor overspecified; otherwise, a poor or incorrect circuit will again result. This is a relatively common error that can be avoided by following the material seen in section 1.5.

Finally, requisite 3 determines that, for hardware implementations, the list of outputs must be exactly the same in all states; otherwise, latches will be inferred, wasting resources and making the time response less predictable. Because this is by far the most common mistake, an example is provided in figure 4.1. In figure 4.1a, y is not specified in state B. If this lack of specification is the result of careless analysis, an incorrect circuit will probably be implemented; otherwise, if the missing specification is because y should keep in state B the same value that it had when the FSM left state A, then $y = y$ should be entered, as depicted in figure 4.1b, making the list of outputs exactly the same in all states and also clarifying what is indeed wanted for y.

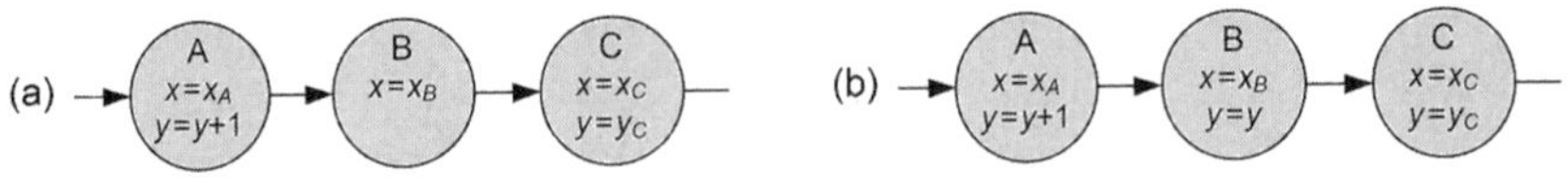
(a)
A
x=xA
y=y+1
B
x=xB
C
x=xC
y=yC
(b)
A
x=xA
y=y+1
B
x=xB
y=y
C
x=xC
y=yC

Figure 4.1
(a) State diagram with incomplete output specifications. (b) Corrected state diagram.

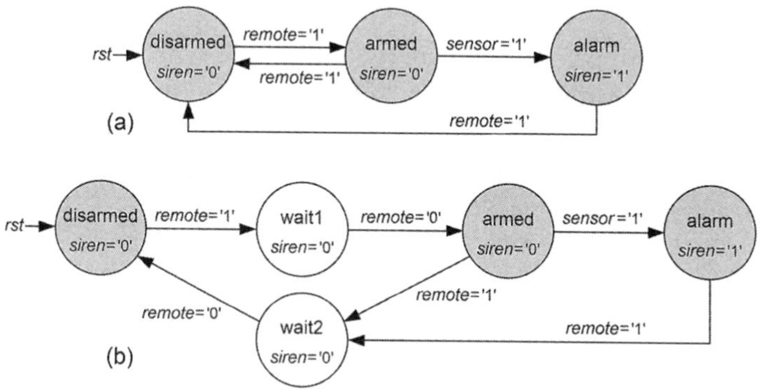
rst
disarmed
siren='0'
remote='1'
remote='1'
armed
siren='0'
sensor='1'
alarm
siren='1'
remote='1'
(a)
rst
disarmed
siren='0'
remote='1'
wait1
siren='0'
remote='0'
armed
siren='0'
sensor='1'
alarm
siren='1'
remote='0'
remote='1'
wait2
siren='0'
remote='1'
(b)

Figure 4.2
Car alarm (a) with and (b) without state bypass.

4.2.4 Existence of State Bypass

The state-bypass problem occurs when the transition conditions for entering a state coincide with the transition conditions for leaving that same state, and such conditions are true during more than one clock cycle.

As an example, consider the car alarm of figure 4.2a. If the alarm is in the *disarmed* state and a command from the remote control (*remote* = '1') is received, the machine passes to the *armed* state, ready to detect any intrusions. However, if the *remote* = '1' command lasts several clock cycles (as is generally the case) and the intrusion sensor is off (*sensor* = '0'), the circuit goes back to *disarmed*, then returns to *armed*, and so on, producing a kind of state bypass (in fact, the states are not exactly bypassed, but rather, the machine remains in each state for just one clock period instead of staying there). Note that in this example state bypass occurs even when *sensor* = '1'.

This problem can be solved with some kind of flag or, more clearly, with wait states, as in figure 4.2b (white circles). Note that the *wait1* and *wait2* states wait until *remote* returns to zero before allowing any other action to take place.

The failure to prevent state bypass can lead to a circuit with occasional malfunctioning that is very difficult to locate later. This is especially true when the state bypass can only occur in very particular situations, which might have been overlooked in the simulations and therefore remained undetected during the design phase.

4.2.5 Lack of Reset

In the design of any FSM the need for reset must always be considered (only few cases are fine without an explicit reset port). Failing to do so can cause incorrect machine initialization or even deadlock. A detailed analysis on the use of reset and its consequences was presented in sections 3.8 and 3.9.

4.2.6 Lack of Synchronizers

Many FSMs have asynchronous inputs, so metastability can occur if synchronizers are not employed. Failing to analyze whether asynchronous signals are involved in the design and the possible consequences of metastability to that particular application can compromise the entire project. Material on the use and construction of synchronizers and their consequences was presented in section 2.3.

4.2.7 Incorrect Timer Construction

Many engineering problems include timed decisions, leading to state machines with time as a transition condition (see figure 1.8). Because timers are just counters (therefore sequential circuits, which can then also be modeled as state machines), one might be tempted to use the FSM approach to design them. There are two main reasons for not doing so in general. The first is that counters are standard circuits, easily designed without the FSM approach. The second is that a counter might have thousands of states and therefore would be impractical to represent as a *regular* state machine.

The recommended approach in such cases is to consider the timer (counter) as an auxiliary circuit, implemented separately and acting as an *input* to the (main) state machine. However, the state machine itself must be responsible for clearing the timer at the proper moments as well as for stopping it or letting it run as needed. Such fine details, sometimes overlooked, are fundamental to attain a correct and optimized design. Such aspects are studied in chapter 8, which deals specifically with timed state machines, and are reinforced in chapters 9 and 10, which show VHDL and SystemVerilog implementions for timed FSMs.

4.2.8 Incomplete VHDL/SystemVerilog Code

Once the state transition diagram has been correctly and completely constructed, we can write a corresponding VHDL or SystemVerilog code to synthesize the circuit. The problem is that here too the coverage of specifications might not be complete, even if the state diagram is complete. Two common mistakes are described below, both related to the combinational logic section of the FSM (more precisely, related to requisites 2 and 3 listed in section 4.2.3).

The first mistake regards incomplete *output* specifications. One might believe that when something was said in a previous state and nothing occurred there is no need to say it again. For example, consider that we are using VHDL and the **case** statement to implement the combinational logic section of an FSM as follows (do not worry about code details for now; they are seen in chapter 6):

```
--Bad:
case pr_state is
   when A =>
```

```
--Good:
case pr_state is
   when A =>
```

```
      output1 <= "0000";
      output2 <= "01";
      nx_state <= B;
   when B =>
      output2 <= "10";
      nx_state <= C;
   when C =>
      output1 <= "1111";
      output2 <= "11";
      nx_state <= A;
end case;
```

```
      output1 <= "0000";
      output2 <= "01";
      nx_state <= B;
   when B =>
      output1 <= "0000";
      output2 <= "10";
      nx_state <= C;
   when C =>
      output1 <= "1111";
      output2 <= "11";
      nx_state <= A;
end case;
```

Note in the code on the left that from state A the machine can only go to state B. If the desired value for *output1* while in B is the same as that in A, one might be tempted to omit it in state B. Recall, however, that the upper section of an FSM is *combinational* (thus memoryless), so there is nothing to prevent its output from changing when the machine leaves a state. For cases like the code above, the compiler will generally infer latches, guessing that the designer wanted the machine to keep the same value that it had in the previous state, which can produce an unsafe behavior because the timing response of latches (built with regular gates) is difficult to predict and is subject to race conditions.

In summary, it is important to remember what was said earlier: the list of outputs must be *exactly the same* in all states (so in this example it must contain *output1* and *output2* in all states, as shown in the code on the right).

The second mistake regards incomplete *transition conditions* specifications. For example, consider again that we are using VHDL and the **case** statement to implement the combinational logic section of an FSM as follows:

```
--Moore machine:
case pr_state is
   when A =>
      output <= <value>;
      if <condition> then
         nx_state <= B;
      elsif...
         ...
      else
         nx_state <= A;
      end if;
   when B =>
      ...
end case;
```

```
--Mealy machine:
case pr_state is
   when A =>
      if <condition> then
         output <= <value>;
         nx_state <= B;
      elsif...
         ...
      else
         output <= <value>;
         nx_state <= A;
      end if;
   when B =>
      ...
end case;
```

Both codes above are correct. Note that in both the **if** statement includes an **else** part, which takes care of *all* remaining options. If this **else** were omitted, an underspecification would occur, and the compiler might again infer unnecessary (and undesirable) latches.

4.2.9 Overregistered VHDL/SystemVerilog Code

This is another common mistake. It is very important to be aware of the code sections that infer registers and close such sections as soon as registers are no longer needed.

An example is shown below, using VHDL. Any signal to which a value is assigned under the **if rising_edge(clk)** statement will be registered, so that **if** must be closed as soon as possible. The code on the left is constructed correctly. Note that the only assignment under the **if rising_edge(clk)** statement is **pr_state <= nx_state**, so only the machine state gets registered. Because the **case** statement used for the upper section is outside that **if** statement, no flip-flops will be inferred for that part of the machine, resulting in a truly combinational circuit for the upper section, which is how it should be.

The code on the right, on the other hand, is an example of an error-prone design. Note that now the **case** statement is inside the **if rising_edge(clk)** statement, so the output will also be registered. As we have already seen, there are cases in which the optional output register is needed, but that is a case-by-case decision, not a forced condition as it is in this code. Probably the worst aspect of this code is that the designer might be completely unaware of what is actually happening. (Note that *pr_state* has no effect in this process' sensitivity list.)

```
--Good:
--lower section of FSM:
process (clk)
begin
   if rising_edge(clk)then
      pr_state <= nx_state;
   end if;
end process;
--upper section of FSM:
process (all)
begin
   case pr_state is
      when A =>
         output <= <value>;
         if <condition> then
            nx_state <= B;
         else
            nx_state <= A;
         end if;
      when B =>
         ...
   end case;
end process;
```

```
--Error prone:
process (clk, pr_state)
begin
   if rising_edge(clk) then
      --lower section of FSM:
      pr_state <= nx_state;
      --upper section of FSM:
      case pr_state is
         when A =>
            output <= <value>;
            if <condition> then
               nx_state <= B;
            else
               nx_state <= A;
            end if;
         when B =>
            ...
      end case;
   end if;
end process;
```

4.3 Design Steps Summary

We close this chapter by summarizing the main steps that should be observed in designing sequential circuits using the FSM approach.

1) *Specifications analysis:* Study the problem specifications carefully. As a final step, decide:

   a) The FSM category (regular, timed, or recursive) to be adopted.
   b) The FSM type (Moore or Mealy) to be used.
   c) Whether the optional output register should be included.

2) *State transition diagram:* Based on your analysis and conclusions above, carefully draw the state transition diagram. The use of a detailed diagram (as in figure 1.4d) is particularly recommended in complex designs or for beginners because it helps visualize and assure that all transition conditions have been completely and correctly covered.

3) *Encoding style and resources usage:* Decide which state-encoding option (e.g., sequential, Gray, Johnson, one-hot) will be employed in the design. After that, the exact number of DFFs that will be needed to build the FSM can be calculated. Do it, so your estimate can be compared later against the actual number reported by the VHDL/SystemVerilog compiler (this is a very important checkpoint).

4) *Reset signal:* Analyze your FSM and decide whether an explicit reset port is needed. Recall that, as seen in sections 3.8 and 3.9, only occasionally is a state machine guaranteed to work properly without a dedicated reset signal.

5) *Input signals:* Two fundamental features must be observed with respect to the input signals. The first regards the aspect of such signals. For example, they might have glitches, or they might be too short or too long, so proper signal conditioning might be required. The second regards synchronism. If any input is asynchronous with respect to the FSM, analyze if metastability (section 2.3) can be critical to the application. If that is the case, and no other part of the circuit is taking care of metastability, add a synchronizer for each asynchronous input from which the machine must be protected. Do not forget to take into account the latency that this will cause.

6) *Code and compilation:* Write the corresponding VHDL/SystemVerilog code and synthesize it (design by hand is viable only for very simple circuits). Compare the number of flip-flops inferred by the compiler against your prediction.

7) *Simulation:* Fully simulate your design (graphically or, preferably, with VHDL/SystemVerilog testbenches). If the simulation is too time consuming, do functional simulation first until the design is debugged; then do timing simulation.

8) *Physical implementation:* Finally, download the resulting FPGA programming file (.pof or .sof) into the physical device in order to program it and proceed to the physical tests.

5 Regular (Category 1) State Machines

5.1 Introduction

We know that, from a hardware perspective, state machines can be classified into two types, based on their *input connections*, as follows.

1) *Moore machines:* The input, if it exists, is connected only to the logic block that computes the next state.
2) *Mealy machines:* The input is connected to both logic blocks, that is, for the next state and for the actual output.

In Section 3.6 we introduced a new classification, also from a hardware point of view, based on the *transition types* and *nature of the outputs*, as follows (see figure 5.1).

1) *Regular (category 1) state machines:* This category, illustrated in figure 5.1a and studied in chapters 5 to 7, consists of machines with only untimed transitions and outputs that do not depend on previous (past) output values.
2) *Timed (category 2) state machines:* This category, illustrated in figure 5.1b and studied in chapters 8 to 10, consists of machines with one or more transitions that depend on time (so they can have all four transition types: conditional, timed, conditional-timed, and unconditional). However, all outputs are still independent from previous (past) output values.
3) *Recursive (category 3) state machines:* This category is illustrated in figure 5.1c and studied in chapters 11 to 13. It can have all four types of transitions, but one or more outputs depend on previous (past) output values. Recall that the outputs are produced by the FSM's *combinational* logic block, so the current output values are "forgotten" after the machine leaves that state; consequently, to implement a recursive (recurrent) machine, some sort of extra memory is needed.

As seen in this and in upcoming chapters, the classifications mentioned above (no other classification is needed) will immensely ease the design of hardware-based

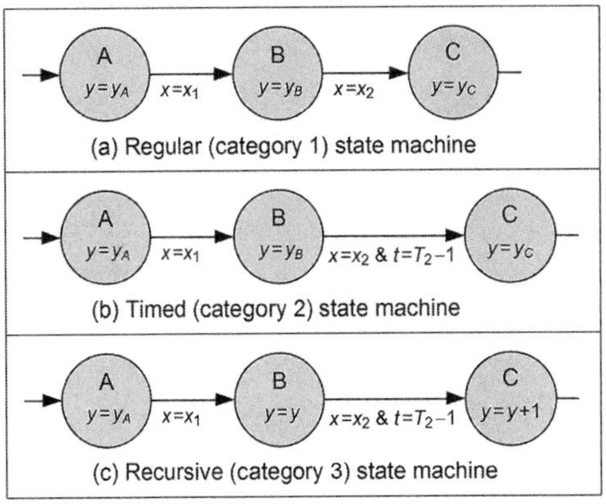
A
$y=y_A$
$x=x_1$
B
$y=y_B$
$x=x_2$
C
$y=y_C$
(a) Regular (category 1) state machine
A
$y=y_A$
$x=x_1$
B
$y=y_B$
$x=x_2$ & $t=T_2-1$
C
$y=y_C$
(b) Timed (category 2) state machine
A
$y=y_A$
$x=x_1$
B
$y=y$
$x=x_2$ & $t=T_2-1$
C
$y=y+1$
(c) Recursive (category 3) state machine

Figure 5.1
State machine categories (from a hardware perspective).

state machines. The two fundamental decisions before starting a design are then the following:

1) Decide the state machine category (regular, timed, or recursive).
2) Next, decide the state machine type (Moore or Mealy).

It is important to recall, however, that regardless of the machine category and type, the state transition diagram must fulfill three fundamental requisites (seen in section 1.3):

1) It must include all possible system states.
2) All state transition conditions must be specified (unless a transition is unconditional) and must be truly complementary.
3) The list of outputs must be exactly the same in all states (standard architecture).

5.2 Architectures for Regular (Category 1) Machines

The architectures for category 1 machines are summarized in figure 5.2. These representations follow the style of figures 3.1b,d, but the style of figures 3.1a,c could be used equivalently. The output register (figure 5.2c) is optional. The four possible constructions, listed in figure 5.2d, are summarized below.

Regular Moore machine (figure 5.2a): In this case, the input (if it exists) is connected only to the logic block for the next state. Consequently, the output depends only on the state in which the machine is (in other words, for each state, the output value in unique), resulting a synchronous behavior (see details in section 3.5). Because modern designs are generally synchronous, this implementation is preferred whenever the application permits.

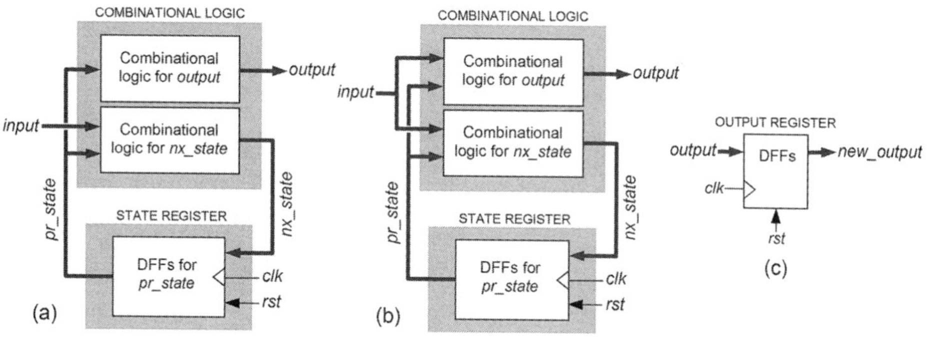
COMBINATIONAL LOGIC
Combinational logic for output
output
input
Combinational logic for nx_state
pr_state
nx_state
STATE REGISTER
DFFs for pr_state
clk
rst
(a)
COMBINATIONAL LOGIC
Combinational logic for output
output
input
Combinational logic for nx_state
pr_state
nx_state
STATE REGISTER
DFFs for pr_state
clk
rst
(b)
OUTPUT REGISTER
output
DFFs
new_output
clk
rst
(c)

(d)

Construction	Resulting category 1 circuit	Behavior
(a) only	Pure Moore machine (preferred)	Order-1 synchronous
(b) only	Pure Mealy machine	Asynchronous
(a) + (c)	Out-registered (pipelined) Moore machine	Order-2 synchronous
(b) + (c)	Out-registered (pipelined) Mealy machine	Order-1 synchronous

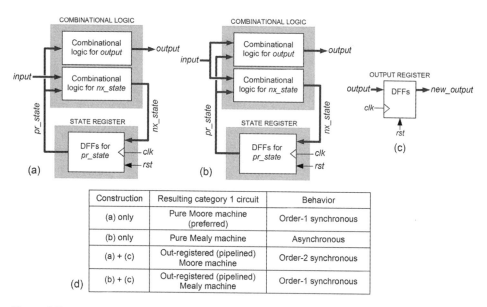

Figure 5.2
Regular (category 1) state machine architectures for (a) Moore and (b) Mealy types. (c) Optional output register. (d) Resulting circuits.

Regular Mealy machine (figure 5.2b): In this case, the input is connected to both logic blocks, so it can affect the output directly, resulting an asynchronous behavior. Therefore, the machine can have more than one output value for the same state (section 3.5).

Out-registered (pipelined) Moore machine: This consists of connecting the register of figure 5.2c to the output of the Moore machine of figure 5.2a. As seen in sections 2.5 and 2.6, two fundamental reasons for doing so are glitch removal and pipelined construction. As a result, the final circuit's output will be delayed with respect to the original machine's output by either one clock period (if the same clock edge is employed in the state register and in the output register) or by one-half of a clock period (if different clock edges are used). Note that the resulting circuit is order-2 synchronous because the original Moore machine was already a registered circuit (in other words, the input–output transfer occurs after two clock edges—see details in section 3.5). If in a given application this extra register is needed but its consequent extra delay is not acceptable, the next alternative can be used.

Out-registered (pipelined) Mealy machine: This consists of connecting the register of figure 5.2c to the output of the Mealy machine of figure 5.2b. The reasons for doing so are the same as for Moore machines. The resulting circuit is order-1 synchronous because the original Mealy machine is asynchronous. Consequently, the overall behavior (with the output register included) is similar to that of a pure Moore machine (without the output register—see details in section 3.5).

5.3 Number of Flip-Flops

In general, and particularly in large designs, it is difficult to estimate the number of logic gates that will be needed to implement the desired solution. However, it is always possible to determine, and *exactly*, the number of flip-flops.

In the case of sequential circuits implemented as category 1 state machines, there are two demands for DFFs, as follows (see state-encoding options in section 3.7).

1) For the state register (see *nx_state* and *pr_state* in figure 5.2a, which are the state memory flip-flops' input and output, respectively; below, M_{FSM} is the number of states):
   For sequential or Gray encoding: $N_{FSM} = \lceil \log_2 M_{FSM} \rceil$. Example: $M_{FSM} = 25 \rightarrow N_{FSM} = 5$.
   For Johnson encoding: $N_{FSM} = \lceil M_{FSM}/2 \rceil$. Example: $M_{FSM} = 25 \rightarrow N_{FSM} = 13$.
   For one-hot encoding: $N_{FSM} = M_{FSM}$. Example: $M_{FSM} = 25 \rightarrow N_{FSM} = 25$.
2) For the output register (figure 5.2c, optional, with b_{output} bits):
   $N_{output} = b_{output}$. Example: $b_{output} = 16 \rightarrow N_{output} = 16$.

Hence, the total is $N_{total} = N_{FSM} + N_{output}$. In the examples that follow, as well as in the actual designs with VHDL and SystemVerilog, the number of flip-flops will be often examined.

5.4 Examples of Regular (Category 1) Machines

A series of regular FSMs are presented next. Several of these examples are designed later using VHDL (chapter 6) and SystemVerilog (chapter 7).

5.4.1 Small Counters

Counters are well-known circuits easily designed without the FSM approach using VHDL or SystemVerilog. Moreover, a counter might have thousands of states, rendering it impractical for representation as a regular state machine. Nevertheless, for designing counters without the help of any EDA tool (as done in sections 3.3 and 3.4), the FSM model can be very helpful, particularly if the counter is not too big and has several control inputs such as enable and up-down. Moreover, the implementation of such counters can be very illustrative of the FSM approach. For these reasons, an example is included in this section.

A 1-to-5 counter with enable and up-down controls is presented in figure 5.3 (just to practice, equivalent detailed and simplified representations are shown—recall figure 1.4). The circuit counts if *ena* = '1', or stops (and holds its last output value) otherwise. If *up* = '1', the circuit counts from 1 to 5, restarting then automatically from 1; oth-

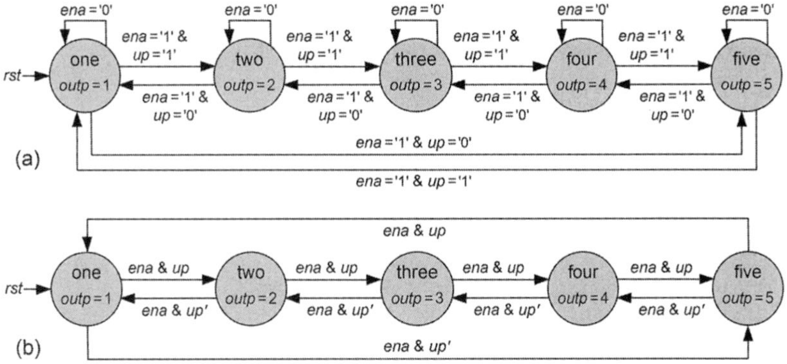
ena='0'
ena='1' & up='1'
ena='1' & up='0'
one
outp=1
two
outp=2
three
outp=3
four
outp=4
five
outp=5
rst
(a)
ena='1' & up='0'
ena='1' & up='1'
ena & up
ena & up'
(b)

Figure 5.3
Detailed (a) and simplified (b) representations for a 1-to-5 counter with enable and up-down controls.

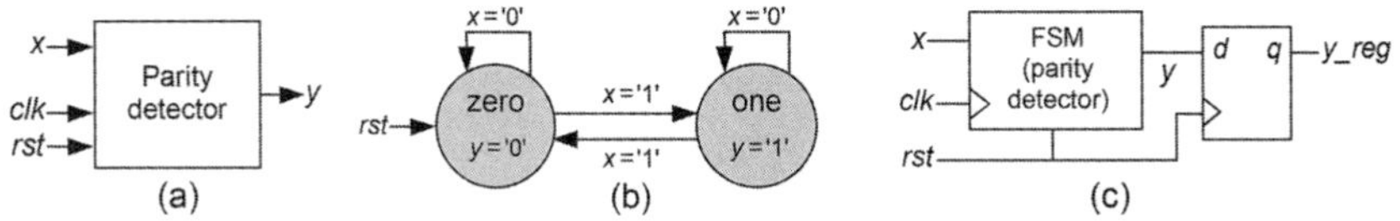
x
clk
rst
Parity detector
y
(a)
x='0'
x='1'
zero
y='0'
one
y='1'
rst
(b)
FSM (parity detector)
d q
y_reg
(c)

Figure 5.4
Parity detector. (a) Circuit ports. (b) State transition diagram. (c) Hardware block diagram.

erwise, it counts from 5 down to 1, restarting then automatically from 5. Because counters are inherently synchronous, the Moore model is the natural choice for their implementations.

Because this machine has $M_{FSM} = 5$ states, and the optional output register is generally not needed in counters, the number of flip-flops required to implement it (see section 5.3) is $N_{FSM} = 3$ if sequential, Gray, or Johnson encoding is used, or 5 for one-hot encoding.

VHDL and SystemVerilog implementations for this counter are presented in sections 6.6 and 7.5, respectively.

5.4.2 Parity Detector

This example concerns a circuit that detects the parity of a serial data stream. As depicted in figure 5.4a, x is the serial data input, and y is the circuit's response. The output must be y = '1' when the number of '1's in x is odd.

A basic solution for the case when a reset pulse is applied before every calculation starts is presented in figure 5.4b. In this case the parity value is the value of y after the last bit has been presented to the circuit (before a new reset pulse is applied). Note

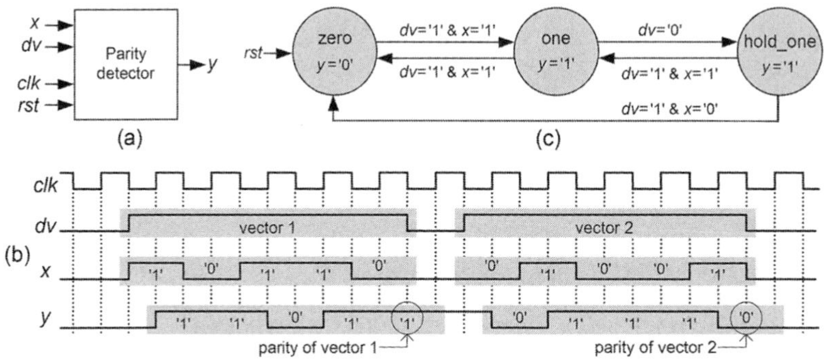
x
dv
clk
rst
Parity detector
y
(a)
rst
zero
y='0'
dv='1' & x='1'
dv='1' & x='1'
one
y='1'
dv='0'
dv='1' & x='1'
hold_one
y='1'
dv='1' & x='0'
(c)
clk
dv
vector 1
vector 2
(b)
x
y
parity of vector 1
parity of vector 2

Figure 5.5
Another parity detector. (a) Circuit ports. (b) Illustrative time behavior. (c) State transition diagram.

the arrangement in figure 5.4c, based on the material seen in section 3.11; when the reset pulse goes up (which subsequently resets the FSM), it causes the value of *y* to be stored in the auxiliary register, producing *y_reg*, which stays stable (constant) until a new calculation is completed (i.e., a new reset pulse occurs).

A slightly different parity detection problem is depicted in figure 5.5, which has to be reset only at power-up (thus a more usual situation). A data-valid (*dv*) bit indicates the extension of the data vector whose parity must be calculated (when *dv* goes up, a new vector begins, finishing when *dv* returns to zero). It is assumed that after a calculation (data stream) is completed, the machine must keep displaying the final parity value until a new vector is presented, as depicted in the illustrative timing diagram of figure 5.5b, which shows two vectors of size 5 bits each, with final parity *y* = '1' for vector 1 and *y* = '0' for vector 2.

A Moore machine that complies with these specifications is presented in figure 5.5c (note that in this example *dv* and *x* are updated at the negative clock edge). Because of *dv*, this machine does not need to be reset before a new calculation starts. Indeed, depending on the encoding scheme (sequential or Gray, for example), this circuit might not need a reset signal at all because deadlock cannot occur (the unused codeword will converge back to one of the machines' states) and *dv* will cause the computations to be correct even if the initial state is arbitrary (see exercise 3.11).

5.4.3 Basic One-Shot Circuit

One-shot circuits are circuits that, when triggered, generate a single voltage or current pulse, possibly with a fixed time duration. This section discusses the particular case in which the time duration of the output is exactly one clock period. In this example it will be considered that the input lasts at least one clock period; generic cases are studied in sections 8.11.8 to 8.11.10, which deal specifically with triggered circuits.

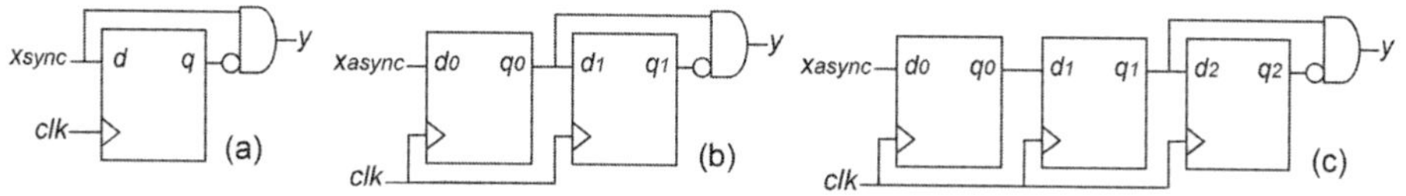
Xsync
d
q
y
clk
(a)
Xasync
d0
q0
d1
q1
y
clk
(b)
Xasync
d0
q0
d1
q1
d2
q2
y
clk
(c)

Figure 5.6
Trivial one-shot circuits. (a) Basic version, for synchronous input only. (b) Preceded by a synchronizing DFF, so the input can be asynchronous. (c) With a two-stage synchronizer.

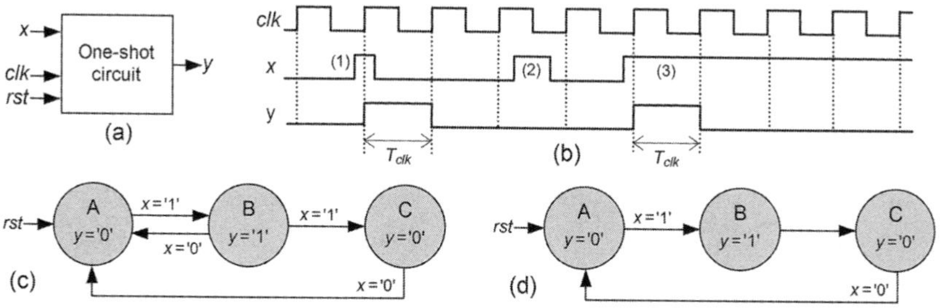
x
clk
rst
One-shot circuit
y
(a)
clk
x
y
(1)
(2)
(3)
Tclk
(b)
Tclk
rst
A
y='0'
x='1'
B
y='1'
x='0'
x='1'
C
y='0'
x='0'
(c)
rst
A
y='0'
x='1'
B
y='1'
C
y='0'
x='0'
(d)

Figure 5.7
One-shot state machine. (a) Circuit ports. (b) Example of expected behavior. (c) State transition diagram. (d) An inferior solution (exercise 5.5).

In fact, a one-shot circuit (not employing the FSM approach) was already seen in chapter 2 (figure 2.10), with its schematic repeated in figure 5.6a. This option, however, is fine only if the triggering input (x) is synchronous; otherwise, the output pulse could last less than T_{clk}. For it to work with asynchronous inputs, another DFF is needed, as shown in figure 5.6b. A version with a full synchronizer (section 2.3) is shown in figure 5.6c.

The general operating principle is illustrated in figure 5.7. The circuit ports are shown in figure 5.7a, where x is the triggering input and y is the one-shot output. An illustrative timing diagram is presented in figure 5.7b, with x having an arbitrary duration and y lasting exactly one clock period. Pulse 1 lasts less than T_{clk} but happened to fall under a positive clock edge, so it was detected. This is obviously not guaranteed to happen, as illustrated for pulse 2. Only if the duration is T_{clk} or longer, as for pulse 3, is the triggering of y guaranteed. Note that x and y are uncorrelated (mutually asynchronous) if x and *clk* are uncorrelated.

A solution using a regular (category 1) Moore machine is presented in figure 5.7c. Note that it stays in state B during only one clock period; because y = '1' occurs only in that state, the desired pulse results. An inferior solution is presented in figure 5.7d (see exercise 5.5).

As a final comment, let us consider the circuit of figure 5.6b, which is a kind of optimized synchronous version of the one-shot circuit. Because the solution in figure 5.7c is also synchronous (all Moore machines are), would you expect the circuit that implements this state machine to be equal or at least similar to that of figure 5.6b? (See exercise 5.5.)

5.4.4 Temperature Controller

Figure 5.8a shows a circuit diagram for a temperature controller of an air conditioning system. In the upper branch, the room temperature is sensed by some type of temperature sensor and converted to digital format by the ADC (analog-to-digital converter), producing the signal T_{room}. In the lower branch, the user, by means of two pushbuttons (*up*, *dn*), selects the reference (desired) temperature, producing the signal T_{ref}. Depending on the values of these two signals, the controller core decides whether to heat the room (h = '1'), to cool it (c = '1'), or to stay in the idle state.

Because mechanical switches are subject to bounces before they finally settle in the proper position, the pushbuttons must be debounced. However, debouncers are timed circuits, thus requiring a timed (category 2) machine to be implemented. Such machines are seen in chapter 8, so for now let us just consider that the proper value is produced for T_{ref} (the design of this block is treated in section 8.11.4). For example, T_{ref} could be selected in the 60°F to 90°F range with an initial value (on power-up, defined by the reset signal) of 73°F, if degrees Fahrenheit are used, or in the 15°C to 30°C range with a default value of 23°C, if degrees centigrade are employed instead.

An important addition to the system is depicted in figure 5.8b, which consists of a display accessed by means of a multiplexer. The display shows the room temperature while the selection pushbutton (*sel*, with no need for debouncing, not shown in the figure) is at rest (*sel* = '0') or the reference temperature while it is pressed (*sel* = '1').

A state machine for the controller core, using the Moore approach, is depicted in figure 5.8c. ΔT represents the system hysteresis, which is generally a fixed circuit

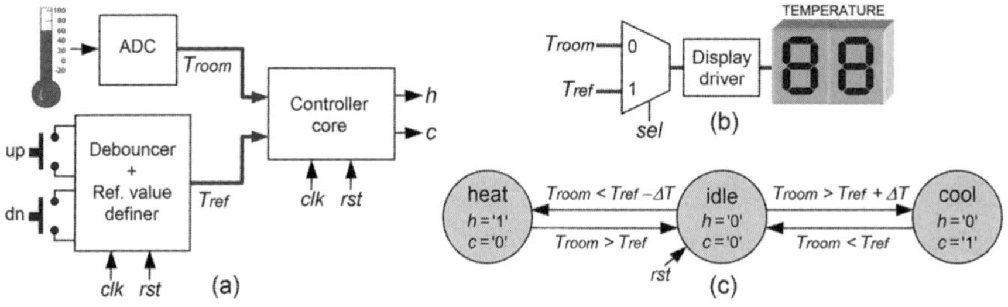
ADC
Troom
Controller core
h
c
up
dn
Debouncer + Ref. value definer
Tref
clk rst
clk rst
(a)
Troom
Tref
0
1
Display driver
sel
(b)
TEMPERATURE
heat
h='1'
c='0'
Troom < Tref − ΔT
Troom > Tref
idle
h='0'
c='0'
rst
Troom > Tref + ΔT
Troom < Tref
cool
h='0'
c='1'
(c)

Figure 5.8
Temperature controller. (a) Overall circuit diagram. (b) Display driver. (c) State machine for the controller core block.

parameter. For example, if ΔT = 1°F, the room temperature will be kept within T_{ref} ± 1°F. By comparing T_{room} to T_{ref} and taking into account the hysteresis, the machine will be able to produce the proper values for *h* and *c*.

Finally, note that the inputs from the pushbuttons are asynchronous with respect to the system clock, which could, in principle, cause metastability (see section 2.3). This, however, is prevented here by the debouncer (section 8.11.3).

5.4.5 Garage Door Controller

This example presents a garage door controller that operates as follows. If the door is completely closed or completely open and the remote is activated, the motor is turned on in the direction to open or close it, respectively. If the door is opening or closing and the remote is activated, the door stops. If the remote is activated again, the motor is turned on to move the door in the opposite direction.

The circuit ports are depicted in figure 5.9a, where *remt* (command from the remote control), *sen1* (door-open sensor), and *sen2* (door-closed sensor) are the inputs (plus the conventional *clk* and *rst* signals), and *ctr* (control) is the output. Note that *ctr* has two bits; *ctr*(1) turns the motor on ('1') or off ('0'), whereas *ctr*(0) defines its direction, opening ('0') or closing ('1') the door (thus the value of the latter does not matter when the former is '0').

A preliminary state diagram is shown in figure 5.9b. The transition control signals are *remt*, *sen1*, and *sen2*. Note that this machine complies with all three requisites of

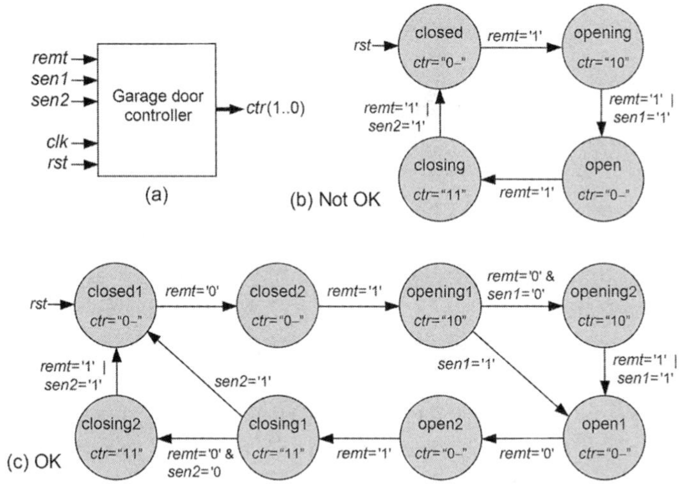
remt
sen1
sen2
clk
rst
Garage door controller
ctr(1..0)
(a)
rst
closed
ctr="0-"
remt='1'
opening
ctr="10"
remt='1' | sen1='1'
remt='1' | sen2='1'
closing
ctr="11"
remt='1'
open
ctr="0-"
(b) Not OK
rst
closed1
ctr="0-"
remt='0'
closed2
ctr="0-"
remt='1'
opening1
ctr="10"
remt='0' & sen1='0'
opening2
ctr="10"
remt='1' | sen2='1'
sen2='1'
sen1='1'
remt='1' | sen1='1'
closing2
ctr="11"
remt='0' & sen2='0
closing1
ctr="11"
remt='1'
open2
ctr="0-"
remt='0'
open1
ctr="0-"
(c) OK

Figure 5.9
Garage door controller. (a) Circuit ports. (b) Bad solution (with state-bypass). (c) Good solution.

section 1.3. However, it exhibits a major problem, which is state bypass (see section 4.2.4). For example, if the door is closed and a long (lasting several clock cycles) *remt* = ‘1’ command is received, the machine goes around the entire loop. Of course, if a one-shot circuit (section 5.4.3) is used to reduce the duration of *remt* to a single clock period, then this machine is fine.

A corrected diagram is presented in figure 5.9c, containing additional states that wait for *remt* to return to zero before proceeding, thus eliminating the state-bypass problem. This is a Moore machine because there is no reason to employ an asynchronous solution in this kind of application. Glitches at the output are not a problem here, so the optional output register is not needed.

A good practice in this kind of application is to include debouncers for the signals coming from the remote control and from the sensors, which not only eliminate the need for synchronizers but also prevent short input glitches (due to lightning or the switching of large electric currents, for example) from activating the machine (in this case, it has to be a full debouncer, like that in section 8.11.3, for example).

Because the machine of figure 5.9c has $M_{FSM} = 8$ states, the required number of DFFs is $N_{FSM} = 3$ if sequential or Gray encoding is used, 4 for Johnson, or 8 for one-hot.

VHDL and SystemVerilog implementations for this garage door controller are presented in sections 6.7 and 7.6, respectively.

5.4.6 Vending Machine Controller

This example deals with a controller for a vending machine. It is assumed that it sells candy bars for the single price of \$0.40, accepting nickel, dime, and quarter coins.

The circuit ports are depicted in figure 5.10a. The inputs *nickel_in*, *dime_in*, and *quarter_in* are generated by the coin collector, informing the type of coin that was deposited by the customer. The inputs *nickel_out* and *dime_out* are generated by the coin dispenser mechanism, informing the type of coin that was returned to the customer. The last nonoperational input is *candy_out*, produced by the candy dispenser mechanism, informing that a candy was delivered to the customer. The outputs *disp_nickel* and *disp_dime* tell the coin dispenser mechanism that a nickel or a dime must be returned to the customer, while the output *disp_candy* tells the candy bar dispenser mechanism that a candy bar must be delivered to the customer.

A corresponding Moore machine is presented in figure 5.10b. To simplify the notation, numbers were used instead of names (see other examples of equivalent state diagram representations in section 1.4). The state names correspond to the accumulated amount (*credit*). The transition conditions refer to the last coin entered, with negative values indicating change returned to the customer. In the coin-return operations it was opted to deliver the largest coins possible. After the machine reaches the state 40 (thick circle), the only way to return to the initial state is by receiving a

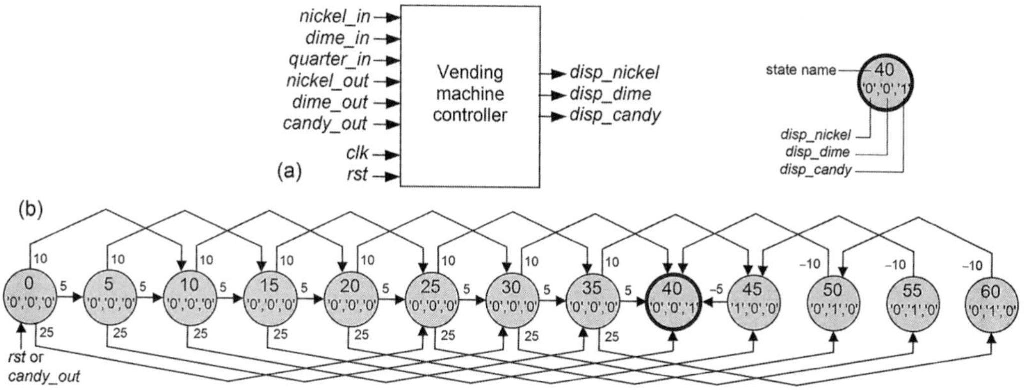

nickel_in
dime_in
quarter_in
nickel_out
dime_out
candy_out
clk
rst
Vending machine controller
disp_nickel
disp_dime
disp_candy
(a)
state name
disp_nickel
disp_dime
disp_candy
(b)
rst or candy_out

Figure 5.10

Controller for a vending machine that sells candy bars for \$0.40, accepting nickels, dimes, and quarters. (a) Circuit ports. (b) Corresponding Moore machine (state-bypass prevention not included).

candy_out = '1' command from the candy-delivering mechanism confirming that a candy bar was dispensed or a reset pulse.

Note that the machine of figure 5.10b is subject to state bypass (section 4.2.4) if the inputs last longer than one clock period (which is generally the case in this kind of application), so wait states (or a flag or one-shot conversion) must be added (exercise 5.11).

Because glitches are definitely not acceptable in this application, the optional output register should be used here. In regard to the inputs, we can assume that they are produced by other circuits that process the actual inputs and hence operate with the same clock as our state machine, dispensing with the use of debouncers and/or synchronizers (although they might be needed at the inputs of preceding circuits).

If we assume that all control inputs to this machine last exactly one clock period (due to one-shot circuits, for example), so state bypass cannot occur and additional states are not needed, the number of DFFs required to build it (with M_{FSM} = 13 states) is N_{FSM} = 4 if sequential or Gray encoding is used, 7 for Johnson, or 13 for one-hot, plus N_{output} = 3 for the output register.

5.4.7 Datapath Control for an Accumulator

Before we examine this example, a review of section 3.13 is suggested.

In this example we assume that the datapath of figure 3.22a must operate as an add-and-accumulate circuit (ACC), accumulating in register A four consecutive values of *inpB*. The data-valid bit (*dv*), when asserted (during just one clock period), will again be responsible for starting the computations, after which the resulting value must remain displayed at *ALUout* until another pulse occurs in *dv*. In summary, the operations are: 0 + B → A, A + B → A, A + B → A, and A + B → A.

Recall that in a datapath-based design the FSM is not responsible for implementing the whole computation but just the *control unit* (shown on the left in figure 3.22a), which controls the datapath. In other words, the FSM must produce the signals *selA* (selects the data source for register A), *wrA* and *wrB* (enable writing into registers A and B), and *ALUop* (produces the ALU opcode, defining its operations, according to the table in figure 3.22b).

An illustrative timing diagram (similar to what was done in figure 3.22c) for an FSM that controls this datapath such that the desired accumulator results is presented in figure 5.11a. Note that the computations take five steps (called *start, acc1, acc2, acc3*, and *acc4*), after which the control unit (FSM) returns to the *idle* state (so the machine has six states). The corresponding state transition diagram, which is a direct translation of the timing diagram (compare the values in the timing diagram against those in the state transition diagram), is exhibited in figure 5.11b. Observe that this control unit is indeed a category 1 machine.

Because this machine has $M_{FSM} = 6$ states, and the optional output register is generally not needed in control units, the number of flip-flops required to implement it (see section 5.3) is $N_{FSM} = 3$ if sequential, Gray, or Johnson encoding is used or 6 for one-hot.

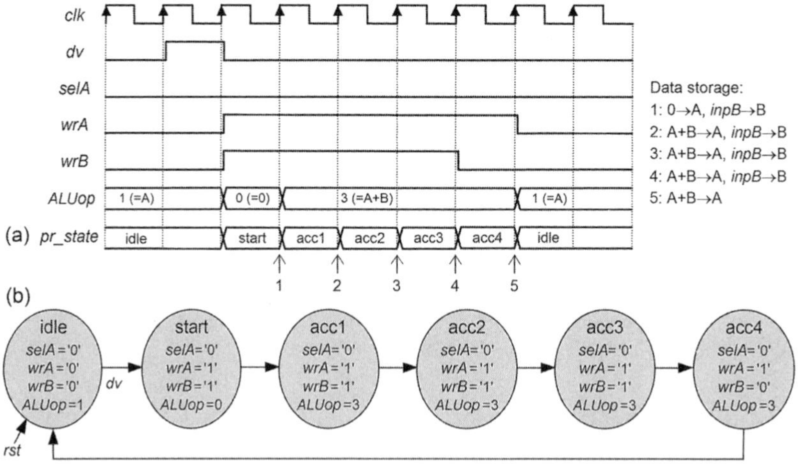
clk
dv
selA
wrA
wrB
ALUop
1 (=A)
0 (=0)
3 (=A+B)
1 (=A)
(a) pr_state
idle
start
acc1
acc2
acc3
acc4
idle
1
2
3
4
5
Data storage:
1: 0→A, inpB→B
2: A+B→A, inpB→B
3: A+B→A, inpB→B
4: A+B→A, inpB→B
5: A+B→A
(b)
idle
selA='0'
wrA='0'
wrB='0'
ALUop=1
dv
start
selA='0'
wrA='1'
wrB='1'
ALUop=0
acc1
selA='0'
wrA='1'
wrB='1'
ALUop=3
acc2
selA='0'
wrA='1'
wrB='1'
ALUop=3
acc3
selA='0'
wrA='1'
wrB='1'
ALUop=3
acc4
selA='0'
wrA='1'
wrB='0'
ALUop=3
rst

Figure 5.11

(a) Illustrative timing diagram for the datapath of figure 3.22a operating as an accumulator. (b) Corresponding Moore machine.

5.4.8 Datapath Control for a Greatest Common Divisor Calculator

Before we examine this example, a review of section 3.13 is suggested. Particular attention should be paid to comment number 4 at the end of that section, which is helpful here.

This section shows another example of a datapath-based circuit. The datapath must compute the GCD (greatest common divisor) between two integers. The corresponding algorithm is shown in figure 5.12; the largest value is substituted with the difference between it and the other value until the values become equal, which is then declared to be the GCD. A corresponding flowchart is also included in figure 5.12. As in the previous example, a *dv* bit, when asserted (during one clock period), must start the computations.

The datapath to be used in this example is depicted in figure 5.13a. The ALU's opcode table is shown in figure 5.13b. The ALU has also an auxiliary output (*sign*) that indicates whether its output (*ALUout*) is zero ("00"), positive ("01"), or negative ("10"), as listed in figure 5.13c.

As shown, the datapath's control signals are *selA* and *selB* (select the data sources for registers A and B), *wrA* and *wrB* (enable writing into registers A and B), and *ALUop* (produces the ALU opcode, defining its operations, according to the table in figure 5.13b). The control unit (FSM) is responsible for generating all control signals.

An illustrative timing diagram for an FSM that controls this datapath such that the desired computations occur is presented in figure 5.13d. Dashed lines indicate "don't care" values. Because *inpA* = 9 and *inpB* = 15 were adopted, the following computations are expected: Iteration 1, 9 → A, 15 → B; Iteration 2, B > A, then 15 – 9 = 6 → B; Iteration 3, A > B, then 9 – 6 = 3 → A; Iteration 4, B > A, so 6 – 3 = 3 → B. Because A = B, GCD = A = 3.

Observe in figure 5.13d that the time slots are identified as *idle* (waiting for a *dv* bit), *load* (*inpA* and *inpB* are stored in A and B), *writeA* (*ALUout* is stored in A), and

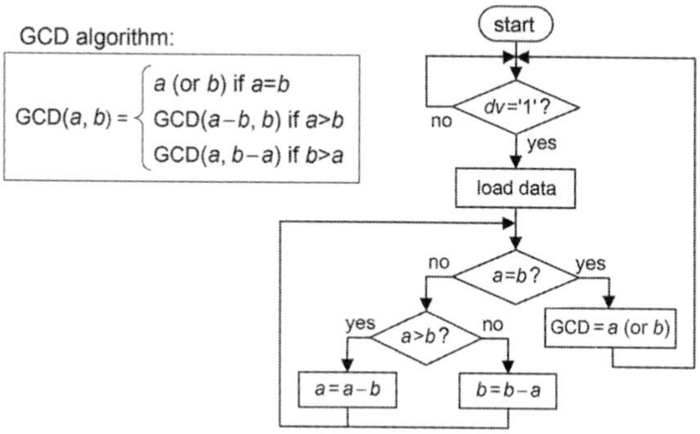
GCD algorithm:
GCD(a, b) =
a (or b) if a=b
GCD(a−b, b) if a>b
GCD(a, b−a) if b>a
start
dv='1'?
no
yes
load data
a=b?
no
yes
a>b?
yes
no
GCD = a (or b)
a = a−b
b = b−a

Figure 5.12
GCD algorithm and flowchart.

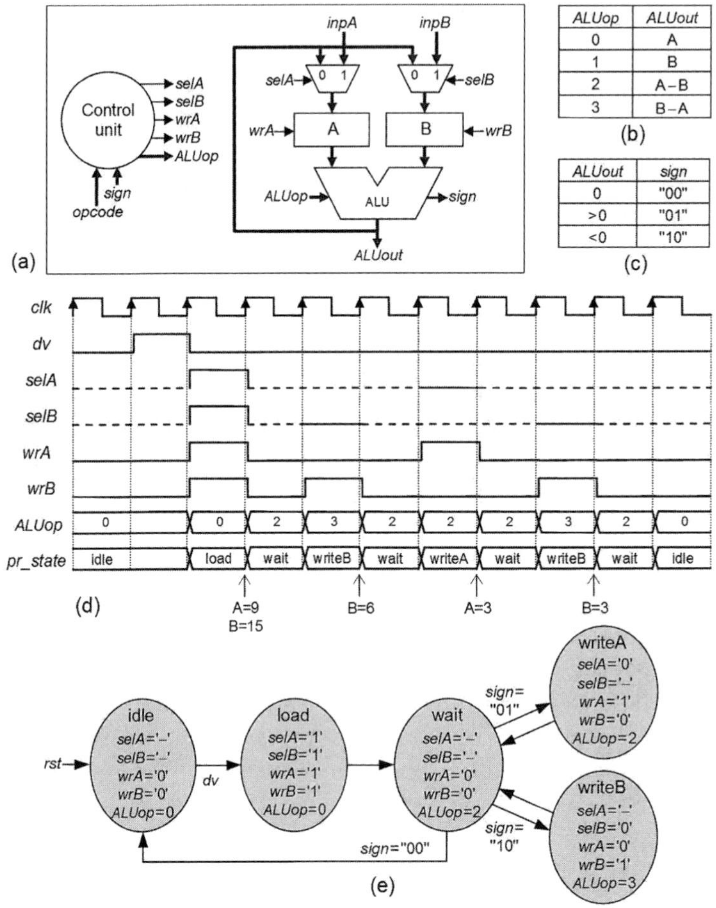
(a)
Control unit
selA
selB
wrA
wrB
ALUop
sign
opcode
inpA
inpB
0 1
A
B
ALU
ALUout
(b)
(c)
clk
dv
selA
selB
wrA
wrB
ALUop
0 0 2 3 2 2 2 3 2 0
pr_state
idle load wait writeB wait writeA wait writeB wait idle
(d)
A=9
B=15
B=6
A=3
B=3
rst
idle
selA='–'
selB='–'
wrA='0'
wrB='0'
ALUop=0
dv
load
selA='1'
selB='1'
wrA='1'
wrB='1'
ALUop=0
wait
selA='–'
selB='–'
wrA='0'
wrB='0'
ALUop=2
sign="01"
writeA
selA='0'
selB='–'
wrA='1'
wrB='0'
ALUop=2
sign="10"
writeB
selA='–'
selB='0'
wrA='0'
wrB='1'
ALUop=3
sign="00"
(e)

ALUop	ALUout
0	A
1	B
2	A–B
3	B–A

(b)

ALUout	sign
0	"00"
>0	"01"
<0	"10"

(c)

Figure 5.13

(a) Datapath and control unit for a GCD calculator. (b) ALU's opcode table. (c) ALU's sign table. (d) Illustrative timing diagram, for *inpA* = 9 and *inpB* = 15. (e) Corresponding state machine.

writeB (*ALUout* is stored in B). Observe also the presence of a *wait* time slot after every data storage, which is needed for the data to be effectively ready for comparison before an actual comparison occurs (recall comment 4 of section 3.13).

A corresponding state transition diagram is presented in figure 5.13e, which is a direct translation of the timing diagram (compare the values in the plots against those in the state transition diagram). Note that after each write-enabling state (*load*, *writeA*, and *writeB*) the machine goes unconditionally to the *wait* state. In the *idle* state, $wrA = wrB =$ '0', so nothing can be written into the registers, and because $ALUop = 0$, the output is $ALUout =$ A, so the computed GCD value is kept unchanged until *dv* is asserted again.

VHDL and SystemVerilog implementations for this control unit are presented in sections 6.8 and 7.7, respectively.

5.4.9 Generic Sequence Detector

This is another interesting example from a conceptual point of view. Say that we want to design a signature detector that searches for the string "*abc*" in a sequential data stream, examining one character at a time (a character here represents a bit vector with any number of bits). So this is exactly the same problem presented in the very first state transition diagram of the book (figure 1.3, repeated in figure 5.14a). In this example it was assumed that $a \neq b \neq c$, so this machine works well. But let us consider now a completely generic situation, in which *a*, *b*, and *c* are *programmable*, so we can no longer assume that they are all different. Will this machine still work?

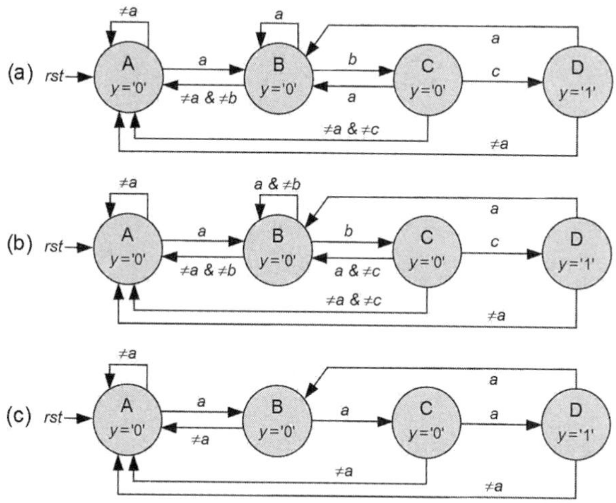
(a)
rst
A
y='0'
B
y='0'
C
y='0'
D
y='1'
≠a
a
a
b
c
≠a & ≠b
a
≠a & ≠c
≠a
(b)
rst
A
y='0'
B
y='0'
C
y='0'
D
y='1'
≠a
a & ≠b
a
b
c
≠a & ≠b
a & ≠c
≠a & ≠c
≠a
(c)
rst
A
y='0'
B
y='0'
C
y='0'
D
y='1'
≠a
a
a
a
≠a
≠a
≠a

Figure 5.14

Generic string detection. (a) Nongeneric case (requires $a \neq b \neq c$). (b) Completely generic implementation due to the inclusion of priorities in the transition conditions. (c) Example for the case of $a = b = c$.

To answer this question, let us assume that $a = b$, so b can be replaced with a in figure 5.14a. Consequently, state B (for example) has the following transition conditions: a in the BB transition; a also in the BC transition; and $\neq a$ & $\neq b = \neq a$ in the BA transition. This shows that state B is now *overspecified* because both BB and BC transitions are governed by the same condition (a). Therefore, this machine is not fine for generic values of a, b, and c.

The new question then is "How do we fix overspecifications?" We do it in the way explained in section 1.5, that is, with the establishment of *priorities*. This is done in figure 5.14b. For state B, the BC transition must have priority over the BB transition, so the transition condition in the former remains just b, while that in the latter becomes a & $\neq b$. Likewise, for state C, the CD transition must have priority over the CB transition; thus, the transition condition in the former remains c, whereas that in the latter becomes a & $\neq c$.

As an example, figure 5.14c shows the extreme case in which $a = b = c$. Then $\neq a$ & $\neq b = \neq a$, $\neq a$ & $\neq c = \neq a$, a & $\neq b$ = null (so the BB transition disappears), and a & $\neq c$ = null (the CB transition also disappears).

The only restriction of this generic string detector is that it detects only nonoverlapping strings.

5.4.10 Transparent Circuits

We close this chapter with the description of a special (although uncommon) type of circuit for FSMs, which consists of sequential circuits that are required to be "transparent" (i.e., the output must "see" the input; in other words, if the input changes, so should the output). If implemented using an FSM, the circuit must provide outputs that are capable of changing when the input changes, even if the machine remains in the same state.

As an example, consider the case in figure 5.15a, with inputs a and b and output y. The output must be $y = a$ during one clock period, $y = a \cdot b$ during the next period, and finally $y = b$ during the third clock cycle, with this sequence repeated indefinitely. Corresponding Moore and Mealy diagrams are included in figures 5.15b,c. Note that because the machine must go to the next state at every clock cycle, its transitions are unconditional.

Because in this case the output depends solely on the machine's state, a Moore machine seems to be the natural choice. However, because the output must change when the input changes, a Mealy machine, being asynchronous, would be recommended. In fact, both are fine.

In the Moore case the transparency problem is circumvented by associating the machine with switches such as the multiplexer in figure 5.15d, in which case the machine plays just the role of mux selector (in this example, the resulting machine is clearly just a 0-to-2 counter), so even though the machine is not transparent, the

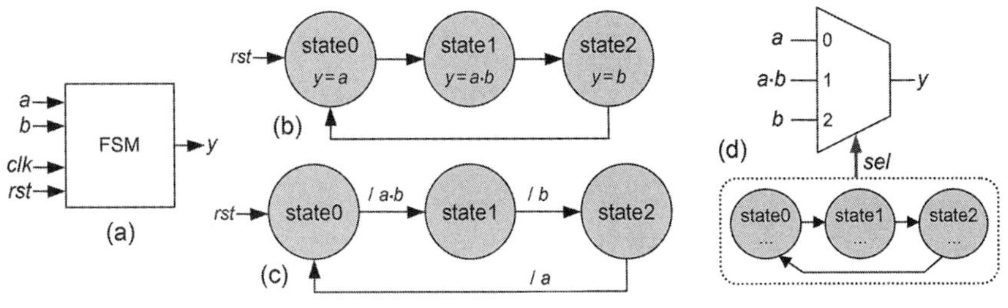
a
b
clk
rst
FSM
y
(a)
rst
state0
y=a
state1
y=a·b
state2
y=b
(b)
rst
state0
/ a·b
state1
/ b
state2
(c)
/ a
(d)
a
a·b
b
0
1
2
y
sel
state0
state1
state2

Figure 5.15
A "transparent" circuit. (a) Circuit ports. (b) Moore and (c) Mealy state transition diagrams. (d) Typical implementation based on the Moore model.

overall circuit is (this is typically what a VHDL/SystemVerilog compiler would do). In the Mealy case the implementation is straightforward, but the output will be one clock cycle ahead of the desired sequence (compare figures 5.15b and 5.15c).

5.4.11 LCD, I²C, and SPI Interfaces

Three special additional design examples are presented in chapter 14, consisting of circuits for interfacing with alphanumeric LCD displays and for implementing I^2C or SPI serial interfaces. Depending on the application, any of the three FSM categories might be needed in these circuits; for instance, in the LCD driver example of section 14.1, a category 1 FSM is employed, whereas in the I^2C and SPI serial interfaces of sections 14.2 and 14.3, categories 2 and 3 are used.

5.5 Exercises

Exercise 5.1: Machine Category and Number of Flip-Flops

a) Why are the state machines in figures 5.3, 5.9c, and 5.13e (among others) said to be of category 1?

b) How many DFFs are needed to implement each of these FSMs using (*i*) sequential encoding, (*ii*) Gray encoding, or (*iii*) one-hot encoding?

Exercise 5.2: Metastability and Synchronizer

a) Solve exercise 2.2 if not done yet.

b) Consider now the garage door controller of figure 5.9. (*i*) Which inputs are asynchronous? (*ii*) If no debouncing circuits (which are synchronous) are adopted for the asynchronous inputs, are synchronizers indispensable in this application?

Exercise 5.3: Need for Reset

a) Solve exercise 3.10 if not done yet.

b) Solve exercise 3.11 if not done yet.

Exercise 5.4: Truly Complementary Transition Conditions

In section 1.5 the importance of having the state transition diagram neither under- nor overspecified was discussed. What happens if, in the garage door controller of figure 5.9c, the condition *sen1* = '0' is removed from the *opening1-opening2* transition, or the condition *sen2* = '0' is removed from the *closing1-closing2* transition?

Exercise 5.5: One-Shot Circuits Analysis

a) It is said in section 5.4.3 that the solution in figure 5.7d is inferior to that in figure 5.7c. Why? (Suggestion: fill in the last two plots of figure 5.16 and you will see the answer.)

b) Is reset indispensable in these two solutions?

c) In order to answer the question posed at the end of section 5.4.3, solve exercise 3.3 if not done yet.

Exercise 5.6: Two-Signal-Triggered One-Shot Circuit

Figure 5.17 shows an illustrative timing diagram for a one-shot circuit that is not triggered by a single signal but rather by a pair of signals. The triggering condition is the following: the one-shot pulse (in *y*) must be generated if the control signal *x* lasts at least as long as the *dv* pulse (this is obviously checked only at positive clock transitions). Note in the figure that only the first pulse of *x* fulfills this requirement, so the one-clock-period pulse in *y* has to be produced only in that case. Draw the state transition diagram for a state machine capable of implementing this circuit.

Exercise 5.7: Arbiter

Arbiters are used to manage access to shared resources. An example is depicted in figure 5.18, which shows three peripherals (P1 to P3) that use a common bus

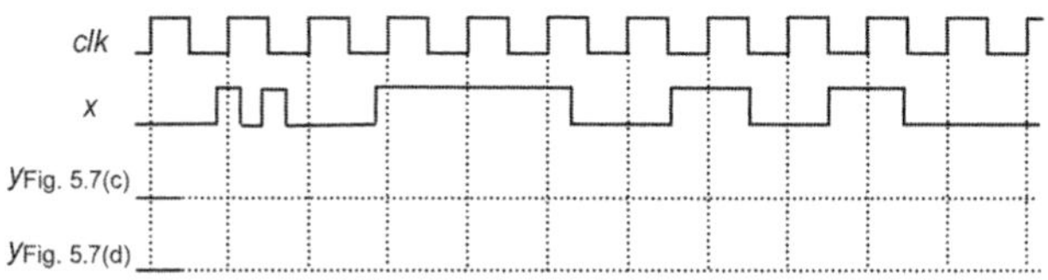
clk
x
yFig. 5.7(c)
yFig. 5.7(d)

Figure 5.16

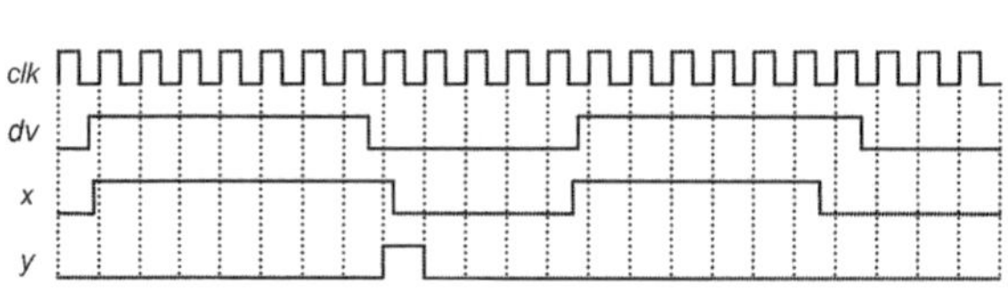
clk
dv
x
y

Figure 5.17

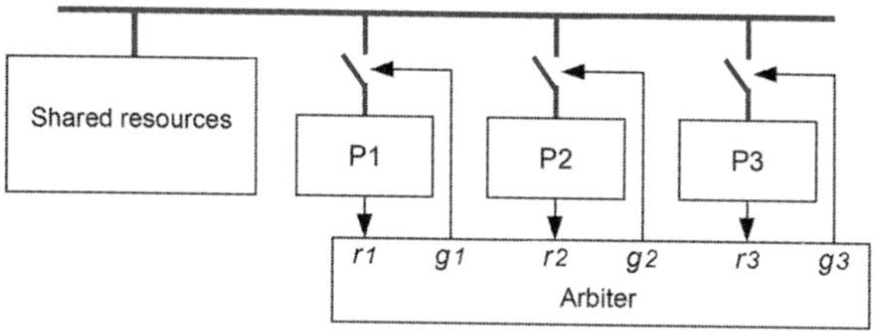
Shared resources
P1
P2
P3
r1
g1
r2
g2
r3
g3
Arbiter

Figure 5.18

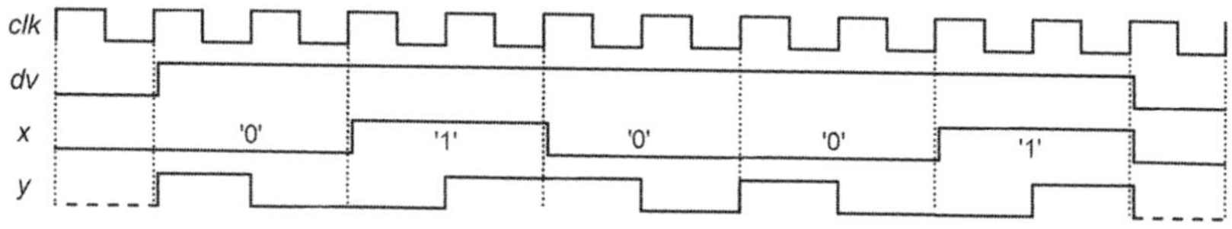
clk
dv
x
y
'0'
'1'
'0'
'0'
'1'

Figure 5.19

to access common resources. Obviously, only one of them can use the bus at a time; for example, if P1 wants to use the bus, it issues a request (r_1 = '1') to the arbiter, which grants (g_1 = '1') access only if the bus is idle at that moment. If multiple requests are received by the arbiter, access is granted based on preestablished priorities. Assuming that the priorities are P1 > P2 > P3, draw a state transition diagram for a machine capable of implementing this arbiter. The machine's input and output are the vectors $r = r_1r_2r_3$ and $g = g_1g_2g_3$, respectively (besides clock and reset, of course).

Exercise 5.8: Manchester Encoder

An IEEE Manchester encoder produces a low-to-high transition when the input is '1' or a high-to-low transition when it is '0', as illustrated in figure 5.19 for the sequence "01001". Note that each input value lasts two clock periods. Observe also the presence of a *dv* bit, which defines the extent of the vector to be encoded (dashed lines in *y* indicate "don't care" values). To be more realistic, *dv* is produced at the same time that the first valid bit is presented; additionally, a small propagation delay is included between clock transitions and corresponding responses. Assume that the machine too must operate at the positive clock edge.

a) Draw a state transition diagram for a Moore machine capable of implementing this encoder.

b) Redraw the illustrative timing diagram of figure 5.19 for your Moore machine, including in it a plot for *pr_state*. Does the Moore circuit behave exactly as in figure 5.19, or is *y* one clock cycle delayed?

c) Redo the design, this time employing a Mealy machine.

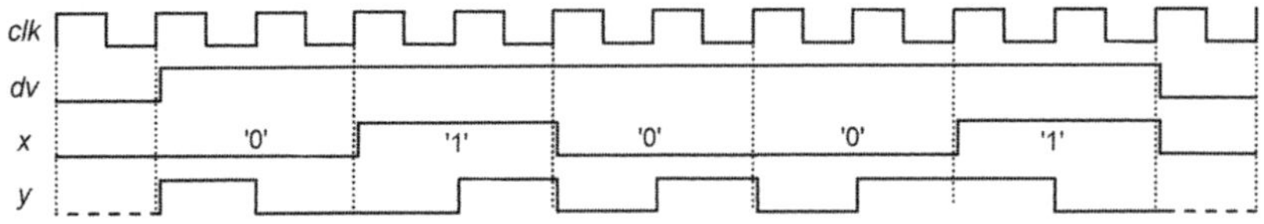
clk
dv
x
y
'0'
'1'
'0'
'0'
'1'

Figure 5.20

d) Repeat part b now for your Mealy solution.

e) Say that we want the output to be completely clean. Are any of the solutions above guaranteed to be glitch-free? If not, how can glitches be removed? What happens then with the time response?

Exercise 5.9: Differential Manchester Encoder

Figure 5.20 illustrates the operation of a differential Manchester encoder for the sequence "01001". Note that the shape of the output pulse remains unchanged when the input is '0' but gets inverted when it is '1'. For example, if the last pulse was a '1'-to-'0' pulse, the next pulse must be '1'-to-'0' if the input is '0' or '0'-to-'1' if it is '1'. Observe the presence of a *dv* bit, which defines the extent of the vector to be encoded (dashed lines in *y* indicate "don't care" values). To be more realistic, *dv* is produced at the same time that the first valid bit is presented; additionally, a small propagation delay has been included between the clock transitions and the corresponding responses. Assume that the machine too must operate at the positive clock edge.

a) Draw a state transition diagram for a Moore machine capable of implementing this encoder.

b) Redraw the illustrative timing diagram of figure 5.20 for your solution, including in it a plot for *pr_state*. Does the Moore circuit behave exactly as in figure 5.20, or is *y* one clock cycle delayed?

Exercise 5.10: Time-Ordered "111" Detector

Draw the state transition diagram for an FSM that detects the sequence *abc* = "111" under the constraint that it must be time ordered; that is, *a* = '1' must occur (and hold), then *b* = '1' must also occur (and hold), and finally, *c* = '1' must happen. The circuit ports are shown in figure 5.21a. The circuit operation is illustrated in figure 5.21b, where *x* = '1' occurs when *abc* = "111", but in a time-ordered fashion.

Exercise 5.11: Vending Machine

It was seen that the vending machine controller of figure 5.10b must be improved to avoid state bypass. Present a solution for this problem. Is it better to include wait

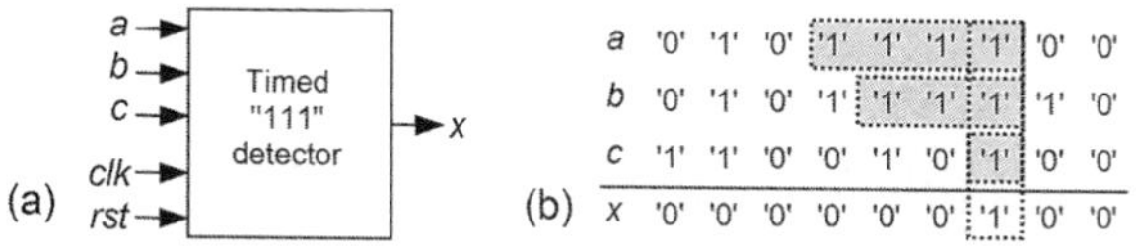
(a)
a
b
c
clk
rst
Timed "111" detector
x
(b)
a '0' '1' '0' '1' '1' '1' '1' '0' '0'
b '0' '1' '0' '1' '1' '1' '1' '1' '0'
c '1' '1' '0' '0' '1' '0' '1' '0' '0'
x '0' '0' '0' '0' '0' '0' '1' '0' '0'

Figure 5.21

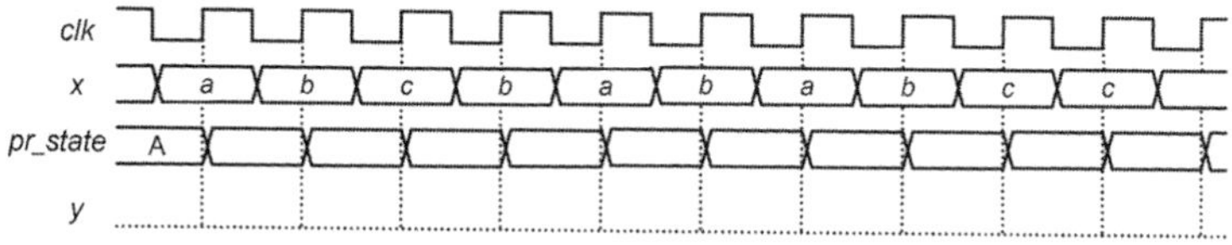
clk
x
a
b
c
b
a
b
a
b
c
c
pr_state
A
y

Figure 5.22

states or a flag or to convert the inputs into one-shot signals with one-clock-period duration?

Exercise 5.12: Time Behavior of a String Detector

Consider the Moore-type state machine of figure 5.14a, which detects the sequence "*abc*" for the case of $a \neq b \neq c$, where x and y represent the input and output, respectively.

a) Complete the timing diagram of figure 5.22 for the given values of x. Note that a little propagation delay was included between the clock transitions and the respective changes in the present state; do the same for y.
b) Does the output go up immediately when the sequence "*abc*" occurs or only at the next (positive) clock edge? Is this result as you expected? (Recall that Moore machines are fully synchronous.)

Exercise 5.13: Generic Overlapping String Detector

We saw in section 5.4.9 a generic approach for the implementation of nonoverlapping string detectors. In that case, if the sequence to be detected were "*aba*", for example, the response to the serial bit stream "*abababab*..." would be "00100010001...", whereas here, because overlaps must be allowed, it should be "0010101...". Can you find a generic solution (with or without a state machine) for this case?

Exercise 5.14: Keypad Encoder

Figure 5.23a shows a 12-key keypad for which we need to design an encoder (and possibly also a debouncer—debouncers are discussed in chapter 8). The actual push-button connections can be seen in figure 5.23b, where $r(3{:}0)$ and $c(2{:}0)$ represent the keypad's rows and columns, respectively. Note that because of the pull-up resistors,

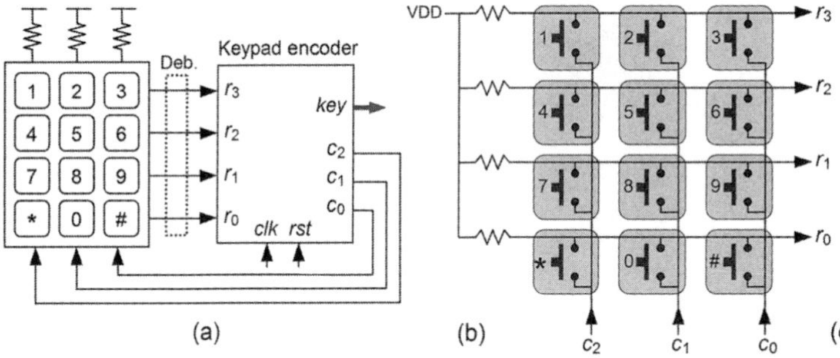
Keypad encoder
Deb.
key
clk rst
VDD
(a)
(b)

(c)

c(2:0)	r(3:0)	key(3:0)
011	0111	0001 (1)
	1011	0100 (4)
	1101	0111 (7)
	1110	1010 (*)
	1111	1111 (n)
101	0111	0010 (2)
	1011	0101 (5)
	1101	1000 (8)
	1110	0000 (0)
	1111	1111 (n)
110	0111	0011 (3)
	1011	0110 (6)
	1101	1001 (9)
	1110	1011 (#)
	1111	1111 (n)

Figure 5.23

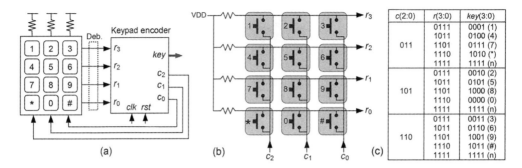

the rows' voltages are all high when no switch is pressed. The keypad encoder must connect the bottom of one column at a time to ground ('0'), then read the resulting row values, converting them into the respective codeword, as listed in figure 5.23c (n stands for "none"); for example, if c = "011", which means that the leftmost column is being inspected, and the reading is r = "1011", then we know that pushbutton 4 is pressed. Present a solution for this encoder. (A possible solution for the debouncer is treated in exercise 8.11.)

Exercise 5.15: Datapath Controller for a Largest-Value Detector

Say that you are given the datapath of figure 5.13a, with *inpB* monitoring a serial data stream, of which the largest value must be determined (placed at the ALU output, *ALUout*). The monitoring should start when a *dv* bit is asserted, ending when *dv* returns to zero.

a) Develop a state transition diagram (as in figure 5.13e) for an FSM capable of implementing the corresponding control unit. Include in it "nop" (no operation) states if necessary to have the number of clock cycles be the same in all iterations.
b) Present an illustrative timing diagram for your machine (as in figure 5.13d), assuming that the values presented to the circuit (while *dv* = '1') are $5 \rightarrow 8 \rightarrow 4 \rightarrow 0$. (If you prefer, do part b before part a.)

Exercise 5.16: Datapath Controller for a Square Root Calculator

To calculate $z = (x^2 + y^2)^{1/2}$, where x, y, and z are unsigned integers, the expression $z = \max(a - a/8 + b/2, a)$ can be used, where $a = \max(x, y)$ and $b = \min(x, y)$. Recall that to divide an integer by 8 or by 2 all that is needed is to shift it to the right three positions or one position, respectively. Make the adjustments that you find necessary in the datapath of figure 5.13a (for example, include a shift-right option in one of the existing registers or in a new register at the ALU output), then devise a state machine that computes the square root above using that datapath.

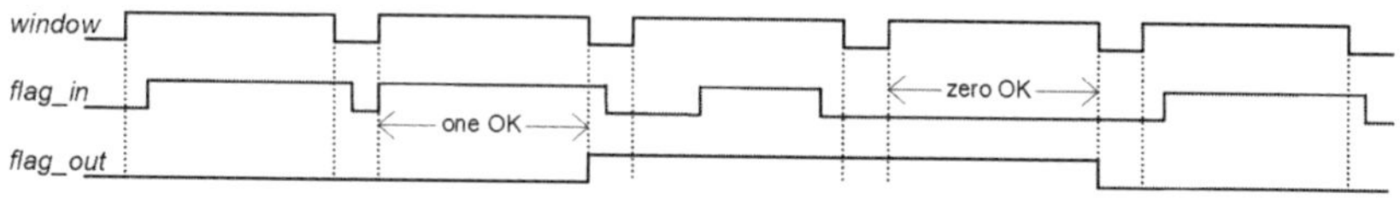
window
flag_in
flag_out
one OK
zero OK

Figure 5.24

Exercise 5.17: Flag Monitor

Develop an FSM for a circuit that monitors a flag such that, if the flag remains constant within a given time window, the output copies the measured (constant) flag value. This is illustrated in figure 5.24; if *flag_in* has no transitions at all while *window* is high, then *flag_out* gets the value of *flag_in*; otherwise, it keeps the same value that it had when the time window started.

6 VHDL Design of Regular (Category 1) State Machines

6.1 Introduction

This chapter presents several VHDL designs of category 1 state machines. It starts by presenting two VHDL templates, for Moore- and Mealy-based implementations, which are used subsequently to develop a series of designs related to the examples introduced in chapter 5.

The codes are always complete (not only partial sketches) and are accompanied by comments and simulation results, illustrating the design's main features. All circuits were synthesized using Quartus II (from Altera) or ISE (from Xilinx). The simulations were performed with Quartus II or ModelSim (from Mentor Graphics). The default encoding scheme for the states of the FSMs was regular sequential encoding (see encoding options in section 3.7; see ways of selecting the encoding scheme at the end of section 6.3).

The same designs will be presented in chapter 7 using SystemVerilog, so the reader can make a direct comparison between the codes.

Note: See suggestions of VHDL books in the bibliography.

6.2 General Structure of VHDL Code

A typical structure of VHDL code for synthesis, with all elements that are needed in this and in coming chapters, is depicted in figure 6.1. It is composed of three fundamental sections, briefly described below.

Library/Package Declarations

As the name says, it contains the libraries and corresponding packages needed in the design. The most common package is *std_logic_1164*, from the IEEE library, which defines the types *std_logic* (for single bit) and *std_logic_vector* (for multiple bits), which are the industry standard.

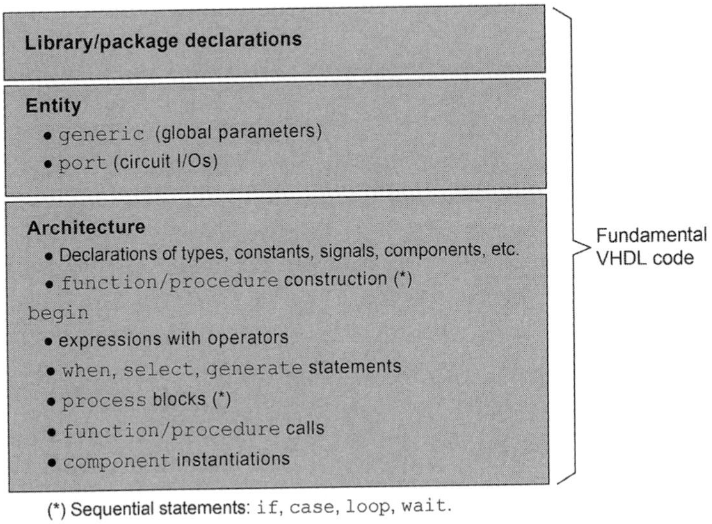
Library/package declarations
Entity
• generic (global parameters)
• port (circuit I/Os)
Architecture
• Declarations of types, constants, signals, components, etc.
• function/procedure construction (*)
begin
• expressions with operators
• when, select, generate statements
• process blocks (*)
• function/procedure calls
• component instantiations
(*) Sequential statements: if, case, loop, wait.
Fundamental VHDL code

Figure 6.1
Typical VHDL code structure for synthesis.

Entity

The entity is divided into two main parts, called **generic** and **port**.

Generic: This portion is optional. It is used for the declaration of global parameters, which can be easily modified to fulfill different system specifications or, more importantly, can be overridden during instantiations (using the **component** construct) into other designs.

Port: This part of the code is mandatory for synthesis. It is just a list with specifications of all circuit ports (I/Os), including their name, mode (**in**, **out**, **inout**, or **buffer**), and type (plus range).

Architecture

The architecture too is divided into two parts, called *declarative part* and *statements part*.

Declarative part: This section precedes the keyword **begin** and is optional. It is used for all sorts of local declarations, including **type**, **signal**, and **component**. It also allows the construction of **function** and **procedure**. These declarations and functions/procedures can also be placed outside the main code, in a **package**.

Statements part: This portion, which starts at the keyword **begin**, constitutes the code proper. As shown in figure 6.1, its main elements (in no particular order) are the following: basic expressions using operators (for simple combinational circuits); expressions using concurrent statements (**when**, **select**, **generate**), generally for simple to midcomplexity combinational circuits; sequential code using **process**, which is constructed using sequential statements (**if**, **case**, **loop**, **wait**), for sequential as well as (complex) combinational circuits; **function/procedure** calls; and, finally, **component** (that is, other design) instantiations, resulting in structural designs.

6.3 VHDL Template for Regular (Category 1) Moore Machines

The template is based on figure 6.2 (derived from figure 5.2), which shows three processes: 1) for the FSM state register; 2) for the FSM combinational logic; and 3) for the optional output register. Note the asterisk on one of the input connections; as we know, if that connection exists it is a Mealy machine, else it is a Moore machine.

There obviously are other ways of breaking the code instead of using the three processes indicated in figure 6.2. For example, the combinational logic section, being not sequential, could be implemented without a process (using purely concurrent code). At the other extreme the combinational logic section could be implemented with two processes, one with the logic for *output*, the other with the logic for *nx_state*.

The VHDL template for the design of category 1 Moore machines, based on figures 6.1 and 6.2, is presented below. Observe the following:

1) To improve readability, the three fundamental code sections (library/package declarations, entity, and architecture) are separated by dashed lines (lines 1, 4, 14, 76).
2) The library/package declarations (lines 2–3) show the package *std_logic_1164*, needed because the types used in the ports of all designs will be *std_logic* and/or *std_logic_vector* (industry standard).
3) The entity, called *circuit*, is in lines 5–13. As seen in figure 6.1, it usually contains two parts: **generic** (optional) and **port** (mandatory for synthesis). The former is employed for the declaration of generic parameters (if they exist), as illustrated in lines 6–8. The latter is a list of all circuit ports, with respective specifications, as illustrated

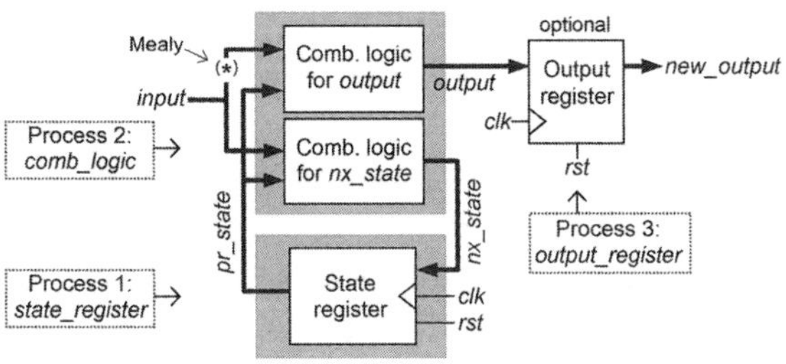
Mealy
(*)
input
Process 2:
comb_logic
Process 1:
state_register
Comb. logic
for output
Comb. logic
for nx_state
pr_state
nx_state
output
State
register
clk
rst
optional
Output
register
new_output
clk
rst
Process 3:
output_register

Figure 6.2
State machine architecture depicting how the VHDL code was broken (three processes).

in lines 9–12. Note that the type used for all ports (lines 10–12) is indeed *std_logic* or *std_logic_vector*.

4) The architecture, called *moore_fsm*, is in lines 15–75. It too is divided into two parts: declarative part (optional) and statements part (code proper, so mandatory).

5) The declarative part of the architecture is in lines 16–19. In lines 16–17 a special enumerated type, called *state*, is created, and then the signals *pr_state* and *nx_state* are declared using that type. In lines 18–19 an optional attribute called *enum_encoding* is shown, which defines the type of encoding desired for the machine's states (e.g., "sequential", "one-hot"). Another related attribute is *fsm_encoding*. See a description for both attributes after the template below. The encoding scheme can also be chosen using the compiler's setup, in which case lines 18–19 can be removed.

6) The statements part (code proper) of the architecture is in lines 20–75 (from **begin** on). In this template it is composed of three **process** blocks, described below.

7) The first process (lines 23–30) implements the state register (process 1 of figure 6.2). Because all of the machine's DFFs are in this section, clock and reset are only connected to this block (plus to the optional output register, of course, but that is not part of the FSM proper). Note that the code for this process is essentially standard, simply copying *nx_state* to *pr_state* at every positive clock transition (thus inferring the DFFs that store the machine's state).

8) The second process (lines 33–61) implements the entire combinational logic section of the FSM (process 2 of figure 6.2). This part must contain all states (A, B, C, . . .), and for each state two things must be declared: the output values/expressions and the next state. Observe, for example, in lines 36–46, relative to state A, the output declarations in lines 37–39 and the next-state declarations in lines 40–46. A very important point to note here is that there is no **if** statement associated with the outputs because in a Moore machine the outputs depend solely on the state in which the machine is, so for a given state each output value/expression is unique.

9) The third and final process (lines 64–73) implements the optional output register (process 3 of figure 6.2). Note that it simply copies each original output to a new output at every positive clock edge (it could also be at the negative edge), thus inferring the extra register. If this register is used, then the names of the new outputs must obviously be the names used in the corresponding port declarations (line 12). If the initial output values do not matter, reset is not required in this register.

10) To conclude, observe the completeness of the code and the correct use of registers (as requested in sections 4.2.8 and 4.2.9, respectively), summarized below.

a) Regarding the use of registers: The circuit is not overregistered. This can be observed in the **elsif rising_edge(clk)** statement of line 27 (responsible for the inference of flip-flops), which is closed in line 29, guaranteeing that only the machine state (line 28) gets registered. The circuit outputs are in the next process, which is purely combinational.

b) Regarding the outputs: The list of outputs (*output1*, *output2*, . . .) is the same in all states (see lines 37–39, 48–50, . . .), and the output values (or expressions) are always declared.

c) Regarding the next state: Again, the coverage is complete because all states (A, B, C, . . .) are included, and in each state the declarations are finalized with an **else** statement (lines 44, 55, . . .), guaranteeing that no condition is left unchecked.

Note 1: See also the comments in sections 6.4, which show some template variations.

Note 2: The VHDL 2008 review of the VHDL standard added the keyword **all** as a replacement for a process' sensitivity list, so **process (all)** is now valid. It also added boolean tests for *std_logic* signals and variables, so **if x='1' then . . .** can be replaced with **if x then. . . .** Both are supported by the current version (12.1) of Altera's Quartus II compiler but not yet by the current version (14.2) of Xilinx's ISE suite (XST compiler).

Note 3: Another implementation approach, for simple FSMs, will be seen in chapter 15.

```
1    ------------------------------------------------------------
2    library ieee;
3    use ieee.std_logic_1164.all;
4    ------------------------------------------------------------
5    entity circuit is
6       generic (
7          param1: std_logic_vector(...) := <value>;
8          param2: std_logic_vector(...) := <value>);
9       port (
10         clk, rst: in std_logic;
11         input1, input2, ...: in std_logic_vector(...);
12         output1, output2, ...: out std_logic_vector(...);
13   end entity;
14   ------------------------------------------------------------
15   architecture moore_fsm of circuit is
16      type state is (A, B, C, ...);
17      signal pr_state, nx_state: state;
18      attribute enum_encoding: string; --optional, see comments
19      attribute enum_encoding of state: type is "sequential";
20   begin
21
22      --FSM state register:
23      process (clk, rst)
24      begin
25         if rst='1' then  --see Note 2 above on boolean tests
26            pr_state <= A;
27         elsif rising_edge(clk) then
28            pr_state <= nx_state;
29         end if;
30      end process;
31
32      --FSM combinational logic:
33      process (all) --see Note 2 above on "all" keyword
34      begin
35         case pr_state is
```

```
36            when A =>
37               output1 <= <value>;
38               output2 <= <value>;
39               ...
40               if <condition> then
41                  nx_state <= B;
42               elsif <condition> then
43                  nx_state <= ...;
44               else
45                  nx_state <= A;
46               end if;
47            when B =>
48               output1 <= <value>;
49               output2 <= <value>;
50               ...
51               if <condition> then
52                  nx_state <= C;
53               elsif <condition> then
54                  nx_state <= ...;
55               else
56                  nx_state <= B;
57               end if;
58            when C =>
59               ...
60         end case;
61      end process;
62
63      --Optional output register:
64      process (clk, rst)
65      begin
66         if rst='1' then   --rst generally optional here
67            new_output1 <= ...;
68            ...
69         elsif rising_edge(clk) then
70            new_output1 <= output1;
71            ...
72         end if;
73      end process;
74
75   end architecture;
76   ------------------------------------------------------------
```

Final Comments

1) On the need for a reset signal: Note in the template above that the sequential portion of the FSM (process of lines 23–30) has a reset signal. As seen in sections 3.8 and 3.9, that is the usual situation. However, as also seen in those sections, if the circuit is implemented in an FPGA (so the flip-flops are automatically reset on power-up) and the codeword assigned to the intended initial (reset) state is the all-zero codeword, then reset will occur automatically.

2) On the *enum_encoding* and *fsm_encoding* attributes: As mentioned earlier, these attributes can be used to select the desired encoding scheme ("sequential", "one-hot", "001 011 010", and others—see options in section 3.7), overriding the compiler's

setup. It is important to mention, however, that support for these attributes varies among synthesis compilers. For example, Altera's Quartus II has full support for *enum_encoding*, so both examples below are fine (where "sequential" can also be "one-hot", "gray", and so on):

```
attribute enum_encoding: string;
attribute enum_encoding of state: type is "sequential";
attribute enum_encoding: string;
attribute enum_encoding of state: type is "001 100 101"; --user defined
```

Xilinx's XST (from the ISE suite), on the other hand, only supports *enum_encoding* for user-defined encoding; for the others ("sequential", "one-hot", etc.), *fsm_encoding* can be used. Two valid examples are shown below:

```
attribute enum_encoding: string;
attribute enum_encoding of state: type is "001 100 101";
attribute fsm_encoding: string;
attribute fsm_encoding of pr_state: signal is "sequential";
```

6.4 Template Variations

The template of section 6.3 can be modified in several ways with little or no effect on the final result. Some options are described below. These modifications are extensible to the Mealy template treated in the next section.

6.4.1 Combinational Logic Separated into Two Processes

A variation sometimes helpful from a didactic point of view is to separate the FSM combinational logic process into two processes: one for the output, another for the next state. Below, the process for the output logic is in lines 33–47, and that for the next state logic is in lines 50–69.

```
32   --FSM combinational logic for output:
33   process (all)
34   begin
35      case pr_state is
36         when A =>
37            output1 <= <value>;
38            output2 <= <value>;
39            ...
40         when B =>
41            output1 <= <value>;
42            output2 <= <value>;
43            ...
44         when C =>
```

```
         ...
      end case;
   end process;

   --FSM combinational logic for next state:
   process (all)
   begin
      case pr_state is
         when A =>
            if <condition> then
               nx_state <= B;
            elsif <condition> then
               nx_state <= ...;
            else
               nx_state <= A;
            end if;
         when B =>
            if <condition> then
               nx_state <= C;
            ...
            end if;
         when C =>
            ...
      end case;
   end process;
```

6.4.2 State Register Plus Output Register in a Single Process

A variation in the other direction (reducing the number of processes from three to two instead of increasing it to four) consists of joining the process for the state register with that for the output register. This is not recommended for three reasons. First, in most projects the optional output register is not needed. Second, having the output register in a separate process helps remind the designer that the need or not for such a register is an important case-by-case decision. Third, one might want to have the output register operating at the other (negative) clock edge, which is better emphasized by using separate processes.

6.4.3 Using Default Values

When the same signal or variable value appears several times inside the *same* process, a default value can be entered at the beginning of the process. An example is shown below for the process of the combinational logic section, with default values for the outputs included in lines 36–38. In lines 40–45 only the values that disagree with these must then be typed in. An example in which default values are used is seen in section 12.4.

```
32   --FSM combinational logic:
33   process (all)
34   begin
```

```
      --Default values:
      output1 <= <value>;
      output2 <= <value>;
      ...
      --Code:
      case pr_state is
         when A =>;
            ...
         when B =>
            ...
      end case;
   end process;
```

6.4.4 A Dangerous Template

A tempting template is shown next. Note that the entire FSM is in a single process (lines 17–43). Its essential point is that the **elsif rising_edge(clk)** statement encloses the whole circuit (it opens in line 21 and only closes in line 42), thus registering it completely (that is, not only the state is stored in flip-flops—this has to be done anyway—but also all the outputs).

This template has several *apparent* advantages. One is that a shorter code results (for instance, we can replace *pr_state* and *nx_state* with a single name—*fsm_state*, for example; also, only one process is needed). Another apparent advantage is that the code will work (no latches inferred) when the list of outputs is not exactly the same in all states. Such features, however, might hide serious problems.

One of the problems is precisely the fact that the outputs are always registered, so the resulting circuit is never the FSM alone but the FSM plus the optional output register of figure 5.2c, which many times is unwanted.

Another problem is that, even if the optional output register were needed, we do not have the freedom to choose in which of the clock edges to operate it because the same edge is used for the FSM and for the output register in this template, reducing the design flexibility.

A third problem is the fact that, because the list of outputs does not need to be the same in all states (because they are registered, latches will not be inferred when an output value is not specified), the designer is prone to overlook the project specifications.

Finally, it is important to remember that VHDL (and SystemVerilog) is not a program but a code, and a shorter code *does not mean* a smaller or better circuit. In fact, longer, better-organized codes tend to ease the compiler's work, helping to optimize the final circuit.

In summary, the template below is a *particular case* of the general template introduced in section 6.3. The general template gets reduced to this one only when all outputs must be registered and the same clock edge must operate both the state register and the output register.

```
--Dangerous template (particular case of the general template)
library ieee;
use ieee.std_logic_1164.all;
---------------------------------------------------------------
entity circuit is
   generic (...);
   port (
      clk, rst: in std_logic;
      input, ...: in std_logic_vector(...);
      output, ...: out std_logic_vector(...);
end entity;
---------------------------------------------------------------
architecture moore_fsm of circuit is
   type state is (A, B, C, ...);
   signal fsm_state: state;
begin
   process (clk, rst)
   begin
      if rst then
         fsm_state <= A;
      elsif rising_edge(clk) then
         case fsm_state is
            when A =>
               output <= <value>;
               if <condition> then
                  fsm_state <= B;
               elsif <condition> then
                  fsm_state <= ...;
               else
                  fsm_state <= A;
               end if;
            when B =>
               output <= <value>;
               if <condition> then
               ...
               else
                  fsm_state <= B;
               end if;
            when C =>
            ...
         end case;
      end if;
   end process;
---------------------------------------------------------------
```

6.5 VHDL Template for Regular (Category 1) Mealy Machines

This template, also based on figures 6.1 and 6.2, is presented below. The only difference with respect to the Moore template just presented is in the process for the combinational logic because the output is specified differently now. Recall that in a Mealy machine the output depends not only on the FSM's state but also on its input, so **if** statements are expected for the output in one or more states because the output values might not be unique. This is achieved by including the output *within* the conditional statements for *nx_state*. For example, observe in lines 20–36, relative to state A, that the output values are now conditional. Compare these lines against lines 36–46 in the template of section 6.3.

Please review the following comments, which can easily be adapted from the Moore case to the Mealy case:

—On the Moore template for category 1, in section 6.3, especially comment 10.
—On the *enum_encoding* and *fsm_encoding* attributes, also in section 6.3.
—On possible code variations, in section 6.4.

```
1    --------------------------------------------------------------
2    library ieee;
3    use ieee.std_logic_1164.all;
4    --------------------------------------------------------------
5    entity circuit is
6       (same as for category 1 Moore, section 6.3)
7    end entity;
8    --------------------------------------------------------------
9    architecture mealy_fsm of circuit IS
10      (same as for category 1 Moore, Section 6.3)
11   begin
12
13      --FSM state register:
14      (same as for category 1 Moore, section 6.3)
15
16      --FSM combinational logic:
17      process (all) --list proc. inputs if "all" not supported
18      begin
19         case pr_state is
20            when A =>
21               if <condition> then
22                  output1 <= <value>;
23                  output2 <= <value>;
24                  ...
25                  nx_state <= B;
26               elsif <condition> then
27                  output1 <= <value>;
28                  output2 <= <value>;
29                  ...
30                  nx_state <= ...;
31               else
32                  output1 <= <value>;
33                  output2 <= <value>;
34                  ...
35                  nx_state <= A;
36               end if;
37            when B =>
38               if <condition> then
39                  output1 <= <value>;
40                  output2 <= <value>;
41                  ...
42                  nx_state <= C;
43               elsif <condition> then
```

```
                output1 <= <value>;
                output2 <= <value>;
                ...
                nx_state <= ...;
             else
                output1 <= <value>;
                output2 <= <value>;
                ...
                nx_state <= B;
             end if;
          when C =>
             ...
       end case;
    end process;

    --Optional output register:
    (same as for category 1 Moore, section 6.3)

end architecture;
--------------------------------------------------------------
```

6.6 Design of a Small Counter

This section presents a VHDL-based design for the 1-to-5 counter with enable and up-down controls introduced in section 5.4.1 (figure 5.3).

Because counters are inherently synchronous, the Moore approach is the natural choice for their implementation, so the VHDL template of section 6.3 is used. Because possible glitches during (positive) clock transitions are generally not a problem in counters, the optional output register shown in the last process of the template is not employed.

The entity, called *counter*, is in lines 5–9. All ports are of type *std_logic* or *std_logic_vector* (industry standard).

The architecture, called *moore_fsm*, is in lines 11–88. As usual, it contains a declarative part (before the keyword **begin**) and a statements part (from **begin** on).

In the declarative part of the architecture (lines 12–13), the enumerated type *state* is created to represent the machine's present and next states. Recall that when neither the *enum_encoding* nor the *fsm_encoding* attribute is used, the encoding scheme must be selected in the compiler's setup.

The first process (lines 17–24) in the statements part implements the state register. As in the template, this is a standard code with clock and reset present only in this process.

The second and final process (lines 27–86) implements the entire combinational logic section. It is just a list of all states, each containing the output value and the next state. Note that in each state the output value is unique because in a Moore machine the output depends only on the state in which the machine is.

Observe the correct use of registers and the completeness of the code, as described in comment 10 of section 6.3. Note in particular the following:

1) Regarding the use of registers: The circuit is not overregistered. This can be observed in the **elsif rising_edge(clk)** statement of line 21 (responsible for the inference of flip-flops), which is closed in line 23, guaranteeing that only the machine state (line 22) gets stored. The output (*outp*) is in the next process, which is purely combinational (thus not registered).
2) Regarding the outputs: The list of outputs (just *outp* in this example) is exactly the same in all states (see lines 31, 42, 53, 64, 75), and the corresponding output values are always properly declared.
3) Regarding the next state: Again, the coverage is complete because all states are included (see lines 30, 41, 52, 63, 74), and in each state the conditional declarations for the next state are always finalized with an **else** statement (lines 38, 49, 60, 71, 82), guaranteeing that no condition is left unchecked.

```
1    ----------------------------------------------------------
2    library ieee;
3    use ieee.std_logic_1164.all;
4    ----------------------------------------------------------
5    entity counter is
6       port (
7          ena, up, clk, rst: in std_logic;
8          outp: out std_logic_vector(2 downto 0));
9    end entity;
10   ----------------------------------------------------------
11   architecture moore_fsm of counter is
12      type state is (one, two, three, four, five);
13      signal pr_state, nx_state: state;
14   begin
15
16      --FSM state register:
17      process (clk, rst)
18      begin
19         if rst='1' then
20            pr_state <= one;
21         elsif rising_edge(clk) then
22            pr_state <= nx_state;
23         end if;
24      end process;
25
26      --FSM combinational logic:
27      process (all) --list proc. inputs if "all" not supported
28      begin
29         case pr_state is
30            when one =>
31               outp <= "001";
32               if ena='1' then
33                  if up='1' then
34                     nx_state <= two;
35                  else
```

```
                  nx_state <= five;
               end if;
            else
               nx_state <= one;
            end if;
         when two =>
            outp <= "010";
            if ena='1' then
               if up='1' then
                  nx_state <= three;
               else
                  nx_state <= one;
               end if;
            else
               nx_state <= two;
            end if;
         when three =>
            outp <= "011";
            if ena='1' then
               if up='1' then
                  nx_state <= four;
               else
                  nx_state <= two;
               end if;
            else
               nx_state <= three;
            end if;
         when four =>
            outp <= "100";
            if ena='1' then
               if up='1' then
                  nx_state <= five;
               else
                  nx_state <= three;
               end if;
            else
               nx_state <= four;
            end if;
         when five =>
            outp <= "101";
            if ena='1' then
               if up='1' then
                  nx_state <= one;
               else
                  nx_state <= four;
               end if;
            else
               nx_state <= five;
            end if;
      end case;
   end process;

end architecture;
--------------------------------------------------------------
```

Synthesis results using the VHDL code above are presented in figure 6.3. The circuit's structure can be seen in the RTL view of figure 6.3a, while the FSM can be seen

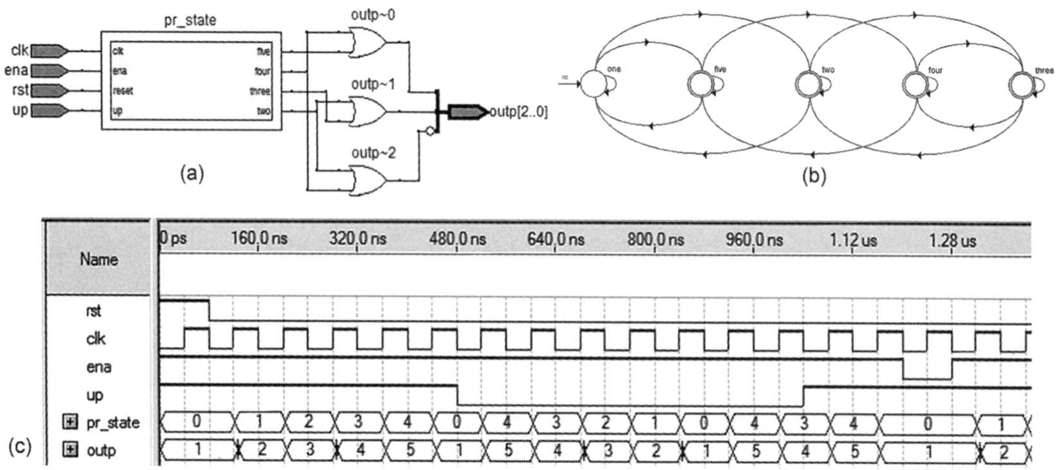
pr_state
clk
ena
rst
up
clk
ena
reset
up
five
four
three
two
outp~0
outp~1
outp~2
outp[2..0]
(a)
one
five
two
four
three
(b)
Name
0 ps
160,0 ns
320,0 ns
480,0 ns
640,0 ns
800,0 ns
960,0 ns
1.12 us
1.28 us
rst
clk
ena
up
pr_state
outp
(c)

Figure 6.3
Results from the VHDL code for the 1-to-5 counter with enable and up-down controls of figure 5.3. (a) RTL view. (b) State machine view. (c) Simulation results.

in figure 6.3b. As expected, the latter coincides with the intended state transition diagram (figure 5.3). Simulation results are exhibited in figure 6.3c. Note that the output changes only at positive clock transitions, counting up when *up* = '1', down when *up* = '0', and stopping if *ena* = '0'.

The number of flip-flops inferred by the compiler after synthesizing the code above was three for sequential, Gray, or Johnson encoding and five for one-hot, matching the predictions made in section 5.4.1.

Note: As smentioned in section 5.4.1, counters can be designed very easily without employing the FSM approach when using VHDL or SystemVerilog. The design above was included, nevertheless, because it illustrates well the construction of VHDL code for category 1 machines. A similar counter, but without the up-down control, results from the code below, where the FSM technique is not employed. Moreover, it is fine for any number of bits.

```
------------------------------------------------------
library ieee;
use ieee.std_logic_1164.all;
use ieee.std_logic_arith.all;
------------------------------------------------------
entity counter is
   generic (
      bits: natural := 3;
      xmin: natural := 1;
      xmax: natural := 5);
   port (
      clk, rst, ena: in std_logic;
```

```
      x_out: out std_logic_vector(bits-1 downto 0));
end entity;
--------------------------------------------------
architecture direct_counter of counter is
   signal x: natural range 0 to xmax;
begin
   process (clk, rst)
   begin
      if rst='1' then
         x <= xmin;
      elsif rising_edge(clk) and ena='1' then
         if x<xmax then
            x <= x + 1;
         else
            x <= xmin;
         end if;
      end if;
   end process;
   x_out <= conv_std_logic_vector(x, bits);
end architecture;
--------------------------------------------------
```

6.7 Design of a Garage Door Controller

This section presents a VHDL-based design for the garage door controller introduced in section 5.4.5. The Moore template of section 6.3 is employed to implement the FSM of figure 5.9c.

The entity, called *garage_door_controller*, is in lines 5–9. All ports are of type *std_logic* or *std_logic_vector* (industry standard).

The architecture, called *moore_fsm*, is in lines 11–94. As usual, it contains a declarative part (before the keyword **begin**) and a statements part (from **begin** on).

In the declarative part of the architecture (lines 12–14), the enumerated type *state* is created to represent the machine's present and next states.

The first process (lines 18–25) in the statements part implements the state register. As in the template, this is a standard code with clock and reset present only in this process.

The second and final process (lines 28–92) implements the entire combinational logic section. It is just a list of all states, each containing the output value and the next state. Note that in each state the output value is unique because in a Moore machine the output depends only on the state in which the machine is.

Observe the correct use of registers and the completeness of the code as described in comment number 10 of section 6.3. Note in particular the following:

1) Regarding the use of registers: The circuit is not overregistered. This can be observed in the **elsif rising_edge(clk)** statement of line 22 (responsible for the inference of flip-flops), which is closed in line 24, guaranteeing that only the machine state (line 23) gets stored. The output (*ctr*) is in the next process, which is purely combinational (thus not registered).

2) Regarding the outputs: The list of outputs (just *ctr* in this example) is exactly the same in all states (see lines 32, 39, 46, ...), and the corresponding output value is always properly declared.

3) Regarding the next state: Again, the coverage is complete because all states are included (see lines 31, 38, 45, ...), and in each state the conditional declarations for the next state are always finalized with an **else** statement (lines 35, 42, 51, ...), guaranteeing that no condition is left unchecked.

```
1    ---------------------------------------------------------
2    library ieee;
3    use ieee.std_logic_1164.all;
4    ---------------------------------------------------------
5    entity garage_door_controller is
6       port (
7          remt, sen1, sen2, clk, rst: in std_logic;
8          ctr: out std_logic_vector(1 downto 0));
9    end entity;
10   ---------------------------------------------------------
11   architecture moore_fsm of garage_door_controller is
12      type state is (closed1, closed2, opening1, opening2,
13         open1, open2, closing1, closing2);
14      signal pr_state, nx_state: state;
15   begin
16
17      --FSM state register:
18      process (clk, rst)
19      begin
20         if rst='1' then
21            pr_state <= closed1;
22         elsif rising_edge(clk) then
23            pr_state <= nx_state;
24         end if;
25      end process;
26
27      --FSM combinational logic:
28      process (all) --or (pr_state, remt, sen1, sen2)
29      begin
30         case pr_state is
31            when closed1 =>
32               ctr <= "0-";
33               if remt='0' then
34                  nx_state <= closed2;
35               else
36                  nx_state <= closed1;
37               end if;
38            when closed2 =>
39               ctr <= "0-";
40               if remt='1' then
41                  nx_state <= opening1;
42               else
43                  nx_state <= closed2;
44               end if;
45            when opening1 =>
```

```
               ctr <= "10";
               if sen1='1' then
                  nx_state <= open1;
               elsif remt='0' then
                  nx_state <= opening2;
               else
                  nx_state <= opening1;
               end if;
            when opening2 =>
               ctr <= "10";
               if remt='1' or sen1='1' then
                  nx_state <= open1;
               else
                  nx_state <= opening2;
               end if;
            when open1 =>
               ctr <= "0-";
               if remt='0' then
                  nx_state <= open2;
               else
                  nx_state <= open1;
               end if;
            when open2 =>
               ctr <= "0-";
               if remt='1' then
                  nx_state <= closing1;
               else
                  nx_state <= open2;
               end if;
            when closing1 =>
               ctr <= "11";
               if sen2='1' then
                  nx_state <= closed1;
               elsif remt='0' then
                  nx_state <= closing2;
               else
                  nx_state <= closing1;
               end if;
            when closing2 =>
               ctr <= "11";
               if remt='1' or sen2='1' then
                  nx_state <= closed1;
               else
                  nx_state <= closing2;
               end if;
         end case;
      end process;

end architecture;
-------------------------------------------------------
```

The number of flip-flops inferred by the compiler after synthesizing the code above was three for sequential or Gray encoding, four for Johnson, and eight for one-hot, matching the predictions made in section 5.4.5.

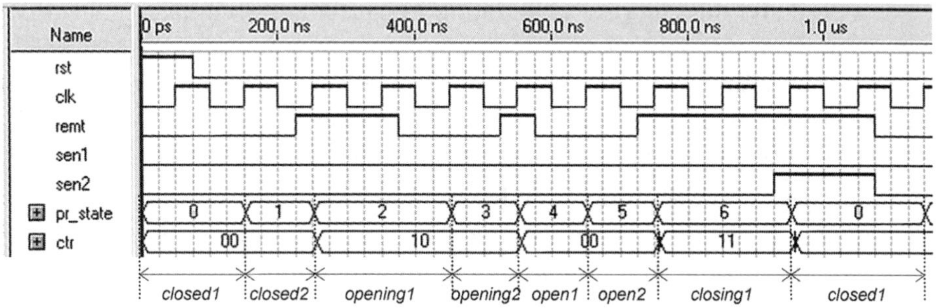
Name
0 ps
200.0 ns
400.0 ns
600.0 ns
800.0 ns
1.0 us
rst
clk
remt
sen1
sen2
pr_state
ctr
0
1
2
3
4
5
6
0
00
10
00
11
closed1
closed2
opening1
opening2
open1
open2
closing1
closed1

Figure 6.4
Simulation results from the VHDL code for the garage door controller of figure 5.9c.

Simulation results are depicted in figure 6.4. The encoding chosen for the states was *sequential* (section 3.7). The states are enumerated from 0 to 7 (there are eight states), in the order in which they were declared in lines 12–13. Be aware, however, that some compilers reserve the value zero for the reset state; because the reset (initial) state in the present example is *closed1* (see lines 20–21), which is the first state in the declaration list, that is not a concern here.

In this simulation the sequence *closed1–closed2–opening1–opening2–open1–open2–closing1–closed1* (see state names in the lower part of figure 6.4) was tested. Note that pulses of various widths were used to illustrate the fact that their width has no effect beyond the first positive clock edge.

6.8 Design of a Datapath Controller for a Greatest Common Divisor Calculator

This section presents a VHDL-based design for the control unit introduced in section 5.4.8, which controls a datapath to produce a greatest common divisor (GCD) calculator. The Moore template of section 6.3 is employed to implement the FSM of figure 5.13e.

The entity, called *control_unit_for_GCD,* is in lines 5–11. All ports are of the type *std_logic* or *std_logic_vector* (industry standard).

The architecture, called *moore_fsm,* is in lines 13–80. As usual, it contains a declarative part (before the keyword **begin**) and a statements part (from **begin** on).

In the declarative part of the architecture (lines 14–15), the enumerated type *state* is created to represent the machine's present and next states.

The first process (lines 19–26) in the statements part implements the state register. As in the template, this is a standard code with clock and reset present only in this process.

The second and final process (lines 29–78) implements the entire combinational logic section. It is just a list of all states, each containing the output values and the next state. Note that in each state the output values are unique because in a Moore machine the outputs depend only on the state in which the machine is.

Observe the correct use of registers and the completeness of the code, as described in comment 10 of section 6.3. Note in particular the following:

1) Regarding the use of registers: The circuit is not overregistered. This can be observed in the **elsif rising_edge(clk)** statement of line 23 (responsible for the inference of flip-flops), which is closed in line 25, guaranteeing that only the machine state (line 24) gets stored. The outputs are in the next process, which is purely combinational (thus not registered).
2) Regarding the outputs: The list of outputs (*selA, selB, wrA, wrB, ALUop*) is exactly the same in all states (see lines 33–37, 44–48, 51–55, . . .), and the corresponding output values are always properly declared.
3) Regarding the next state: Again, the coverage is complete because all states are included (see lines 32, 43, 50, . . .), and in each state any conditional declarations for the next state are finalized with an **else** statement (lines 40 and 60), guaranteeing that no condition is left unchecked.

```
1     -----------------------------------------------------
2     library ieee;
3     use ieee.std_logic_1164.all;
4     -----------------------------------------------------
5     entity control_unit_for_GCD is
6        port (
7           dv, clk, rst: in std_logic;
8           sign: in std_logic_vector(1 downto 0)
9           selA, selB, wrA, wrB: out std_logic;
10          ALUop: out std_logic_vector(1 downto 0));
11    end entity;
12    -----------------------------------------------------
13    architecture moore_fsm of control_unit_for_GCD is
14       type state is (idle, load, waitt, writeA, writeB);
15       signal pr_state, nx_state: state;
16    begin
17
18       --FSM state register:
19       process (clk, rst)
20       begin
21          if rst='1' then
22             pr_state <= idle;
23          elsif rising_edge(clk) then
24             pr_state <= nx_state;
25          end if;
26       end process;
27
28       --FSM combinational logic:
29       process (all)  --or (pr_state, dv, sign)
```

```
   begin
      case pr_state is
         when idle =>
            selA <= '-';
            selB <= '-';
            wrA <= '0';
            wrB <= '0';
            ALUop <= "00";
            if dv='1' then
               nx_state <= load;
            else
               nx_state <= idle;
            end if;
         when load =>
            selA <= '1';
            selB <= '1';
            wrA <= '1';
            wrB <= '1';
            ALUop <= "00";
            nx_state <= waitt;
         when waitt =>
            selA <= '-';
            selB <= '-';
            wrA <= '0';
            wrB <= '0';
            ALUop <= "10";
            if sign="01" then
               nx_state <= writeA;
            elsif sign="10" then
               nx_state <= writeB;
            else
               nx_state <= idle;
            end if;
         when writeA =>
            selA <= '0';
            selB <= '-';
            wrA <= '1';
            wrB <= '0';
            ALUop <= "10";
            nx_state <= waitt;
         when writeB =>
            selA <= '-';
            selB <= '0';
            wrA <= '0';
            wrB <= '1';
            ALUop <= "11";
            nx_state <= waitt;
      end case;
   end process;

end architecture;
----------------------------------------------------
```

Simulation results are presented in figure 6.5. The encoding chosen for the states was *sequential* (section 3.7). The states are enumerated from 0 to 4 (there are five states)

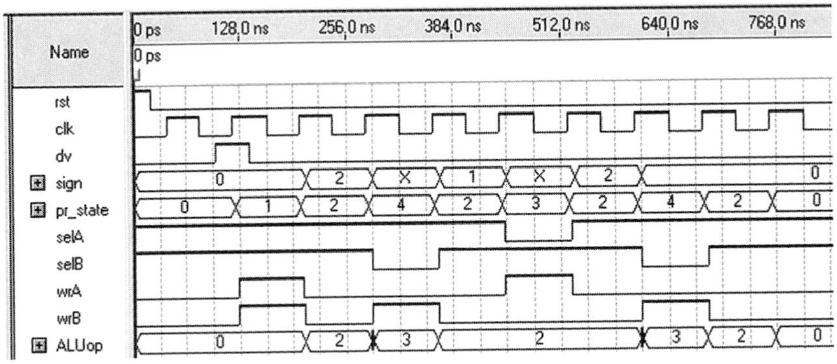
Name
0 ps
128,0 ns
256,0 ns
384,0 ns
512,0 ns
640,0 ns
768,0 ns
rst
clk
dv
sign
pr_state
selA
selB
wrA
wrB
ALUop

Figure 6.5
Simulation results from the VHDL code for the control unit of figure 5.13e, which controls a datapath for GCD calculation.

in the order in which they were declared in line 14 (be aware, however, that some compilers reserve the value zero for the reset state). The stimuli are exactly as in figure 5.13d (GCD for 9 and 15). The reader is invited to inspect these results and compare them against the waveforms in figure 5.13d.

6.9 Exercises

Exercise 6.1: Parity Detector

This exercise concerns the parity detector of figure 5.5c.

a) How many flip-flops are needed to implement it with sequential and one-hot encoding?
b) Implement it using VHDL. Check whether the number of DFFs inferred by the compiler matches each of your predictions.
c) Simulate it using the same stimuli of figure 5.5b and check if the same waveform results for *y*.

Exercise 6.2: One-Shot Circuits

This exercise concerns the one-shot circuits of figures 5.7c,d.

a) Solve exercise 5.5 if not done yet.
b) How many flip-flops are needed to implement each FSM with sequential encoding?
c) Implement both circuits using VHDL. Check whether the number of DFFs inferred by the compiler matches each of your predictions.
d) Simulate each circuit using the same stimuli of exercise 5.5 (figure 5.16) and check whether the same results are obtained here.

Exercise 6.3: Manchester Encoder

This exercise concerns the Manchester encoder treated in exercise 5.8.

a) Solve exercise 5.8 if not done yet.
b) Implement the Moore machine relative to part a of that exercise using VHDL. Simulate it using the same stimuli of part b, checking if the results match.
c) Implement the Mealy machine relative to part c of that exercise using VHDL. Simulate it using the same stimuli of part d, checking if the results match.

Exercise 6.4: Differential Manchester Encoder

This exercise concerns the differential Manchester encoder treated in exercise 5.9.

a) Solve exercise 5.9 if not done yet.
b) Implement the FSM relative to part a of that exercise using VHDL. Simulate it using the same waveforms of part b, checking if the results match.

Exercise 6.5: String Detector

This exercise concerns the string detector of figure 5.14a, which detects the sequence "*abc*".

a) Solve exercise 5.12 if not done yet.
b) Implement the FSM of figure 5.14a using VHDL. Simulate it using the same stimuli of exercise 5.12, checking if the same results are obtained here.

Exercise 6.6: Generic String Detector

This exercise concerns the generic string detector of figure 5.14b. Implement it using VHDL and simulate it for the following cases:

a) To detect the sequence "*abc*".
b) To detect the sequence "*aab*".
c) To detect the sequence "*aaa*".

Exercise 6.7: Keypad Encoder

This exercise concerns the keypad encoder treated in exercise 5.14. It is repeated in figure 6.6, with a seven-segment display (SSD—see figure 8.13) at the output, which must display the last key pressed (use the characters "A" and "b" for * and #, respectively). (A deboucer is generally needed in this kind of design; see exercise 8.9.)

a) Solve exercise 5.14 if not done yet.
b) Implement the FSM obtained above using VHDL. Instead of encoding *r*(3:0) according to the table in figure 5.23c, encode it as an SSD driver, using the table in figure 8.13d (so *key* is now a 7-bit signal).

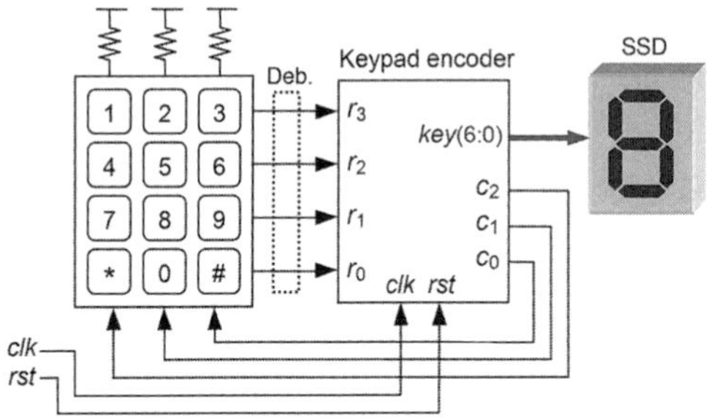
SSD
Keypad encoder
Deb.
1
2
3
4
5
6
7
8
9
*
0
#
r3
r2
r1
r0
key(6:0)
c2
c1
c0
clk rst
clk
rst

Figure 6.6

c) Physically test your design by connecting an actual keypad (or an arrangement of pushbuttons) to the FPGA in your development board, with *key* displayed by one of the board's SSDs.

Exercise 6.8: Datapath Controller for a Largest-Value Detector

This exercise concerns the control unit treated in exercise 5.15.

a) Solve exercise 5.15 if not done yet.

b) Implement the FSM obtained above using VHDL. Present meaningful simulation results.

7 SystemVerilog Design of Regular (Category 1) State Machines

7.1 Introduction

This chapter presents several SystemVerilog designs of category 1 state machines. It starts by presenting two SystemVerilog templates, for Moore- and Mealy-based implementations, which are used subsequently to develop a series of designs related to the examples introduced in chapter 5.

The codes are always complete (not only partial sketches) and are accompanied by comments and simulation results illustrating the design's main features. All circuits were synthesized using Quartus II (from Altera) or ISE (from Xilinx). The simulations were performed with Quartus II or ModelSim (from Mentor Graphics). The default encoding scheme for the states of the FSMs was regular sequential encoding (see encoding options in section 3.7).

The same designs were developed in chapter 6 using VHDL, so the reader can make a direct comparison between the codes.

Note: See suggestions of SystemVerilog books in the bibliography.

7.2 General Structure of SystemVerilog Code

A typical structure of SystemVerilog code for synthesis, with all elements that will be needed in this and in coming chapters, is depicted in figure 7.1. It is composed of three fundamental sections, briefly described below.

Module Header

The **module** header is similar to **entity** in VHDL (section 6.2), also divided into two parts, called *parameter declarations* and *port declarations*.

Parameter declarations: This portion, similar to **generic** in VHDL, is optional. It is used for the declaration of global parameters, which can be easily modified to fulfill different system specifications or, more importantly, can be overridden during instantiations into other designs (structural code).

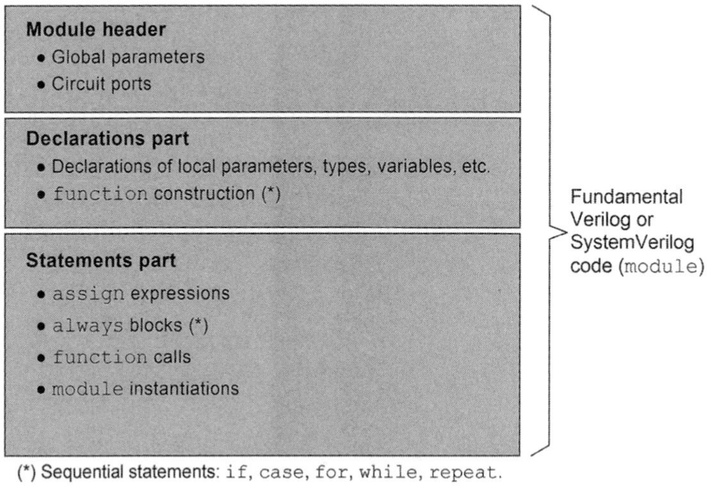
Module header
• Global parameters
• Circuit ports
Declarations part
• Declarations of local parameters, types, variables, etc.
• function construction (*)
Statements part
• assign expressions
• always blocks (*)
• function calls
• module instantiations
Fundamental Verilog or SystemVerilog code (module)
(*) Sequential statements: if, case, for, while, repeat.

Figure 7.1
Typical SystemVerilog code structure for synthesis.

Port declarations: This portion, similar to **port** in VHDL, is mandatory for synthesis. It is just a list with specifications of all circuit ports (I/Os), including their mode (**input**, **output**, or **inout**), type (plus range), and name.

Declarations Part

The declarations part of a SystemVerilog code is similar to the declarative part of **architecture** in VHDL (section 6.2). It too is optional and allows all sorts of local declarations (e.g., local parameters, data types, variables) as well as **function** (and **task**) constructions.

Statements Part

The statements part of a SystemVerilog code is similar to the statements part of **architecture** in VHDL (section 6.2). As shown in figure 7.1, its main elements (in no particular order) are the following: **assign** statements, normally using operators, for simple combinational circuits; **always** blocks, constructed using sequential statements (**if**, **case**, **for**, **while**, **repeat**), for both sequential as well as (complex) combinational circuits; **function** (and **task**) calls; and, finally, **module** (that is, other design) instantiations.

7.3 SystemVerilog Template for Regular (Category 1) Moore Machines

The template is based on figure 7.2 (derived from figure 5.2), which shows three **always** blocks: 1) for the FSM state register; 2) for the FSM combinational logic; and

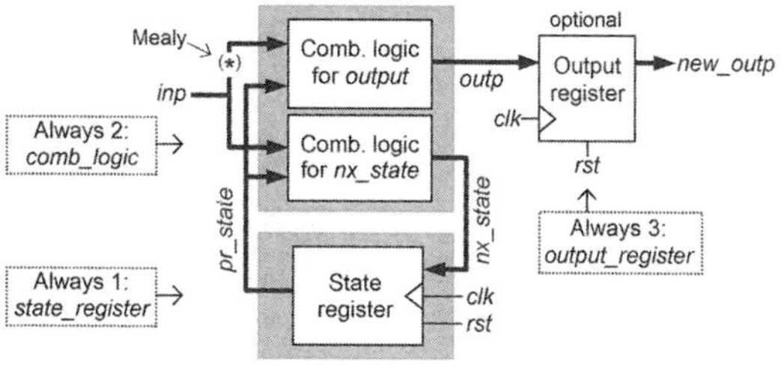
Mealy
(*)
inp
Always 2: comb_logic
Comb. logic for output
Comb. logic for nx_state
pr_state
nx_state
Always 1: state_register
State register
clk
rst
optional
outp
Output register
new_outp
clk
rst
Always 3: output_register

Figure 7.2
State machine architecture depicting how the SystemVerilog code was broken (three **always** blocks).

3) for the optional output register. Note the asterisk on one of the input connections; as we know, if that connection exists it is a Mealy machine, else it is a Moore machine.

There obviously are other ways of breaking the code instead of using the three **always** blocks indicated in figure 7.2. For example, the combinational logic section could be implemented with two **always** blocks, one with the logic for *output*, the other with the logic for *nx_state*.

The SystemVerilog template for the design of category 1 Moore machines is presented below. Observe the following:

1) To improve readability, the three fundamental code sections were separated by dashed lines (lines 1, 11, 17, 61).

2) The first part of the code is the module header, in lines 1–9. It contains two sections: global parameter declarations (optional, lines 3–5) and circuit ports (mandatory for synthesis, lines 7–9). Note that all ports are of type **logic**, with one or more bits.

3) The second part of the code is the declarations part, in lines 11–15. A special enumerated type, called *state*, is created in line 14, then the signals *pr_state* and *nx_state* are declared using that type in line 15.

4) The third part of the code is the statements part (code proper), in lines 17–60. In this template, it contains three **always** blocks, described next.

5) The first **always** block (lines 20–22) is an **always_ff** because we want flip-flops to be inferred. It implements the machine's state register (always 1 block of figure 7.2). This register is reset when *rst*='1' occurs; if *rst*='0', the input is copied to the output at every positive clock edge.

6) The second **always** block (lines 25–47) is an **always_comb** because we want a purely combinational circuit to be inferred (see always 2 block in figure 7.2). This part must contain all states (A, B, C, . . .), and for each state two things must be declared: the output values/expressions and the next state. Note, for example, in lines 27–34, relative to state A, the output declarations in lines 28–30 and the next state declarations in lines 31–33. A very important point to observe here is that there is no **if**

statement associated with the outputs because in a Moore machine the outputs depend solely on the state in which the machine is, so for a given state each output value/ expression is unique.

7) The third and final **always** block (lines 50–58) implements the optional output register (always 3 block of figure 7.2). Note that it simply copies each original output to a new output at every positive clock edge (it could also be at the negative edge), thus inferring the extra register. If this register is used, then the names of the new outputs must obviously be the names used in the corresponding port declarations (line 9). If the initial output values do not matter, reset is not required in this register.

8) To conclude, observe the completeness of the code and the correct use of registers (as requested in sections 4.2.8 and 4.2.9, respectively), summarized below.

a) Regarding the use of registers: The circuit is not overregistered. This can be observed in the **always_ff** statement of line 20 (responsible for the inference of flip-flops), which is closed in line 22, guaranteeing that only the machine state (line 22) gets registered. The output is in the **always_comb** block, which is purely combinational.

b) Regarding the outputs: The list of outputs (*outp1*, *outp2*, ...) is exactly the same in all states (see lines 28–30, 36–38, . . .), and the output values/expressions are always declared.

c) Regarding the next state: Again, the coverage is complete because all states (A, B, C, . . .) are included and in each state the conditional declarations are finalized with an **else** statement (lines 33, 41, . . .), guaranteeing that no condition is left unchecked.

Note: Another implementation approach, for simple FSMs, will be seen in chapter 15.

```
1    //Part 1: Module header:----------------------------
2    module module_name
3       #(parameter
4       param1 = <value>,
5       param2 = <value>)
6       (
7       input logic clk, rst, ...
8       input logic [7:0] inp1, inp2, ...
9       output logic [15:0] outp1, outp2, ...);
10
11   //Part 2: Declarations:-----------------------------
12
13      //FSM states type:
14      typedef enum logic [2:0] {A, B, C, ...} state;
15      state pr_state, nx_state;
16
17   //Part 3: Statements:-------------------------------
18
19      //FSM state register:
20      always_ff @(posedge clk, posedge rst)
21         if (rst) pr_state <= A;
22         else pr_state <= nx_state;
23
```

```
24    //FSM combinational logic:
25    always_comb
26       case (pr_state)
27          A: begin
28             outp1 <= <value>;
29             outp2 <= <value>;
30             ...
31             if (condition) nx_state <= B;
32             else if (condition) nx_state <= ...;
33             else nx_state <= A;
34          end
35          B: begin
36             outp1 <= <value>;
37             outp2 <= <value>;
38             ...
39             if (condition) nx_state <= C;
40             else if (condition) nx_state <= ...;
41             else nx_state <= B;
42          end
43          C: begin
44             ...
45          end
46          ...
47       endcase
48
49    //Optional output register:
50    always_ff @(posedge clk, posedge rst)
51       if (rst) begin   //rst might be not needed here
52          new_outp1 <= ...;
53          new_outp2 <= ...; ...
54       end
55       else begin
56          new_outp1 <= outp1;
57          new_outp2 <= outp2; ...
58       end
59
60 endmodule
61 //-----------------------------------------------------
```

7.4 SystemVerilog Template for Regular (Category 1) Mealy Machines

This template, also based on figures 7.1 and 7.2, is presented below. The only difference with respect to the Moore template just presented is in the **always_comb** block for the combinational logic because the output is specified differently now. Recall that in a Mealy machine the output depends not only on the FSM's state but also on its input, so **if** statements are expected for the output in one or more states because the output values might not be unique. This is achieved by including the output *within* the conditional statements for *nx_state*. For example, observe in lines 15–33, relative to state A, that the output values are now conditional. Compare these lines against lines 27–34 in the previous template.

Please read all comments made for the Moore template in section 7.3 because, except for the difference mentioned above, they all apply to the Mealy template below as well. Particular attention should be paid to the recommendations in comment 8, which can be easily adapted from the Moore case to the Mealy case.

```
//Part 1: Module header:---------------------------------
   (same as for category 1 Moore, section 7.3)

//Part 2: Declarations:----------------------------------
   (same as for category 1 Moore, section 7.3)

//Part 3: Statements:------------------------------------

   //FSM state register:
   (same as for category 1 Moore, section 7.3)

   //FSM combinational logic:
   always_comb
      case (pr_state)
         A:
            if (condition) begin
               outp1 <= <value>;
               outp2 <= <value>;
               ...
               nx_state <= B;
            end
            else if (condition) begin
               outp1 <= <value>;
               outp2 <= <value>;
               ...
               nx_state <= ...;
            end
            else begin
               outp1 <= <value>;
               outp2 <= <value>;
               ...
               nx_state <= A;
            end
         B:
            if (condition) begin
               outp1 <= <value>;
               outp2 <= <value>;
               ...
               nx_state <= C;
            end
            else if (condition) begin
               outp1 <= <value>;
               outp2 <= <value>;
               ...
               nx_state <= ...;
            end
            else begin
               outp1 <= <value>;
               outp2 <= <value>;
               ...
               nx_state <= B;
            end
```

```
            C: ...
            ...
         endcase

      //Optional output register:
      (same as for category 1 Moore, section 7.3)

   endmodule
   //-----------------------------------------------------
```

7.5 Design of a Small Counter

This section presents a SystemVerilog-based design for the 1-to-5 counter with enable and up-down controls introduced in section 5.4.1 (figure 5.3).

Because counters are inherently synchronous, the Moore approach is the natural choice for their implementation, so the SystemVerilog template of section 7.3 is used. Because possible glitches at (positive) clock transitions are generally not a problem in counters, the optional output register shown in the final portion of the template is not employed.

The first part of the code (*module header*) is in lines 1–4. The module's name is *counter*. Note that all ports are of type **logic**, with one bit for each input and three bits for the output.

The second part of the code (*declarations*) is in lines 6–9. The enumerated type *state* is created in it to represent the machine's present and next states.

The third and final part of the code (*statements*) is in lines 11–57. It contains two **always** blocks, described next.

The first **always** block (lines 13–15) is an **always_ff**, which implements the machine's state register. This is a standard code, similar to the template.

The second **always** block (lines 18–55) is an **always_comb**, which implements the entire combinational logic section. It is just a list of all states, each containing the output value and the next state. Note that in each state the output value is unique because in a Moore machine the output depends only on the state in which the machine is.

Finally, and very importantly, observe the correct use of registers and the completeness of the code, as described in comment 8 of section 7.3. Observe in particular the following: 1) all states are included; 2) the list of outputs (only *outp* in this case) is exactly the same in all states, and the corresponding values are always included; 3) the specifications for *nx_state* are always finalized with an **else** statement, so no condition is left unchecked.

```
1   //Module header:------------------------------------------
2   module counter (
3      input logic up, ena, clk, rst,
```

```
   output logic [2:0] outp);

//Declarations:------------------------------------------------
   //FSM states type:
   typedef enum logic [2:0] {one, two, three, four, five} state;
   state pr_state, nx_state;

//Statements:--------------------------------------------------
   //FSM state register:
   always_ff @(posedge clk, posedge rst)
      if (rst) pr_state <= one;
      else pr_state <= nx_state;

   //FSM combinational logic:
   always_comb
      case (pr_state)
         one: begin
            outp <= 1;
            if (ena)
               if (up) nx_state <= two;
               else nx_state <= five;
            else nx_state <= one;
         end
         two: begin
            outp <= 2;
            if (ena)
               if (up) nx_state <= three;
               else nx_state <= one;
            else nx_state <= two;
         end
         three: begin
            outp <= 3;
            if (ena)
               if (up) nx_state <= four;
               else nx_state <= two;
            else nx_state <= three;
         end
         four: begin
            outp <= 4;
            if (ena)
               if (up) nx_state <= five;
               else nx_state <= three;
            else nx_state <= four;
         end
         five: begin
            outp <= 5;
            if (ena)
               if (up) nx_state <= one;
               else nx_state <= four;
            else nx_state <= five;
         end
      endcase

endmodule
//-------------------------------------------------------------
```

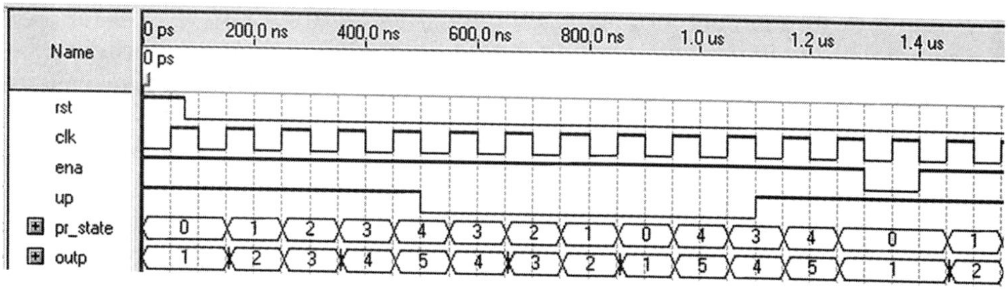
Name
0 ps
200.0 ns
400.0 ns
600.0 ns
800.0 ns
1.0 us
1.2 us
1.4 us
rst
clk
ena
up
pr_state
outp
0 1 2 3 4 3 2 1 0 4 3 4 0 1
1 2 3 4 5 4 3 2 1 5 4 5 1 2

Figure 7.3
Simulation results from the SystemVerilog code for the 1-to-5 counter with enable and up-down controls of figure 5.3.

Simulation results from the code above are exhibited in figure 7.3. Note that the output changes only at positive clock transitions, counting up when *up*='1', down when *up*='0', and stopping if *ena*='0'.

The number of flip-flops inferred by the compiler was three for sequential, Gray, or Johnson encoding and five for one-hot, matching the predictions made in section 5.4.1.

7.6 Design of a Garage Door Controller

This section presents a SystemVerilog-based design for the garage door controller introduced in section 5.4.5. The Moore template of section 7.3 is employed to implement the FSM of figure 5.9c.

The first part of the code (*module header*) is in lines 1–4. The module's name is *garage_door_controller*. Note that all ports are of type **logic**.

The second part of the code (*declarations*) is in lines 6–10. The enumerated type *state* is created in it to represent the machine's present and next states.

The third and final part of the code (*statements*) is in lines 12–65. It contains two **always** blocks, described next.

The first **always** block (lines 14–16) is an **always_ff**, which implements the machine's state register. This is a standard code, similar to the template.

The second **always** block (lines 19–63) is an **always_comb**, which implements the entire combinational logic section. It is just a list of all states, each containing the output value and the next state. Note that in each state the output value is unique because in a Moore machine the output depends only on the state in which the machine is.

Finally, and very importantly, observe the correct use of registers and the completeness of the code, as described in comment 8 of section 7.3. Observe in particular the following: 1) all states are included; 2) the list of outputs (only *ctr* in this case) is exactly the same in all states, and the corresponding values are always included; 3) the specifications for *nx_state* are always finalized with an **else** statement, so no condition is left unchecked.

```
//Module header:---------------------------------------
module garage_door_controller (
   input logic remt, sen1, sen2, clk, rst,
   output logic [1:0] ctr);

//Declarations:----------------------------------------
   //FSM states type:
   typedef enum logic [2:0] {closed1, closed2, opening1,
      opening2, open1, open2, closing1, closing2} state;
   state pr_state, nx_state;

//Statements:------------------------------------------
   //FSM state register:
   always_ff @(posedge clk, posedge rst)
      if (rst) pr_state <= closed1;
      else pr_state <= nx_state;

   //FSM combinational logic:
   always_comb
      case (pr_state)
         closed1: begin
            ctr <= 2'b0x;
            if (~remt) nx_state <= closed2;
            else nx_state <= closed1;
         end
         closed2: begin
            ctr <= 2'b0x;
            if (remt) nx_state <= opening1;
            else nx_state <= closed2;
         end
         opening1: begin
            ctr <= 2'b10;
            if (sen1) nx_state <= open1;
            else if (~remt) nx_state <= opening2;
            else nx_state <= opening1;
         end
         opening2: begin
            ctr <= 2'b10;
            if (remt | sen1) nx_state <= open1;
            else nx_state <= opening2;
         end
         open1: begin
            ctr <= 2'b0x;
            if (~remt) nx_state <= open2;
            else nx_state <= open1;
         end
         open2: begin
            ctr <= 2'b0x;
            if (remt) nx_state <= closing1;
```

```
50                else nx_state <= open2;
51             end
52             closing1: begin
53                ctr <= 2'b11;
54                if (sen2) nx_state <= closed1;
55                else if (~remt) nx_state <= closing2;
56                else nx_state <= closing1;
57             end
58             closing2: begin
59                ctr <= 2'b11;
60                if (remt | sen2) nx_state <= closed1;
61                else nx_state <= closing2;
62             end
63          endcase
64
65    endmodule
66    //-------------------------------------------------------------
```

The number of flip-flops inferred by the compiler after synthesizing the code above was three for sequential or Gray encoding, four for Johnson, and eight for one-hot, matching the predictions made in section 5.4.5.

Simulation results are depicted in figure 7.4. The encoding chosen for the states was *sequential* (section 3.7). The states are enumerated from 0 to 7 (there are eight states) in the order in which they were declared in lines 8–9. Be aware, however, that some compilers reserve the value zero for the reset state; because the reset (initial) state in the present example is *closed1* (see line 15), which is the first state in the declaration list, that is not a concern here.

In this simulation the sequence *closed1—closed2—opening1—opening2—open1—open2—closing1—closed1* (see state names in the lower part of figure 7.4) was tested. Note that pulses of various widths were used to illustrate the fact that their width has no effect beyond the first positive clock edge.

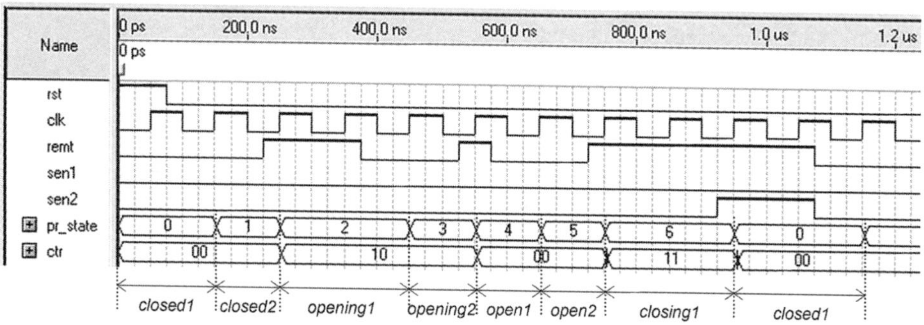
Name
0 ps
200,0 ns
400,0 ns
600,0 ns
800,0 ns
1.0 us
1.2 us
rst
clk
remt
sen1
sen2
pr_state
ctr
0
1
2
3
4
5
6
0
00
10
00
11
00
closed1
closed2
opening1
opening2
open1
open2
closing1
closed1

Figure 7.4
Simulation results from the SystemVerilog code for the garage door controller of figure 5.9c.

7.7 Design of a Datapath Controller for a Greatest Common Divisor Calculator

This section presents a SystemVerilog-based design for the control unit introduced in section 5.4.8, which controls a datapath to produce a greatest common divisor (GCD) calculator. The Moore template of section 7.3 is employed to implement the FSM of figure 5.13e.

The first part of the code (*module header*) is in lines 1–6. The module's name is *control_unit_for_GCD*. Note that all ports are of type **logic**.

The second part of the code (*declarations*) is in lines 8–11. The enumerated type *state* is created in it to represent the machine's present and next states.

The third and final part of the code (*statements*) is in lines 13–67. It contains two **always** blocks, described next.

The first **always** block (lines 15–17) is an **always_ff**, which implements the machine's state register. This is a standard code, similar to the template.

The second **always** block (lines 20–65) is an **always_comb**, which implements the entire combinational logic section. It is just a list of all states, each containing the output values and the next state. Note that in each state the output values are unique because in a Moore machine the outputs depend only on the state in which the machine is.

Finally, and very importantly, observe the correct use of registers and the completeness of the code, as described in comment 8 of section 7.3. Observe in particular the following: 1) all states are included; 2) the list of outputs is exactly the same in all states, and the corresponding values are always included; 3) the conditional specifications for *nx_state* are always finalized with an **else** statement, so no condition is left unchecked.

```
1    //Module header:-----------------------------------------
2    module control_unit_for_GCD (
3       input logic dv, clk, rst,
4       input logic [1:0] sign,
5       output logic selA, selB, wrA, wrB,
6       output logic [1:0] ALUop);
7
8    //Declarations:------------------------------------------
9       //FSM states type:
10      typedef enum logic [2:0] {idle, load, waitt, writeA, writeB} state;
11      state pr_state, nx_state;
12
13   //Statements:--------------------------------------------
14      //FSM state register:
15      always_ff @(posedge clk, posedge rst)
16         if (rst) pr_state <= idle;
17         else pr_state <= nx_state;
18
19      //FSM combinational logic:
```

```
    always_comb
       case (pr_state)
          idle: begin
             selA <= 1'bx;
             selB <= 1'bx;
             wrA <= 1'b0;
             wrB <= 1'b0;
             ALUop <= 0;
             if (dv) nx_state <= load;
             else nx_state <= idle;
          end
          load: begin
             selA <= 1'b1;
             selB <= 1'b1;
             wrA <= 1'b1;
             wrB <= 1'b1;
             ALUop <= 0;
             nx_state <= waitt;
          end
          waitt: begin
             selA <= 1'bx;
             selB <= 1'bx;
             wrA <= 1'b0;
             wrB <= 1'b0;
             ALUop <= 2;
             if (sign==1) nx_state <= writeA;
             else if (sign==2) nx_state <= writeB;
             else nx_state <= idle;
          end
          writeA: begin
             selA <= 1'b0;
             selB <= 1'bx;
             wrA <= 1'b1;
             wrB <= 1'b0;
             ALUop <= 2;
             nx_state <= waitt;
          end
          writeB: begin
             selA <= 1'bx;
             selB <= 1'b0;
             wrA <= 1'b0;
             wrB <= 1'b1;
             ALUop <= 3;
             nx_state <= waitt;
          end
       endcase

endmodule
//-------------------------------------------------------------------
```

7.8 Exercises

Exercise 7.1: Parity Detector

Solve exercise 6.1 using SystemVerilog instead of VHDL.

Exercise 7.2: One-Shot Circuits

Solve exercise 6.2 using SystemVerilog instead of VHDL.

Exercise 7.3: Manchester Encoder

Solve exercise 6.3 using SystemVerilog instead of VHDL.

Exercise 7.4: Differential Manchester Encoder

Solve exercise 6.4 using SystemVerilog instead of VHDL.

Exercise 7.5: String Detector

Solve exercise 6.5 using SystemVerilog instead of VHDL.

Exercise 7.6: Generic String Detector

Solve exercise 6.6 using SystemVerilog instead of VHDL.

Exercise 7.7: Keypad Encoder

Solve exercise 6.7 using SystemVerilog instead of VHDL.

Exercise 7.8: Datapath Controller for a Largest-Value Detector

Solve exercise 6.8 using SystemVerilog instead of VHDL.

8 Timed (Category 2) State Machines

8.1 Introduction

We know that state machines can be classified into two types, based on their *input connections*, as follows.

1) *Moore machines*: The input, if it exists, is connected only to the logic block that computes the next state.
2) *Mealy machines*: The input is connected to both logic blocks, that is, for the next state and for the actual output.

In section 3.6 we introduced a new, additional classification, also from a hardware point of view, based on the *transition types* and *nature of the outputs*, as follows (see figure 8.1).

1) *Regular (category 1) state machines*: This category, illustrated in figure 8.1a and studied in chapters 5 to 7, consists of machines with only untimed transitions and outputs that do not depend on previous (past) values, so none of the outputs need to be registered for the machine to function.
2) *Timed (category 2) state machines*: This category, illustrated in figure 8.1b and studied in chapters 8 to 10, consists of machines with one or more transitions that depend on time (so they can have all four transition types: conditional, timed, conditional-timed, and unconditional). However, all outputs are still independent from previous (past) values.
3) *Recursive (category 3) state machines*: This category is illustrated in figure 8.1c and studied in chapters 11 to 13. It can have all four types of transitions, but one or more outputs depend on previous (past) values, so such outputs must be stored in auxiliary registers for the machine to function.

The two fundamental decisions before starting a design in hardware are then the following:

1) The state machine category (regular, timed, or recursive).

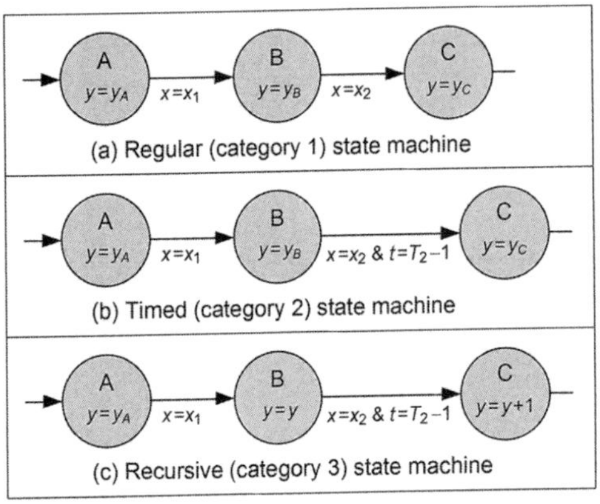
A
y=y_A
x=x_1
B
y=y_B
x=x_2
C
y=y_C
(a) Regular (category 1) state machine
A
y=y_A
x=x_1
B
y=y_B
x=x_2 & t=T_2−1
C
y=y_C
(b) Timed (category 2) state machine
A
y=y_A
x=x_1
B
y=y
x=x_2 & t=T_2−1
C
y=y+1
(c) Recursive (category 3) state machine

Figure 8.1
State machine categories (from a hardware perspective).

2) The state machine type (Moore or Mealy).

It is important to recall, however, that regardless of the machine category and type, the state transition diagram must fulfill three fundamental requisites (seen in section 1.3):

1) It must include all possible system states.
2) All state transition conditions must be specified (unless a transition is unconditional) and must be truly complementary.
3) The list of outputs must be exactly the same in all states (standard architecture).

8.2 Architectures for Timed (Category 2) Machines

The general architecture for category 2 machines is summarized in figure 8.2a. This representation follows the style of figures 3.1b and 3.1d, but the style of figures 3.1a and 3.1c could be used equivalently. The output register (figure 8.2b) is still optional, but the timer (in figure 8.2a) is compulsory.

Note that the timer operates as an auxiliary circuit, producing the signal t, needed by the state machine. However, the FSM itself is responsible for controlling the timer, as represented symbolically by the control signal *ctr* in the figure. In other words, the machine is who decides when the timer should run or stop and when it should be zeroed.

The four possible constructions, listed in figure 8.2c, are summarized below.

Timed Moore machine: The circuit of figure 8.2a is used with the input (if it exists) connected only to the logic block for the next state, as in figure 5.2a. Consequently, it behaves exactly as a pure Moore machine, just with an auxiliary timer operating as

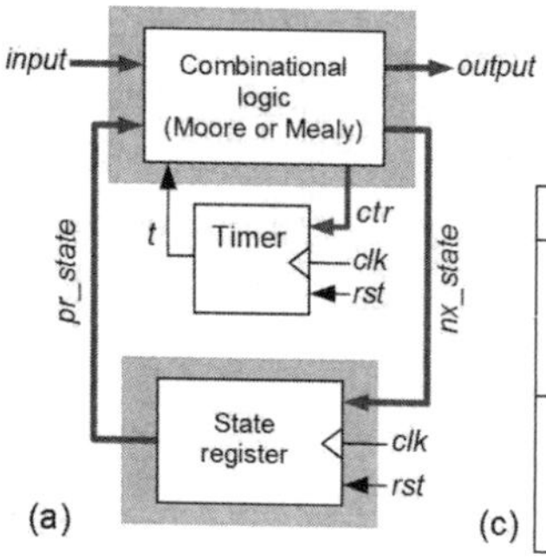
input
Combinational logic (Moore or Mealy)
output
pr_state
nx_state
t
Timer
ctr
clk
rst
State register
clk
rst
(a)

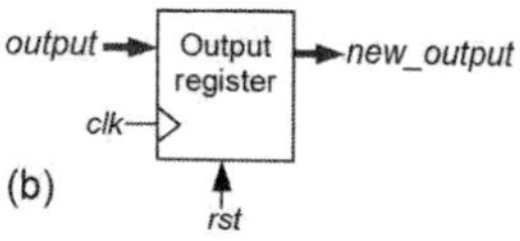
output
Output register
new_output
clk
rst
(b)

(c)

Construction	Resulting category 1 circuit	Behavior
(a) only	Timed Moore machine (preferred)	Order-1 synchronous
	Timed Mealy machine	Asynchronous
(a) + (b)	Out-registered (pipelined) timed Moore machine	Order-2 synchronous
	Out-registered (pipelined) timed Mealy machine	Order-1 synchronous

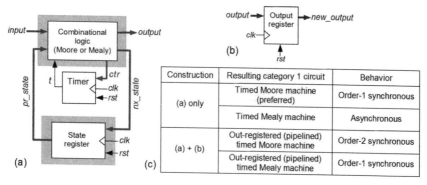

Figure 8.2
Timed (category 2) state machine architectures. (a) Moore or Mealy type (depending on input connections). (b) Optional output register. (c) Resulting circuits.

an extra input. Because the output depends only on the state in which the machine is, this circuit is synchronous (see details in section 3.5). Because modern designs are generally synchronous, this option is preferred over any other timed implementation whenever the application permits.

Timed Mealy machine: Again, the circuit of figure 8.2a is used, but this time with the input connected to both logic blocks (for output and for next state), as in figure 5.2b. Consequently, it behaves exactly as a pure Mealy machine, just with an auxiliary timer operating as an extra input. Because the input–output transfer is asynchronous, this machine can have more than one output value for the same state (see details in section 3.5).

Out-registered (pipelined) timed Moore machine: The extra register of figure 8.2b is connected to the output of the timed Moore machine. As seen in sections 2.5 and 2.6, two fundamental reasons for doing so are glitch removal and pipelined construction. The new output will be one or one-half of a clock cycle (depending on the selected clock edge) behind the original output. The resulting circuit is order-2 synchronous because the original Moore machine was already a registered circuit (in other words, the input–output transfer occurs after two clock edges—see details in section 3.5). If in a given application this extra register is needed but its consequent extra delay is not acceptable, the next alternative can be considered.

Out-registered (pipelined) timed Mealy machine: The extra register of figure 8.2b is connected to the output of the timed Mealy machine. The reasons for doing so are the same as for Moore machines. The resulting circuit is order-1 synchronous because the input–output relationship in the original Mealy machine can be asynchronous. Consequently, the overall behavior (with the output register included) is similar to that of a timed Moore machine without the output register (see details in section 3.5).

8.3 Timer Interpretation

It is very important to interpret the timer correctly. The analysis below and that in the section that follows are based on the state machine of figure 8.3, where x is the actual input, t is an auxiliary input generated by a timer (see the timer in figure 8.2a), and y is the actual output. Note that this FSM contains all four possible transition types (see section 1.6).

8.3.1 Time Measurement Unit

The time in timed machines (t and T in figure 8.3, for example) is not expressed in seconds but rather in "number of clock cycles." For example, if we want the machine to stay in a certain state during t_{state} = 2 ms, and the clock frequency is f_{clk} = 50 MHz, we simply adopt $T = t_{state} \times f_{clk} = 2{\cdot}10^{-3} \times 50{\cdot}10^{6}$ = 100,000 clock cycles.

8.3.2 Timer Range

If a regular sequential counter with initial value zero is used to build the timer, the counter's range for the timer to span T clock periods is then from $t = 0$ to $t = T - 1$ (so $t_{max} = T - 1$).

If the machine has multiple timed transitions, requiring it to stay T_1 clock cycles in state S_1, T_2 clock cycles in state S_2, and so on, then the value of T can be determined using the expression $T = \max\{T_1, T_2, \ldots\}$. The same is true if multiple values of T are required in the same state.

Note that indeed a counter running up to any value above t_{max} would also do. For example, one could choose to use a timer that runs up to the next power-of-two, in which case only the counter's MSB would need to be monitored, simplifying the circuit construction (at the expense of an extra DFF; also, the transition conditions should be changed from $t = t_{max}$ to $t \geq t_{max}$ in the conditional-timed cases).

8.3.3 Number of Bits

The number of bits needed to implement the timer is $N = \lceil \log_2 T \rceil$. In other words, N must satisfy $T \leq 2^N$. For example, if we want $T_1 = 25$ and $T_2 = 8$, $T = \max\{25, 8\} = 25$

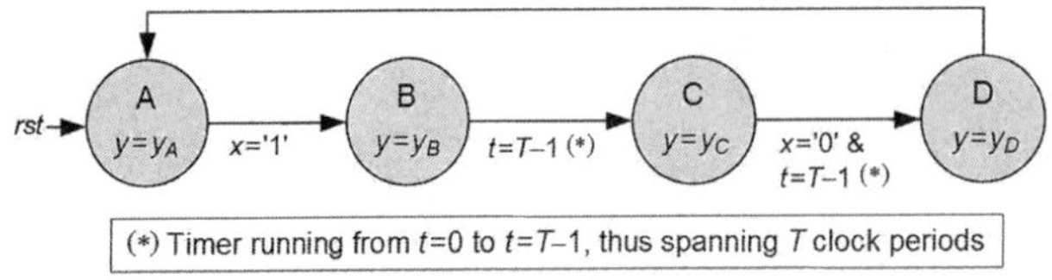
A
y=yA
B
y=yB
C
y=yC
D
y=yD
rst
x='1'
t=T–1 (*)
x='0' &
t=T–1 (*)
(*) Timer running from t=0 to t=T–1, thus spanning T clock periods

Figure 8.3
State machine with all four possible transition types.

results, so a five-bit counter is needed to build the timer (thus able to run from 0 up to 31). The ranges of interest in this case are 0-to-24 in state S_1 and 0-to-7 in state S_2.

8.4 Transition Types and Timer Usage

The state machine of figure 8.3, which contains all four possible transition types, is again used in the analysis that follows.

Transition AB is time independent, so the timer is not needed. Consequently, we can let the timer run freely (for example, from 0 to $2^N - 1$, restarting then automatically from 0), or let it run up to a certain value and then stop it, or simply keep it stopped (at zero, for example). Keeping the timer stopped saves power but can increase the complexity of the comparator. However, if the timer runs up to a certain value and then stops (remaining so until the machine changes its state), the additional power consumption will generally be negligible. In case one decides to keep the timer stopped at zero, $T = 1$ should be used (timer running from 0 to $t_{max} = T - 1 = 0$).

Transition BC is timed, so the timer is needed. The machine must stay in state B during *exactly* T clock periods, moving then to state C. Consequently, we can stop the timer when the monitored value ($t_{max} = T - 1$) is reached or we can simply let it run freely (for example, from 0 to $2^N - 1$, restarting then automatically from 0) because the machine will change its state anyway after $t = t_{max}$ occurs.

Transition CD is conditional-timed, so the timer is again needed. The machine must move to state D at the first (positive) clock edge that finds x = '0' after staying in state C during T clock periods (so it will stay in C during *at least* T clock periods). In this case we cannot let the timer run freely because then if x = '0' is not satisfied when the timer reaches the monitored value ($t_{max} = T - 1$) the condition x = '0' will only be effective again when the timer passes through that value once more. A possible solution here is to stop the timer when the monitored value is reached (indeed, any value $\geq t_{max}$ would do—see comments in section 8.3.2).

Finally, transition DA is unconditional, so the same comments made for transition AB apply here.

In the next section, the possible timer usages described above will be considered in order to develop systematic strategies for designing the timer.

8.5 Timer Control Strategies

We can now develop systematic strategies for controlling the timer. Figure 8.4 is used to illustrate the discussions that follow. Note that all four machines are timed. The timed states (states that need the timer) are represented with a darker shade of gray. A simplified representation was employed for the transition conditions; for example,

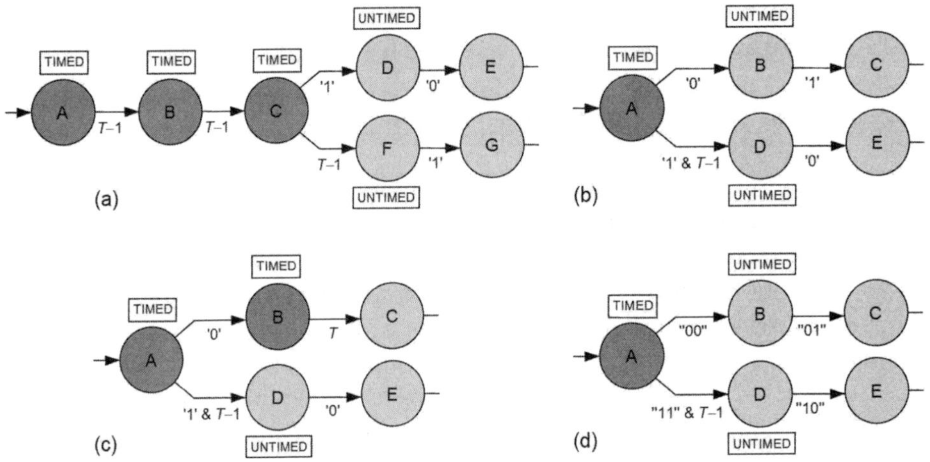
UNTIMED
TIMED
TIMED
TIMED
A
B
C
D
E
F
G
T–1
T–1
'1'
'0'
T–1
'1'
UNTIMED
(a)
UNTIMED
TIMED
A
B
C
D
E
'0'
'1'
'1' & T–1
'0'
UNTIMED
(b)
TIMED
TIMED
A
B
C
D
E
'0'
T
'1' & T–1
'0'
UNTIMED
(c)
UNTIMED
TIMED
A
B
C
D
E
"00"
"01"
"11" & T–1
"10"
UNTIMED
(d)

Figure 8.4

Four timed machines. (a) With only conditional and timed transitions. (b) With conditional and conditional-timed transitions but with state B untimed. (c) Same as b but with state B timed. (d) Same as b but with conditional values ("01" and "10") that might require the machine to remain in state A longer than T clock periods.

'1' means x = '1' and $T - 1$ means $t = T - 1$. As usual, it is assumed that a regular sequential counter running from $t = 0$ up to $t = T - 1$ is employed to build the timer.

8.5.1 Preliminary Analysis

A "tentative" strategy is assumed in this preliminary analysis, which consists of zeroing the timer after it reaches the monitored value ($t_{max} = T - 1$), with $t_{max} = 0$ adopted in the untimed states.

The machine in figure 8.4a has only conditional and timed transitions, so the timer always runs exactly up to t_{max}, after which the machine changes its state. Since it is assumed here that the timer is always zeroed after t_{max} occurs, the timer will always be cleared when the FSM enters a new state, causing it to work properly.

The machine in figure 8.4b has a conditional-timed transition. If x = '0' occurs before $t = t_{max}$, the machine moves from A to B with the timer at an unknown ($<t_{max}$) value. Consequently, the timer will not be zeroed here. However, because state B is untimed, so $t_{max} = 0$, the timer will be zeroed at the end of the first clock period after entering state B. As a result, the timer will be ready to operate properly even if state C is timed.

The case in figure 8.4c is similar to that in figure 8.4b, but state B is now timed. Because the machine will enter state B with $t < t_{max}$, the timer will span in state B only the number of clock cycles needed to complete state B's t_{max}. In summary, our tentative timer control strategy is not appropriate for this machine.

The case in figure 8.4d is also similar to that in figure 8.4b, with state B again untimed. However, note that there are values of x ("01" and "10") that might cause the machine to stay in state A even if t_{max} is reached. Because the timer is zeroed after t_{max} occurs, our tentative strategy does not work here either. A possible solution in this case is to stop the timer at t_{max}, zeroing it only when the machine changes state.

Based on the analysis above and that in section 8.4, two timer control strategies are proposed next.

8.5.2 Timer Control Strategy #1 (Generic)

A strategy that complies with all conditions described in section 8.4 and, consequently, with all conditions in the examples of figure 8.4, is summarized below.

For stopping the timer: Stop the timer when it reaches the monitored value (or a predefined value above that). Keep it so until the machine changes its state.
For zeroing the timer: Zero the timer whenever the machine changes state.

To apply the timer-zeroing technique above, we can compare *pr_state* to *nx_state*. If they are different, it means that the FSM will change its state at the next clock edge, so a flip-flop clearing command can be produced to zero the timer when such a transition occurs.

The advantages of this strategy are that it is generic, simple to understand, and simple to implement. The construction of state transition diagrams using it is simple and direct as well. Additionally, the timer does not need to be controlled in the untimed states because it will run only up to a certain value and will stop anyway, so power consumption is generally not a problem. Also, if one wants, a value greater than t_{max} can be employed (see comments in section 8.3.2), which can simplify the t-to-t_{max} comparator (recall that this comparator can be large; for example, to produce a 1 s delay from a 100 MHz clock, a 27-bit counter is needed); for instance, if T (= t_{max} + 1) is a power of 2, only a single bit (the MSB) needs to be monitored.

Its main disadvantage is that the *pr_state*-to-*nx_state* comparator can be a large circuit, because the number of bits in these two signals can be large, particularly when the number of states is high and one-hot encoding is employed (sequential or gray encoding is suggested when using strategy #1).

The following procedure is recommended: Use strategy #1, which is generic, to draw the state transition diagram. After completing it, check whether it complies with condition 1 or 2 described below for strategy #2. If it does, strategy #2 too can be used to build the timer.

There are only few cases in which strategy #1 cannot be applied completely, but the required adjustments are simple to handle. Such cases will be illustrated in sections 8.7 and 8.11.8.

When using VHDL or SystemVerilog, one of the following codes can be used to implement the timer using strategy #1. Note the use of $t \neq t_{max}$, which can be a slightly smaller comparator circuit than $t < t_{max}$, but either one is fine.

```
--Timer for strategy #1-----------------------

--VHDL----------------------------------------
process (clk, rst)
begin
   if rst='1' then
      t <= 0;
   elsif rising_edge(clk) then
      if pr_state /= nx_state then
         t <= 0;
      elsif t /= tmax then --see comment
         t <= t + 1;
      end if;
   end if;
end process;
----------------------------------------------

--SystemVerilog-------------------------------
always_ff @(posedge clk, posedge rst)
   if (rst) t <= 0;
   else if (pr_state != nx_state) t <= 0;
   else if (t != tmax) t <= t + 1; --see comment
----------------------------------------------
```

8.5.3 Timer Control Strategy #2 (Nongeneric)

This strategy is not generic because it cannot be employed in any timed machine. For example, it only works properly for machines a and b of figure 8.4. The procedure is summarized below.

For stopping the timer: Do not stop the timer.
For zeroing the timer: Zero the timer after it reaches $t_{max} = T - 1$. In the untimed states, adopt $t_{max} = 0$ (timer stopped at zero).

This strategy can be applied in the following cases:

1) To any timed machine without conditional-timed transitions (figure 8.4a, for example).
2) To timed machines with conditional-timed transitions but only if no state has more than one value for T, if no state can last longer than T clock periods, and if any transition that might last less than T clock cycles goes to an untimed state (figure 8.4b, for example).

The advantage of this strategy is that it avoids the *pr_state*-to-*nx_state* comparator, which can be a large circuit.

The disadvantages are that it is not generic and that the resulting circuit is not guaranteed to be smaller than that for strategy #1. Because here the value of t_{max} must be specified in all states (with $t_{max} = 0$ in the untimed states) when the machine has conditional-timed transitions, the t-to-t_{max} comparator (which also can be large) is more complex.

Since strategy #2 is not generic, the suggested procedure is to draw the state transition diagram using strategy #1, checking next if it complies with condition 1 or 2 above in order to determine whether strategy #2 can be used as well.

When using VHDL or SystemVerilog, one of the codes below can be used to implement the timer for strategy #2. Note the use of $t < t_{max}$ instead of $t \neq t_{max}$, needed to guarantee that the timer will be zeroed if the FSM leaves a timed state before the timer has reached t_{max} (entering therefore an untimed state). However, such a situation can only occur if the machine has conditional-timed transitions (figure 8.4b, for example); if the machine does not have conditional-timed transitions (figure 8.4a, for example), then $t \neq t_{max}$ is fine, too.

```
--Timer for strategy #2-----------------------

--VHDL---------------------------------------
process (clk, rst)
begin
   if rst='1' then
      t <= 0;
   elsif rising_edge(clk) then
      if t < tmax then --see comment
         t <= t + 1;
      else
         t <= 0;
      end if;
   end if;
end process;
---------------------------------------------

--SystemVerilog------------------------------
always_ff @(posedge clk, posedge rst)
   if (rst) t <= 0;
   else if (t < tmax) t <= t + 1; --see comment
   else t <= 0;
---------------------------------------------
```

8.5.4 Time Behavior of Strategies #1 and #2

Figure 8.5 shows an example of FSM that in spite of having a conditional-timed transition can be implemented using any of the timer control strategies proposed above (note that this machine falls in the category depicted in figure 8.4b). The purpose of this example is to illustrate the differences in terms of time behavior between strategies #1 and #2.

The machine must implement a triggered circuit with input x and output y. The intended behavior is depicted in figure 8.5a. In this case, y must go up as soon as (i.e., at the next positive clock edge) x goes up, returning to zero T clock cycles (more precisely, T clock *edges*) after x returns to zero. Observe in the final part of the plot that when x comes down but goes up again before the time T has ended the circuit is retriggered.

A solution is shown in figure 8.5b. $t_{max} = T - 2$ is used in the CA transition, thus spanning $T - 1$ clock periods (in fact, $T - 1$ clock edges, as indicated by the black dots), because one period is spent in the BC transition.

When strategy #1 is used the overall behavior is objective and very simple to understand, as can be observed in figure 8.5c, which shows an illustrative timing diagram for $T = 4$. It is assumed that the timer stops as soon as t_{max} is reached (though not required for strategy #1, $t_{max} = 0$ was used in the untimed states).

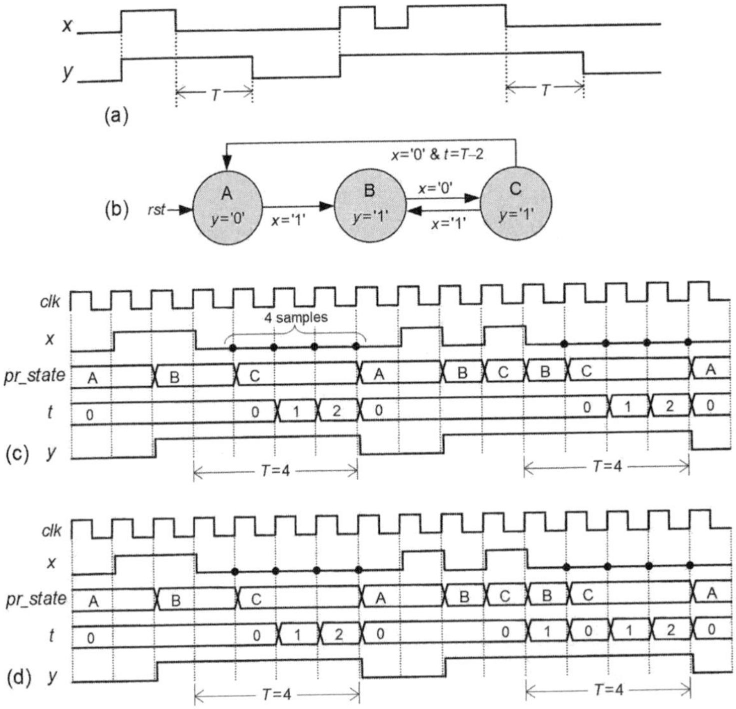

x
y
T
T
(a)
x='0' & t=T–2
A
y='0'
B
y='1'
C
y='1'
x='0'
x='1'
x='1'
(b)
rst
clk
x
4 samples
pr_state
t
(c) y
T=4
T=4
clk
x
pr_state
t
(d) y
T=4
T=4

Figure 8.5
(a) Desired circuit behavior for a triggered circuit. (b) A solution, which can be implemented directly with either strategy #1 or #2. (c) Timing diagram for strategy #1. (d) Timing diagram for strategy #2.

Strategy #2 is a little more difficult to examine. This is due to the CB transition, which can only happen with $t < t_{max}$, causing the FSM to enter state B without any command to zero the timer at the next clock edge. Observe in the B-to-C transition in the timing diagram of figure 8.5d that t is still incremented when the FSM enters state B, being only zeroed at the next clock pulse. However, in spite of this detail, the machine operates adequately, as can be seen in the plot for y, which is exactly the same as that in figure 8.5c.

The reader is invited to examine these two timing diagrams carefully to fully understand and appreciate the differences between these two timer control strategies.

8.6 Truly Complementary Time-Based Transition Conditions

As discussed in Section 3.8, when a circuit does not have any sort of reset mechanism, the initial state (either '0' or '1') of its flip-flops upon power-up might be undetermined. Say that that is the case and that our machine has a timed transition that must span 10 clock periods, thus requiring a 4-bit counter, where 0-to-9 is the range of interest. Since a 4-bit counter is capable of counting from 0 to 15, the initial (random) state might fall in the 10-to-15 range. Recall from section 1.5 that the outward transition conditions in any state must be *truly complementary* (i.e., they must include all possible combinations of the transition control signals, and obviously all just once), so the 10-to-15 range must also be considered.

Figure 8.6a shows an example of timed machine with under-specified transition conditions, which falls in the situation described above because the $t > T - 1$ range is not covered. The problem in fixed in figure 8.6b by assigning that range to the AB transition. Another alternative is presented in figure 8.6c, with the missing range assigned to the AA transition. Either one of the last two options should be used. The decision between one or the other depends on the application; more specifically, it depends on where we want the machine to be in case $t > T - 1$ happens (at power-up, for example).

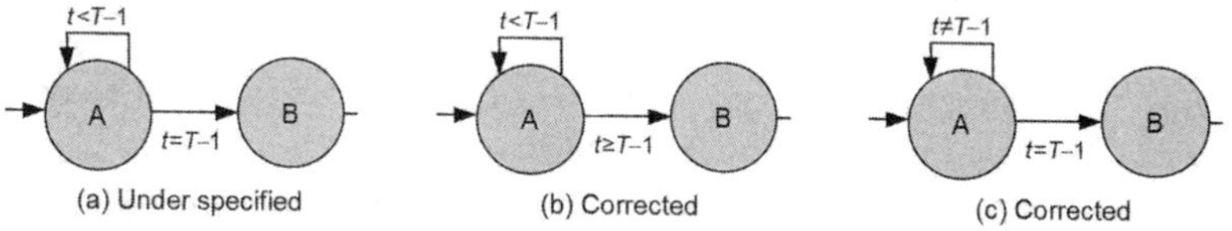
t<T–1
A
B
t=T–1
(a) Under specified
t<T–1
A
B
t≥T–1
(b) Corrected
t≠T–1
A
B
t=T–1
(c) Corrected

Figure 8.6

(a) Under-specified transition conditions ($t > T - 1$ range not covered). (b) $t > T - 1$ range assigned to the AB transition. (c) $t > T - 1$ range assigned to the AA transition.

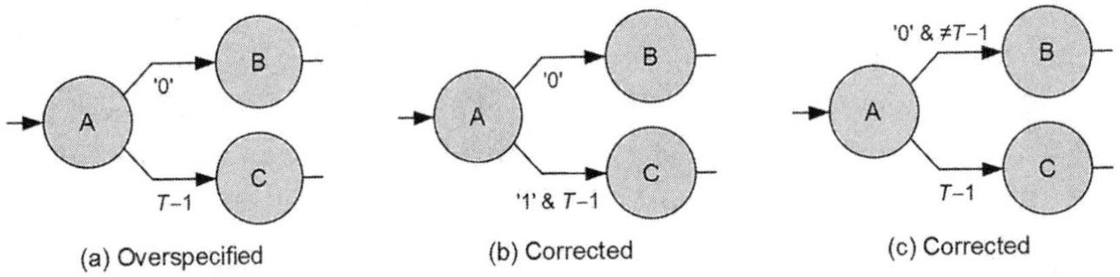
'0'
A
B
C
T−1
(a) Overspecified
'0'
A
B
C
'1' & T−1
(b) Corrected
'0' & ≠T−1
A
B
C
T−1
(c) Corrected

Figure 8.7
(a) Overspecified transition conditions ('0' and $T-1$ can happen at the same time). (b) Solution with priority given to transition AB. (c) Solution with priority given to transition AC.

Note, however, that the problem described above can only happen before the first run of the timer, after which the FSM has full control over the timer.

Another timed machine with incorrect transition conditions is shown in figure 8.7. However, contrary to the previous example, this machine is over-specified, because more than one transition can be true at the same time ('0' and $T-1$ can occur at the same time). As seen in section 1.5, overspecification can be resolved by establishing priorities. AB was considered to have priority over AC in the solution of figure 8.7b, whereas the opposite was assumed in figure 8.7c.

8.7 Repetitively Looped State Machines

This section discusses the particular case of repetitively looped state machines, found, for example, in serial data communications circuits (serial data receiver/transmitter, I^2C interface, SPI interface, etc.). An equivalent implementation will be seen in section 11.5 using the category 3 approach.

The first case is shown in figure 8.8a, where a pair of states is repeated T times (this kind of problem was in fact introduced in section 3.12). If T is large, it is obviously impractical to represent this circuit as a regular FSM.

An equivalent representation for this problem is shown in figure 8.8b, with a loop replacing the repeated states. This loop must be repeated T times in the AB direction (in the BA direction the total is $T-1$ times). Consequently, by converting the category 1 machine of figure 8.8a into the category 2 machine of figure 8.8b, the FSM representation becomes viable and the problem can be easily solved.

A possible implementation is depicted in figure 8.8c, with the timer incremented in both directions (AB and BA), therefore being not zeroed in any of them (note the thick circles and the different arrows; a thick circle means that there is at least one transition into that state in which the timer should not be zeroed, while the different arrow with a dot at its origin identifies which that transition is).

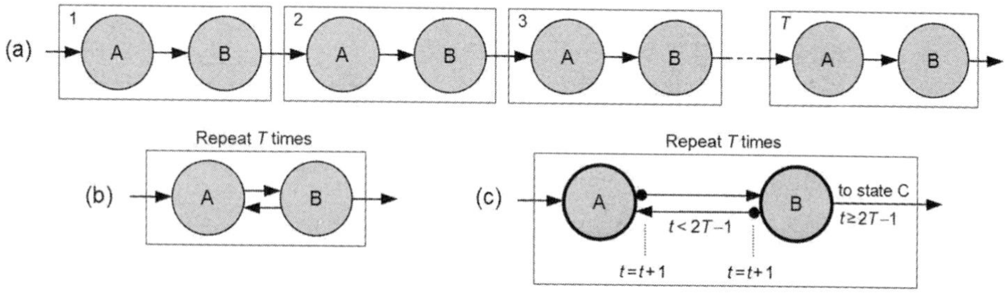
(a)
1
2
3
T
A
B
Repeat T times
(b)
Repeat T times
(c)
to state C
t≥2T−1
t<2T−1
t=t+1
t=t+1

Figure 8.8
(a) FSM with a pair of states repeated T times. (b) Equivalent looped representation. (c) An alternative for the timer, counting in both directions, thus being not zeroed in any of them.

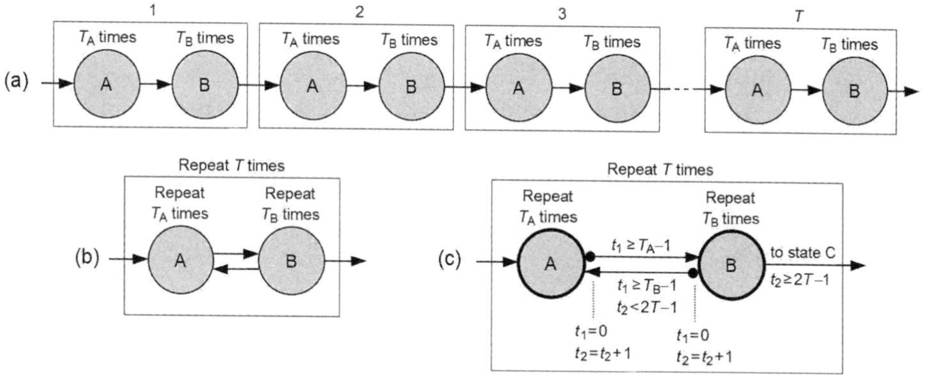
1
2
3
T
TA times
TB times
(a)
A
B
Repeat T times
Repeat TA times
Repeat TB times
(b)
(c)
t1 ≥ TA−1
t1 ≥ TB−1
t2<2T−1
to state C
t2≥2T−1
t1=0
t2=t2+1
t1=0
t2=t2+1

Figure 8.9
Generalization of the case seen in figure 8.8. Not only the loop is repeated T times but also the machine stays T_A clock periods in A and T_B clock periods in B.

A more general case is presented in figure 8.9. Here, not only the loop must be repeated T times, but the machine must also stay T_A clock periods in A and T_B clock periods in B. The problem is stated in figure 8.9a, with an equivalent representation shown in figure 8.9b. A possible solution is shown in figure 8.9c, using two timers. While timer t_1 controls the time the machine stays in state A or state B, timer t_2 measures the number of loop repetitions. Consequently, only timer t_2 is not zeroed in the state transitions.

8.8 Time Behavior of Timed Moore Machines

In section 3.5 an analysis of the general time behavior of Moore and Mealy machines was presented. This section and the next present extensions to that analysis for the case when timed transitions are also involved.

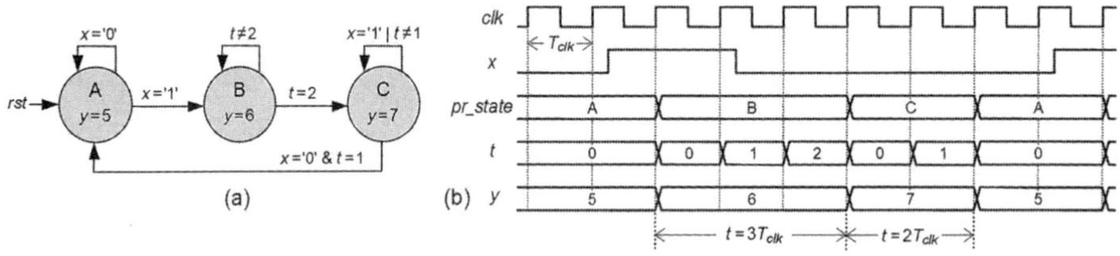
x='0'
t≠2
x='1' | t≠1
rst
A
y=5
x='1'
B
y=6
t=2
C
y=7
x='0' & t=1
(a)
clk
Tclk
x
pr_state
t
(b) y
t=3Tclk
t=2Tclk

Figure 8.10
(a) A Moore machine and (b) a corresponding timing diagram.

The Moore machine of figure 8.10a, which includes three transition types, is used to illustrate the analysis. It is assumed that the timer control strategy #1 is adopted to build the timer. Observe the following in the accompanying timing diagram of figure 8.10b:

1) $T_B = 3$ and $T_C = 2$ clock cycles.
2) When x changes, the output does not change. This is expected because in a Moore machine the output is synchronous, thus changing only when the state changes.
3) The stay in state A depends only on x, so the machine moves to state B at the first (positive) clock edge that finds x = '1'.
4) Because $T_B = 3$, state B lasts exactly three clock cycles (the timer counts from 0 to 2).
5) Because $T_C = 2$ but the CA transition is conditional-timed, state C lasts at least two clock cycles (the timer counts from 0 to 1). The "at least" restriction is due to the x = '0' condition, which might not be true when the timer reaches the monitored ($T_C - 1 = 1$) value. In this example x = '0' was already available, so state C did last only two clock periods.
6) In the states where the timer is not needed (only state A in this example), the timer was kept stopped at zero.

In conclusion, in a Moore machine the output and the state are in perfect sync, changing at the same time. Each output value then has the same duration as its associated state.

8.9 Time Behavior of Timed Mealy Machines

The Mealy machine of figure 8.11a, which is the Mealy counterpart of the Moore machine of figure 8.10a, is used to illustrate the analysis. Observe the following in the accompanying timing diagram of figure 8.11b:

1) $T_B = 3$ and $T_C = 2$ clock cycles.
2) Contrary to the Moore case, y can change when x changes. This is expected because Mealy machines are asynchronous.

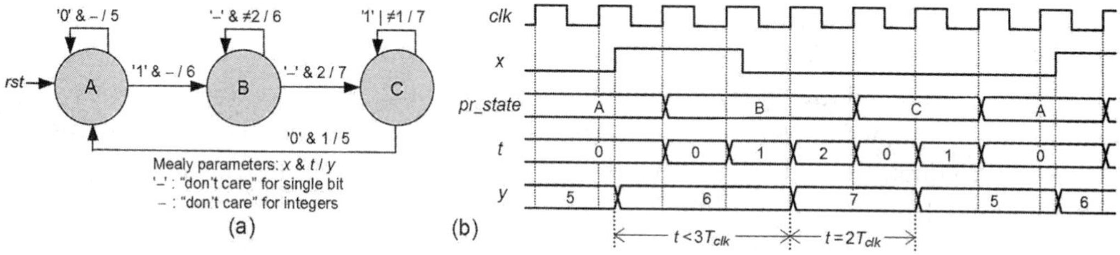
'0' & – / 5
'–' & ≠2 / 6
'1' | ≠1 / 7
rst
A
'1' & – / 6
B
'–' & 2 / 7
C
'0' & 1 / 5
Mealy parameters: x & t / y
'–' : "don't care" for single bit
– : "don't care" for integers
(a)
(b)
clk
x
pr_state
t
y
A
B
C
A
0
0
1
2
0
1
0
5
6
7
5
6
t<3Tclk
t=2Tclk

Figure 8.11

(a) Mealy counterpart of the Moore machine of figure 8.10a. (b) A corresponding timing diagram.

3) The output value while in state A depends on x, so two different values can occur: $y = 5$ if $x =$ '0' or $y = 6$ if $x =$ '1'.

4) Because $T_B = 3$, state B lasts exactly three clock cycles (the timer counts from 0 to 2).

5) Because $T_C = 2$ but the CA transition is conditional-timed, state C lasts at least two clock cycles (the timer counts from 0 to 1). The "at least" restriction is due to the $x =$ '0' condition, which might not be true when the timer reaches the monitored ($T_C - 1 = 1$) value. In this example $x =$ '0' was already available, so state C did last only two clock periods.

6) In the states where the timer is not needed (only state A in this example), the timer was kept stopped at zero.

7) The states and the timer operate exactly as in the Moore case, but that is not true for the output.

8) As already seen, contrary to the Moore case, the output value is not unique in all states.

9) Contrary to the Moore case, the output does not change together with the state. It changes earlier.

10) Contrary to the Moore case, the output values do not necessarily last as long as the states. They last less than the associated state if the transition condition into that state is asynchronous, or they last exactly the same as the associated state if the transition condition into that state is synchronous (a timed transition, for example). Note that the output value $y = 6$ lasts less than three clock periods (the transition control signal into state B is x, which is asynchronous), but $y = 7$ lasts exactly two clock periods (the transition control signal into state C is t, which is synchronous).

In conclusion, in a Mealy machine the output changes earlier than the state, either by a fraction of a clock period (if the transition condition into that state is asynchronous) or by a full clock period (for synchronous conditions, such as time). The duration of each output value can then be different from that of its associated state.

8.10 Number of Flip-Flops

Having understood the timer, we pass now to the last analysis before the presentation of timed (category 2) FSM examples. The analysis regards the number of flip-flops needed to implement the intended circuit. As mentioned earlier, in general, and particularly in large designs, it is difficult to estimate the number of logic gates that will be needed to implement the desired solution, but it is always possible to determine, and exactly, the number of flip-flops.

In the particular case of sequential circuits implemented as category 2 state machines, there are three demands for DFFs, as follows.

1) For the state register (see *nx_state* and *pr_state* in figure 8.2a, which are the state memory flip-flops' input and output, respectively; below, M_{FSM} is the number of states):
   For sequential or Gray encoding, $N_{FSM} = \lceil \log_2 M_{FSM} \rceil$. For example, $M_{FSM} = 25 \rightarrow N_{FSM} = 5$.
   For Johnson encoding, $N_{FSM} = \lceil M_{FSM}/2 \rceil$. For example, $M_{FSM} = 25 \rightarrow N_{FSM} = 13$.
   For one-hot encoding, $N_{FSM} = M_{FSM}$. For example, $M_{FSM} = 25 \rightarrow N_{FSM} = 25$.
2) For the output register (figure 8.2b, optional, with b_{output} bits):
   $N_{output} = b_{output}$. For example, $b_{output} = 16 \rightarrow N_{output} = 16$.
3) To build the timer (figure 8.2a, compulsory):
   $N_{timer} = \lceil \log_2 T_{max} \rceil$, where T_{max} is the largest transition time, expressed in "number of clock cycles," that is, $T_{max} = t_{state\_max} \times f_{clk}$, where t_{state_max} is the largest transition time, in seconds, and f_{clk} is the clock frequency, in hertz. For example, for the machine to be able to stay $t_{state\_max} = 8$ µs in the state with longest duration, and assuming that $f_{clk} = 50$ MHz, $T_{max} = 8 \cdot 10^{-6} \times 50 \cdot 10^{6} = 400$ clock cycles must be used, from which $N_{timer} = 9$ results.

Therefore, the total number of DFFs is $N_{total} = N_{FSM} + N_{output} + N_{timer}$. In the examples that follow, as well as in the actual designs with VHDL and SystemVerilog, the number of flip-flops will be often examined.

8.11 Examples of Timed (Category 2) Machines

A series of timed FSMs are presented next. To draw the corresponding state transition diagrams, strategy #1 (section 8.5.2) is considered as the default strategy for controlling the timer. If the resulting machine fulfills condition 1 or 2 for strategy #2 (section 8.5.3), then that strategy too can be used to control the timer.

Several of the examples described in this chapter will be implemented later using VHDL (chapter 9) and SystemVerilog (chapter 10).

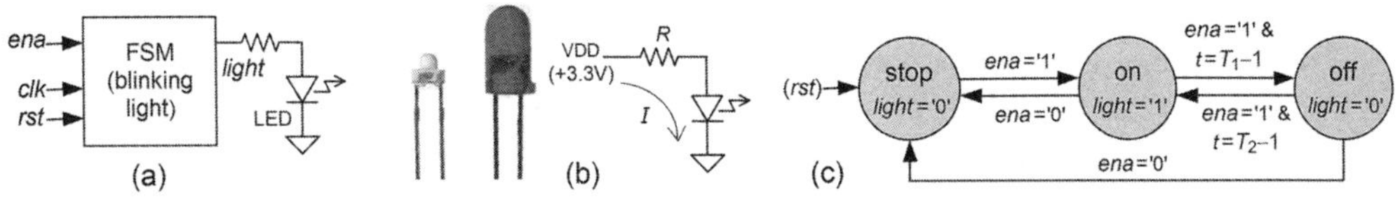
ena
clk
rst
FSM
(blinking
light)
light
LED
(a)
VDD
(+3.3V)
R
I
(b)
(rst)
stop
light='0'
ena='1'
ena='0'
on
light='1'
ena='1' &
t=T1−1
ena='1' &
t=T2−1
off
light='0'
ena='0'
(c)

Figure 8.12
Circuit that feeds a blinking light (an LED, in this case). (a) Circuit ports. (b) Commercial LEDs and typical usage. (c) Moore-type solution.

8.11.1 Blinking Light

This example is simple enough and yet very illustrative of the general behavior of timed FSMs. It concerns a circuit that must turn a light on and off, remaining on during T_1 clock periods and off during T_2 clock periods. An important desired feature is that when the circuit is enabled it must start from a state with the light on (to prevent the user from thinking that the circuit is not working when large transition times are involved).

The circuit ports are depicted in figure 8.12a. The input comes from a switch called *ena*, which enables the circuit when asserted. The output is a port called *light* that feeds a light-emitting diode (LED).

Figure 8.12b shows examples of commercial LEDs and their typical usage. An LED consists of a PN junction fabricated, for example, with gallium arsenide (GaAs). It emits light when forward biased, as shown on the right of figure 8.12b, which also shows a current-limiting resistor, *R*. The emitted radiation can be in the infrared spectrum (used, for example, in remote controls) or in the visible spectrum (used, for example, in alphanumeric displays and as signaling lamps in all sorts of equipment). To radiate in the visible spectrum, other materials must be added to GaAs, such as aluminum, in AlGaAs (red light), or phosphorus, in GaAsP (red or yellow light). Other materials are used to obtain radiations at higher frequencies, such as zinc selenide (ZnSe) for blue LEDs.

A Moore-type solution is presented in figure 8.12c. It has three states, called *stop* (not blinking, with light off), *on* (circuit enabled, with light on), and *off* (circuit enabled, with light off). Note that while the switch is closed (*ena* = '1') the circuit flips back and forth between states *on* and *off*, staying T_1 clock periods in the former and T_2 clock periods in the latter. Note also that when *ena* is asserted the machine moves immediately (at the next positive clock edge) from state *stop* to state *on*; likewise, when enable is turned off, the machine moves immediately to state *stop*.

This machine falls in the situation depicted in figure 8.4b, so it can be implemented with either timer control strategy (#1, section 8.5.2, or #2, section 8.5.3).

An interesting aspect of this machine is that it might not need a reset signal at all, depending on the chosen encoding scheme (for example, sequential or Gray) and on the equations used to implement the next state (see exercise 8.5). Also, despite *ena* being an asynchronous input and being produced by a mechanical switch, neither a synchronizer (section 2.3) nor a debouncer (section 8.11.3) is needed because of the nature of this application.

Based on section 8.10, the number of flip-flops needed to implement this circuit is as follows. For the state register: $M_{FSM} = 3$ states; therefore, $N_{FSM} = 2$ if sequential, Gray, or Johnson encoding is used, or 3 for one-hot. For the optional output register: not needed in this application, so $N_{output} = 0$. For the timer: assuming 0.5 s for T_1 and 1 s for T_2, with $f_{clk} = 50$ MHz, $T_{max} = 5{\cdot}10^7$ clock cycles results, so $N_{timer} = 26$. Therefore, $N_{total} = 28$ or 29.

8.11.2 Light Rotator

This example shows another simple and yet very illustrative application for timed machines. It consists of a circuit that produces a rotating movement in a seven-segment display (SSD).

SSDs are just special seven-LED arrangements (eight if a decimal point is also included). This kind of device is illustrated in figure 8.13. In figure 8.13a an example of a commercial SSD device (two digits with decimal points) is shown. In figure 8.13b a typical notation for the segment names (*abcdefg*) is shown. In figure 8.13c the common-anode configuration is presented, in which a '0' turns a segment on and a '1' turns it off. Finally, the table in figure 8.13d shows the logic values that must be used to obtain the traditional 0-to-F hexadecimal characters.

The circuit ports for the current example are presented in figure 8.14a. The inputs are a stop switch (*stp*), clock (*clk*), and reset (*rst*), and the output is a seven-bit signal (*ssd*) that feeds the seven segments of the SSD.

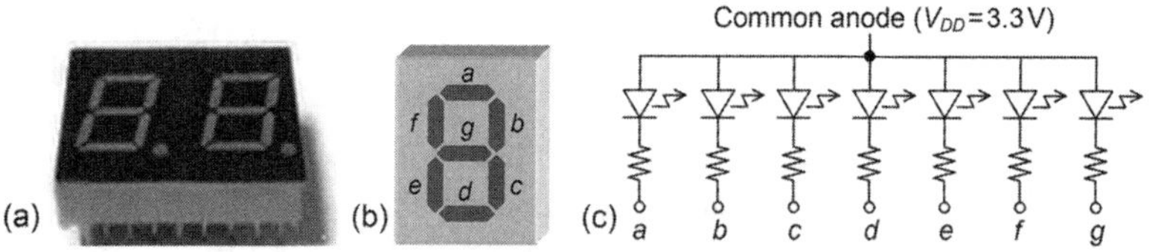
(a)
(b)
a
f
g
b
e
d
c
(c)
Common anode (V_{DD}=3.3V)
a
b
c
d
e
f
g

(d)

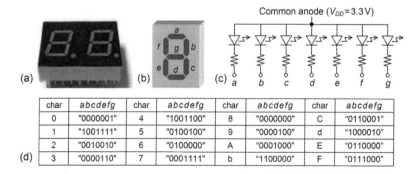

char	abcdefg	char	abcdefg	char	abcdefg	char	abcdefg
0	"0000001"	4	"1001100"	8	"0000000"	C	"0110001"
1	"1001111"	5	"0100100"	9	"0000100"	d	"1000010"
2	"0010010"	6	"0100000"	A	"0001000"	E	"0110000"
3	"0000110"	7	"0001111"	b	"1100000"	F	"0111000"

Figure 8.13
(a) A commercial seven-segment display (SSD) device. (b) Segment names (*abcdefg*). (c) Common-anode configuration (a '0' lights a segment, whereas a '1' turns it off). (d) Logic values to obtain the 0-to-F hexadecimal characters.

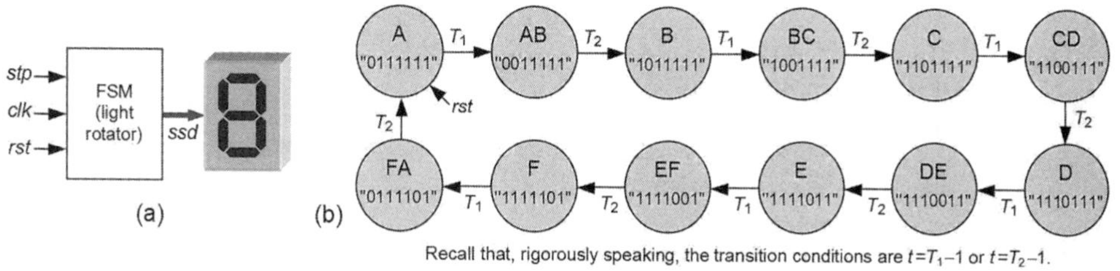
stp
clk
rst
FSM
(light
rotator)
ssd
(a)
(b)
A
"0111111"
AB
"0011111"
B
"1011111"
BC
"1001111"
C
"1101111"
CD
"1100111"
D
"1110111"
DE
"1110011"
E
"1111011"
EF
"1111001"
F
"1111101"
FA
"0111101"
rst
Recall that, rigorously speaking, the transition conditions are t=T1–1 or t=T2–1.

Figure 8.14
SSD rotator. (a) Circuit ports. (b) Corresponding (simplified) Moore-type state diagram.

The segments must be lit sequentially in the clockwise direction, with overlaps used to provide a smoother rotation. In other words, segment *a* must be lit for some time, then *a* and *b* should be lit together, then *b* only, next *b* and *c*, then *c* only, and so on. If the stop switch is asserted (*stp* = '1'), the movement must stop, resuming exactly from the same position when *stp* returns to '0'. If reset is asserted (*rst* = '1'), the circuit must return asynchronously to the initial state (only segment *a* lit).

A corresponding Moore solution is presented in figure 8.14b, where each state name denotes which SSD segments are lit while the machine is in that state. Note that it is a simplified diagram, with only the output value (no output name) shown inside each state circle and only the time values (no stop conditions) specified on the arrows. Moreover, note the relaxed use of T_1 and T_2 alone on the arrows (instead of $t = T_1 - 1$ and $t = T_2 - 1$), indicating simply the total number of clock periods that the machine must spend in each state.

As in the previous example, the input (*stp*) is asynchronous and produced by a mechanical switch. However, because of the nature of the application, again neither a synchronizer (section 2.3) nor a debouncer (section 8.11.3) is needed.

Based on section 8.10, the number of flip-flops needed to implement this circuit is as follows. For the state register: $M_{FSM} = 12$ states; therefore, $N_{FSM} = 4$ if sequential or Gray encoding is used, 6 for Johnson, or 12 for one-hot. For the optional output register: not needed in this application, so $N_{output} = 0$. For the timer: assuming 120 ms for T_1 and 35 ms for T_2, with $f_{clk} = 50$ MHz, $T_{max} = 6 \cdot 10^6$ clock cycles results, so $N_{timer} = 23$. Therefore, $N_{total} = 27$, 29, or 35.

VHDL and SystemVerilog implementations for this light rotator are presented in sections 9.4 and 10.4, respectively. Because this machine falls in the situation depicted in figure 8.4a, it can be implemented with either timer control strategy (#1, section 8.5.2, or #2, section 8.5.3). Strategy #2 was adopted there, but both strategies are explored and compared in exercises 9.1 and 10.1.

8.11.3 Switch Debouncer

Figure 8.15a shows a switch that produces x = '0' when open or x = '1' when closed. The problem with mechanical switches is that they might bounce a few times before

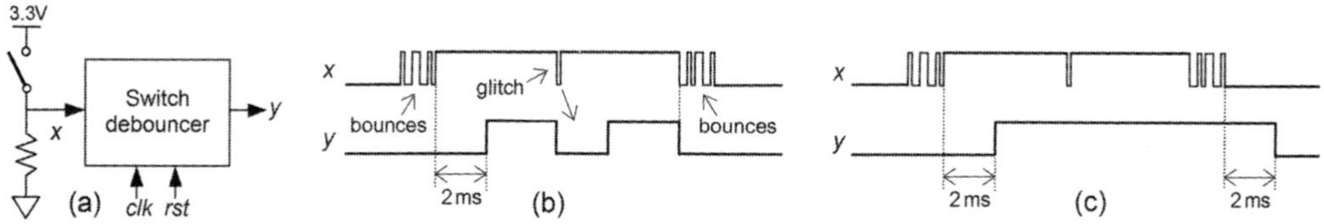
3.3V
Switch debouncer
x
y
(a)
clk
rst
x
y
glitch
bounces
bounces
2ms
(b)
x
y
2ms
(c)
2ms

Figure 8.15
(a) Mechanical switch and debouncer ports. (b) Bounces processed by a one-sided (low-to-high) debouncer. (c) With a two-sided debouncer.

finally settling in the proper position, as illustrated in figures 8.15b,c. Depending on the switch characteristics, such bounces can last from a fraction of a millisecond up to several milliseconds.

Switch bounces are not acceptable in several applications. A disastrous example is when the signal produced by the switch must act as a clock to some process because the corresponding flip-flops will understand the bounces as several clock pulses.

Two debouncing approaches are depicted in figures 8.15b,c. The first is one-sided (only the low-to-high transition is debounced), whereas the second is two-sided (both transitions are debounced). The debouncing strategy here consists of checking the input permanently (at every clock cycle) and accepting a new value only after it has remained fixed for a certain amount of time. For example, if the debouncing time is 2 ms and the clock frequency is 50 MHz, the same result must occur $2 \cdot 10^{-3} \times 50 \cdot 10^{6} = 100{,}000$ consecutive times to be considered valid.

Note that in figure 8.15b the one-sided debouncer automatically filters the other transition, but it does not protect the circuit against unexpected input transients/glitches (caused, for example, by the switching of large current loads onto the same power supply or by lightning).

In debouncers, glitches at the output are generally undesired because providing a safe, clean signal is precisely the purpose of this circuit, so the optional output register of figure 8.2b should be employed unless *y* comes directly from a DFF (this depends on the encoding scheme and can be checked in the compilation report equations).

A flowchart for the two-sided debouncer of figure 8.15c is presented in figure 8.16a.

An initial (bad) solution is presented in figure 8.16b. The problem here is that it only checks the condition x = '1' (or '0') *at the end* of T clock cycles. Consequently, to obtain a full debouncer, each transition of figure 8.16b must be replaced with three pure transitions, resulting in the FSM of figure 8.16c. Although the "–2" factor in the timed $t = T - 2$ condition does not matter in this application, it was kept as a reminder of the precise value.

Even though the input (x) is asynchronous, a synchronizer (section 2.3) is not needed because y can change its value only after a time T, which is a synchronous condition (the timer operates with the same clock as the FSM).

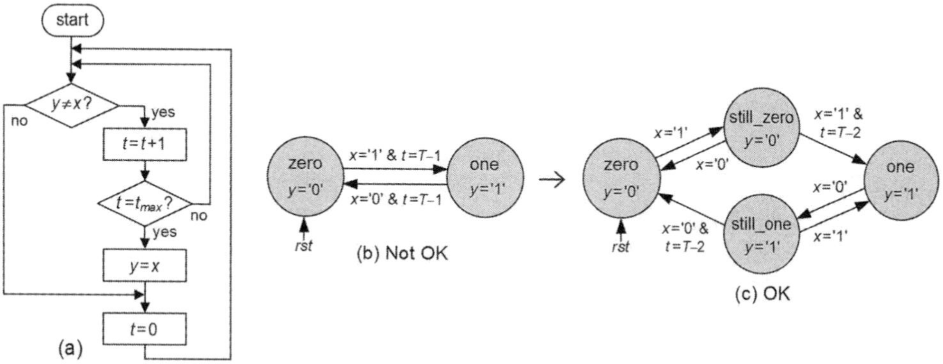
start
y≠x?
no
yes
t=t+1
t=tmax?
no
yes
y=x
t=0
(a)
zero
y='0'
x='1' & t=T-1
x='0' & t=T-1
one
y='1'
rst
(b) Not OK
zero
y='0'
x='1'
x='0'
still_zero
y='0'
x='1' &
t=T-2
one
y='1'
x='0'
x='0' &
t=T-2
still_one
y='1'
x='1'
rst
(c) OK

Figure 8.16
Full switch debouncer. (a) Flowchart. (b) Bad and (c) good solutions.

To discuss the need for an explicit reset signal (see sections 3.8 and 3.9), let us divide the problem into two cases. The first regards implementation in FPGAs whose flip-flops are automatically reset to '0' at power-up. In this case, reset is not needed if any of the encoding schemes described in section 3.7 is used, except for one-hot, but fine for the modified version of one-hot seen in figure 3.10b, with the only restriction that *zero* must be declared as the initial (reset) state. The second case regards implementation in devices whose DFFs' initial state is arbitrary. If sequential or Gray encoding is used, all two-bit codewords will be consumed to encode the machine, so the initial state will fall necessarily inside the machine, and deadlock cannot occur. Consequently, we only need to consider the consequences of having the machine start from a state other than state *zero*. It is clear from figure 8.16c that the value of y will adjust itself automatically to the value of x after at most T clock periods; therefore, reset is required only if having y = '1' during such a short time period might be enough to turn on a critical application (a factory machine, for example).

Based on section 8.10, the number of flip-flops needed to implement this circuit is as follows. For the state register: $M_{FSM} = 4$ states; therefore, $N_{FSM} = 2$ if sequential, Gray, or Johnson encoding is used, or 4 for one-hot. For the optional output register: assuming that y comes directly from a DFF, $N_{output} = 0$. For the timer: with $t_{state\_max} = 2$ ms and $f_{clk} = 50$ MHz, $T_{max} = 10^5$ clock cycles results, so $N_{timer} = 17$. Therefore, $N_{total} = 19$ or 21.

8.11.4 Reference-Value Definer

This section deals with a problem that is common in control applications. It consists of a circuit that sets a reference value. For example, a temperature controller for an air conditioning system must have a way of letting the user choose the desired (*reference*) room temperature (see section 5.4.4).

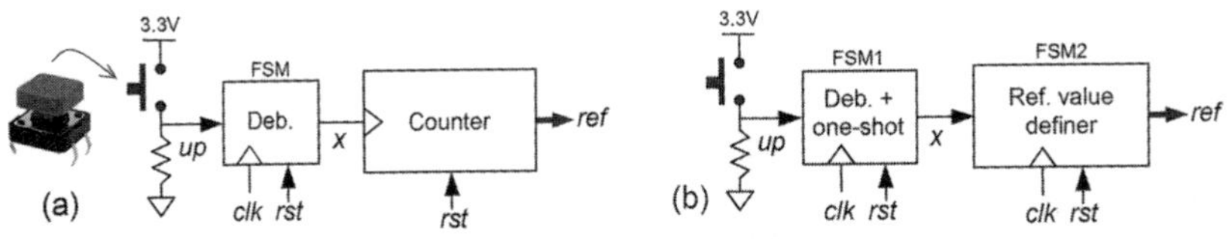
3.3V
FSM
Deb.
up
x
Counter
ref
clk rst
rst
(a)
3.3V
FSM1
Deb. +
one-shot
up
x
FSM2
Ref. value
definer
ref
clk rst
clk rst
(b)

Figure 8.17
Reference-value definer implemented (a) with a counter and (b) with a state machine.

An example is presented in figure 8.17, in which the reference value is set by a pushbutton. Two cases are considered; in both a debouncer is needed, but in figure 8.17a the reference value is produced by a common counter (note that x plays the role of clock), whereas in figure 8.17b it is produced by an FSM (the actual system clock plays the role of clock). Even though a counter too is an FSM, the reference here is to the fact that only the latter is implemented using the FSM approach.

The case in figure 8.17a is advantageous when the reference values are regular (next = present + constant). For example, if we want to set the desired room temperature for the air conditioning system mentioned above, the counter can be incremented by one unit (1°F or 1°C) every time the pushbutton is pressed (and released, of course), going from the minimum to the maximum reference value, returning then to the minimum value and restarting from there. Another advantage of this alternative is that the number of reference values can be arbitrarily large.

The case in figure 8.17b is advantageous when the reference values are irregular and the number of reference values is small. Note that we have made a little modification in the debouncer (FSM1), embedding in it a one-shot converter (see sections 2.4 and 5.4.3), which converts the (long) debounced signal into a pulse with duration equal to one clock period.

A solution for the case of figure 8.17b is presented in figure 8.18. The time behavior of FSM1 is illustrated in figure 8.18a, and the corresponding state diagram is shown in figure 8.18b. It is a '0'-to-'1' debouncer, so it requires three states; just one extra state is needed to include the one-shot conversion, thus totaling four states. The state diagram for FSM2 is depicted in figure 8.18c, for a total of 10 states with arbitrary reference values r_1, r_2, …, r_{10}. The one-shot modification of FSM1 is important because it eliminates the need for FSM2 to check the return of x to '0', which would double its number of states.

In many applications both up and down controls are needed. This kind of situation is illustrated in figures 8.19a and 8.19b, which are generalizations of the cases in figures 8.17a and 8.17b, respectively (i.e., in the former the reference value is set by a counter, whereas in the latter it is set by a state machine).

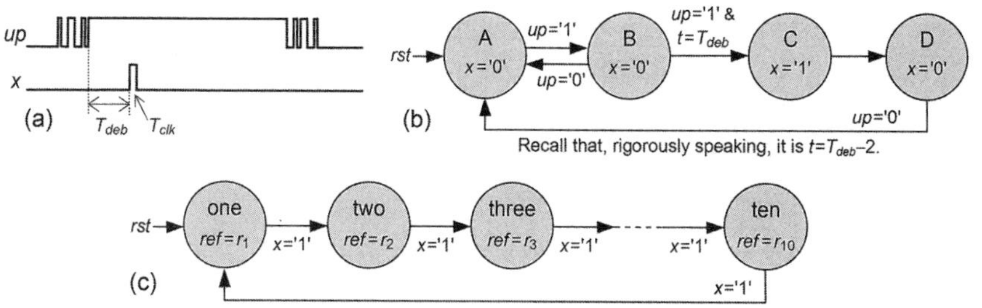
up
x
(a)
T_{deb}
T_{clk}
rst
(b)
A
x='0'
up='1'
up='0'
B
x='0'
up='1' & t=T_{deb}
C
x='1'
D
x='0'
up='0'
Recall that, rigorously speaking, it is t=T_{deb}–2.
rst
one
ref=r_1
x='1'
two
ref=r_2
x='1'
three
ref=r_3
x='1'
x='1'
ten
ref=r_{10}
x='1'
(c)

Figure 8.18

Solution for the circuit of figure 8.17b. (a) Time behavior and (b) FSM for the pushbutton debouncer with embedded one-shot conversion. (c) FSM for the reference-value definer, with 10 arbitrary values.

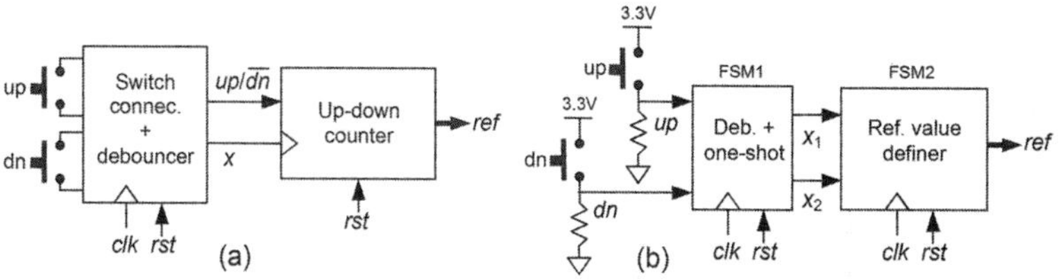
up
dn
Switch connec. + debouncer
up/$\overline{dn}$
x
Up-down counter
ref
clk rst
rst
(a)
3.3V
up
3.3V
dn
up
dn
FSM1
Deb. + one-shot
x_1
x_2
FSM2
Ref. value definer
ref
clk rst
clk rst
(b)

Figure 8.19

Reference-value definer with up and down controls implemented (a) with a counter and (b) with a state machine.

The circuit of figure 8.19a consists of a block with switch connections plus a debouncer, followed by a regular counter with up-down control (it counts upward if *up/dn'* = '1' or downward otherwise). Note that *x* acts as the clock to the counter, so *x* must be debounced. When a switch is pressed, it must not only generate a pulse in *x* but also define the value of *up/dn'* (see exercise 8.12).

The circuit of figure 8.19b also consists of two blocks. Note that it is similar to that in figure 8.17b, but with two control inputs. The first block contains a debouncer plus a one-shot conversion circuit, producing two short (one-clock-period duration) pulses at x_1 and x_2, which are not expected to happen simultaneously. The second block is a reference-value-definer state machine, moving up if x_1 = '1' or down if x_2 = '1'. The construction of this circuit can be based on figure 8.18. As in that case, it might be advantageous to build the first block with a single FSM that combines the debouncer and the one-shot circuit. Additionally, because x_1 and x_2 are not expected to be active at the same time, it might be advantageous to build a "combined" debouncer for both *up* and *dn* signals (see exercises 8.9 and 8.10).

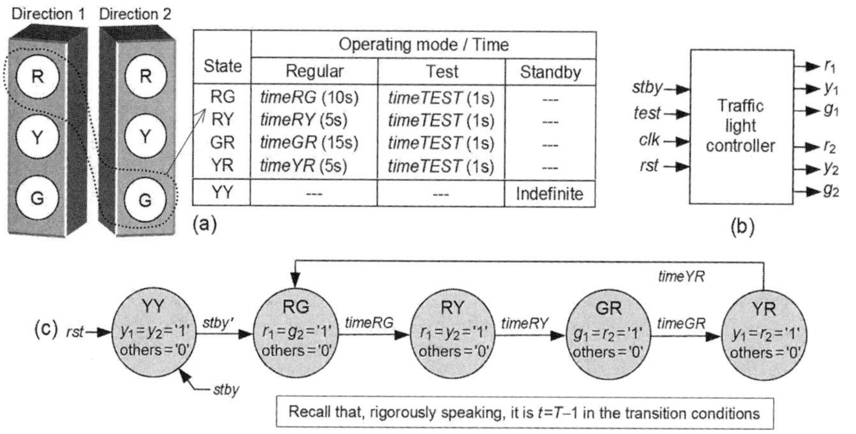
Direction 1
Direction 2
R
Y
G
R
Y
G
State
Operating mode / Time
Regular
Test
Standby
RG
RY
GR
YR
YY
timeRG (10s)
timeRY (5s)
timeGR (15s)
timeYR (5s)
timeTEST (1s)
timeTEST (1s)
timeTEST (1s)
timeTEST (1s)
Indefinite
(a)
stby
test
clk
rst
Traffic light controller
r_1
y_1
g_1
r_2
y_2
g_2
(b)
(c) rst
YY
$y_1=y_2=$'1'
others='0'
stby'
stby
RG
$r_1=g_2=$'1'
others='0'
timeRG
RY
$r_1=y_2=$'1'
others='0'
timeRY
GR
$g_1=r_2=$'1'
others='0'
timeGR
YR
$y_1=r_2=$'1'
others='0'
timeYR
Recall that, rigorously speaking, it is $t=T-1$ in the transition conditions

Figure 8.20

Traffic light controller. (a) Time-related specifications. (b) Circuit ports. (c) Corresponding (simplified) Moore machine.

8.11.5 Traffic Light Controller

A classical timed application is described in this example, which consists of a traffic light controller.

The overall specifications, summarized in figure 8.20a, are as follows:

a) It must have three operating modes: *regular*, *test*, and *standby*.
b) Regular mode consists of four states of operation, called *RG* (red in direction 1 and green in direction 2 turned on), *RY* (red in direction 1 and yellow in direction 2 turned on), *GR* (green in direction 1 and red in direction 2 turned on), and *YR* (yellow in direction 1 and red in direction 2 turned on), each with an independent time duration.
c) Test mode must allow all preprogrammed times to be overwritten (by activating a manual switch) with a small value (1 s per state), such that the system can be easily tested during maintenance.
d) Standby mode, if set (by a sensor accusing malfunctioning, for example, or by a manual switch), must have the system activate the yellow lights in both directions, remaining so while the standby signal is active.

The circuit ports are shown in figure 8.20b. The inputs are two switches, called *stby* and *test*, plus clock and reset. The standby switch selects between the regular mode (*stby* = '0') and the standby mode (*stby* = '1'), and the test switch forces the system into test mode when asserted (*test* = '1'). The output consists of six signals that control the six traffic lights (RYG in direction 1 plus RYG in direction 2).

A corresponding Moore solution is presented in figure 8.20c. If the system is in standby mode, the machine remains in state YY; otherwise, it circulates through states

RG → RY → GR → YR → RG, and onward. The time values shown in the figure (*timeRG*, *timeRY*, and so on) are for the regular operation mode, which must change to *timeTEST* if the system is switched to the test mode (not included in the state diagram for the sake of simplicity). Note that, due to the nature of this application, *stby* can operate in a way similar to reset (after proper synchronization/glitch removal).

Based on section 8.10, the number of flip-flops needed to implement this circuit is as follows. For the state register: $M_{FSM} = 5$ states; therefore, $N_{FSM} = 3$ if sequential, Gray, or Johnson encoding is used, or 5 for one-hot. For the optional output register: not needed in this application, so $N_{output} = 0$. For the timer: knowing that $t_{state\_max} = 15$ s (table of figure 8.20a) and assuming $f_{clk} = 50$ MHz, $T_{max} = 75{\cdot}10^7$ clock cycles results, so $N_{timer} = 30$. Therefore, $N_{total} = 33$ or 35.

The analysis on the need for reset and synchronizers is left as an exercise (exercise 8.13).

8.11.6 Car Alarm (with Chirps)

A car alarm was presented in section 4.2.4. The example shown here is an extension to that one, now with chirps included to announce when the alarm is turned on (one chirp) or off (two chirps). Because the chirps are brief siren activations, a timed machine is now needed.

The circuit ports are shown in figure 8.21a. The inputs are *remt* (command from the remote control) and *sen* (from sensors indicating intrusion) plus clock and reset.

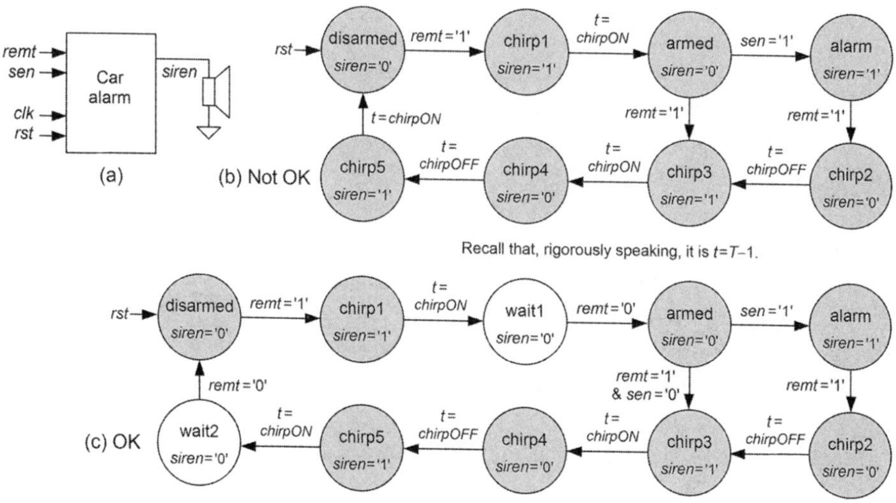
remt
sen
clk
rst
Car alarm
siren
(a)
rst
disarmed
siren='0'
remt='1'
chirp1
siren='1'
t=chirpON
armed
siren='0'
sen='1'
alarm
siren='1'
remt='1'
remt='1'
t=chirpON
chirp5
siren='1'
t=chirpOFF
chirp4
siren='0'
t=chirpON
chirp3
siren='1'
t=chirpOFF
chirp2
siren='0'
(b) Not OK
Recall that, rigorously speaking, it is t=T–1.
rst
disarmed
siren='0'
remt='1'
chirp1
siren='1'
t=chirpON
wait1
siren='0'
remt='0'
armed
siren='0'
sen='1'
alarm
siren='1'
remt='0'
remt='1' & sen='0'
remt='1'
(c) OK
wait2
siren='0'
t=chirpON
chirp5
siren='1'
t=chirpOFF
chirp4
siren='0'
t=chirpON
chirp3
siren='1'
t=chirpOFF
chirp2
siren='0'

Figure 8.21

Car alarm. (a) Circuit ports. (b) Bad (with state bypass and non–true complementarity) and (c) good solutions.

The output is *siren*, which must be turned on (= '1') when an intrusion occurs or during the chirps.

A corresponding Moore solution is presented in figure 8.21b, with *disarmed, armed,* and *alarm* as the fundamental states and *chirp1* to *chirp5* as the chirp-generating states. Note the timed transitions. The time duration of a chirp is *chirpON* clock cycles, and the time interval between two siren activations is *chirpOFF* clock periods. Observe, however, that this machine exhibits the state-bypass problem described in section 4.2.4, which occurs when a long *remt* = '1' command is received because then the circuit simply circulates in the loop *disarmed* → *chirp1* → *armed* → *chirp3* → *chirp4* → *chirp5* → *disarmed*, and so on. An additional (minor) problem is that not all transition conditions are truly complementary (section 1.5); for example, observe in state *armed* that there is no priority definition in case *remt* = '1' and *sen* = '1' occur simultaneously.

A corrected machine is presented in figure 8.21c, in which two wait states (white circles) were added to eliminate state bypass. Of course, if a one-shot circuit (section 5.4.3) were added to the previous solution to reduce the duration of *remt* to a single clock period, and noncomplementarity were corrected, then that machine would work well too. Note that in the presented solution the alarm can be turned on with *sen* = '1'; if that is not wanted, all that is needed is to use the condition "*remt* = '1' & *sen* = '0'" in the *disarmed*-to-*chirp1* transition.

As mentioned in a similar application in section 5.4.5 (garage door controller), a good practice in this kind of application is to include debouncers for the signals coming from the remote control and sensors, which not only eliminate the need for synchronizers but also prevent short glitches (due to lightning, for example) from (de) activating the alarm (they have to be full debouncers, similar to that in section 8.11.3).

VHDL and SystemVerilog implementations for this car alarm are presented in sections 9.5 and 10.5, respectively. The analysis of the number of flip-flops and the need for reset and synchronizers is treated in exercise 8.14.

8.11.7 Password Detector

This section describes a password detector, used, for example, in password-protected door locks like that in figure 8.22a.

The circuit ports are depicted in figure 8.22b. The inputs are *key* (which represents the keypad pushbutton pressed by the user) plus the traditional clock and reset signals; *key* is composed of four bits, so it can encode a keypad with up to 15 pushbuttons (one codeword is reserved for the no-button-pressed case). In the development below it is assumed that the bits of *key* are already debounced and encoded according to the table in exercise 5.14. The outputs are *led1* (turned on when the system is in the *idle* state) and *led2* (turned on for a few seconds when the correct password has been typed in).

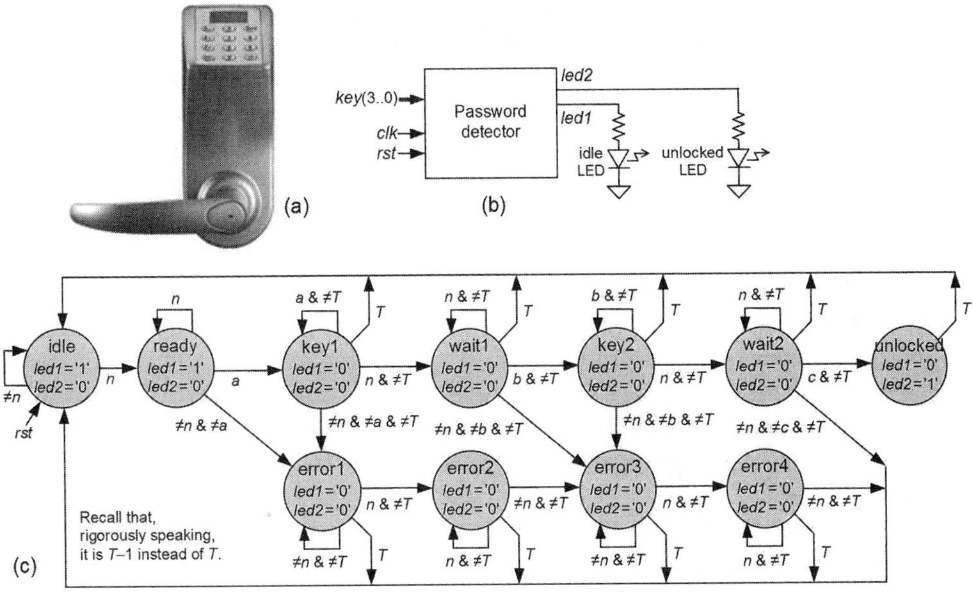
(a)
key(3..0)
clk
rst
Password detector
led2
led1
idle LED
unlocked LED
(b)
idle
led1='1'
led2='0'
≠n
rst
n
ready
led1='1'
led2='0'
a
≠n & ≠a
a & ≠T
T
key1
led1='0'
led2='0'
n & ≠T
≠n & ≠a & ≠T
wait1
led1='0'
led2='0'
b & ≠T
≠n & ≠b & ≠T
b & ≠T
key2
led1='0'
led2='0'
n & ≠T
≠n & ≠b & ≠T
wait2
led1='0'
led2='0'
c & ≠T
≠n & ≠c & ≠T
unlocked
led1='0'
led2='1'
error1
led1='0'
led2='0'
≠n & ≠T
error2
led1='0'
led2='0'
≠n & ≠T
n & ≠T
error3
led1='0'
led2='0'
error4
led1='0'
led2='0'
Recall that, rigorously speaking, it is T–1 instead of T.
(c)

Figure 8.22
Password detector. (a) A password-protected door lock. (b) Circuit ports. (c) A Moore-type solution.

The desired circuit features are the following (where T corresponds to 3 s):

a) When the system is in the *idle* state, LED *led1* (*idle*) must be on and LED *led2* (*unlocked*) must be off.
b) During the time a password is being entered both LEDs must be off.
c) If the correct password is entered, *led2* must be turned on for a time T, with *led1* still off, after which the system must return to *idle* (during that time interval a new password must not be accepted).
d) If the time interval during which a key is kept pressed or between two key presses is longer than T, it must be considered an error, so the machine should return to *idle*.
e) Passwords with repeated digits must be allowed.

A Moore-type solution for this problem is presented in figure 8.22c. The three digits that comprise the password are called a, b, and c; n means *none*, which is the character corresponding to none of the keys pressed ("1111"—see the table in exercise 5.14). Note that both LEDs remain off during the process. To keep the diagram as clean as possible, a slightly simplified representation was used (for example, the a and T conditions on the arrows mean $key = a$ and $t = T - 1$, respectively). The time during which *led2* stays on is the time that the user has to open the door in a corresponding physical implementation.

Based on section 8.10, the number of flip-flops needed to implement this circuit is as follows. For the state register: M_{FSM} = 11 states; therefore, N_{FSM} = 4 if sequential or Gray encoding is used, 6 for Johnson, or 11 for one-hot. For the optional output register: glitches are generally not a problem in this kind of application, so $N_{output} = 0$. For the timer: because t_{state_max} = 3 s, and assuming f_{clk} = 50 MHz, $T_{max} = 15 \cdot 10^7$ clock cycles results, so N_{timer} = 28. Therefore, N_{total} = 32, 34, or 39.

The analysis of the need for reset and synchronizers is left as an exercise (exercise 8.15).

8.11.8 Triggered Circuits

This section shows FSMs for triggered circuits with both bistable and monostable behavior. The former can hold any logic level ('0' or '1') forever, whereas the latter (also called *one-shot*) always returns to the initial value ('0', for example) after a finite time interval. The input (triggering signal) is denoted by x, and the output (response) is called y.

The input can be either synchronous (generated by a circuit operating with the same clock as the triggered circuit) or asynchronous. However, the circuit itself is always synchronous, so the output goes up or down only at the proper clock edge. For example, if we say that y goes to '1' when x goes to '1', it means that y goes to '1' at the first (positive) clock transition after x goes to '1'.

Two signals produced by bistable circuits are depicted in figure 8.23. Note that y does not return to '0' (initial value) automatically. The signal in figure 8.23a is triggered by the condition "x = '1' during T clock cycles" and detriggered by x = '0'. The signal in figure 8.23b is triggered by x = '1' and detriggered by the condition "x = '0' during T clock cycles."

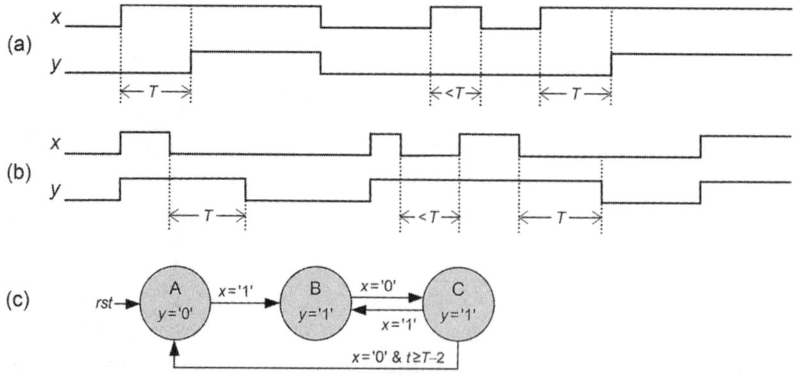
(a)
x
y
T
<T
T
(b)
x
y
T
<T
T
(c)
rst
A
y='0'
x='1'
B
y='1'
x='0'
x='1'
C
y='1'
x='0' & t≥T–2

Figure 8.23

(a, b) Signals produced by triggered bistable circuits (note that the output does not return to zero automatically). (c) Solution for the case in b.

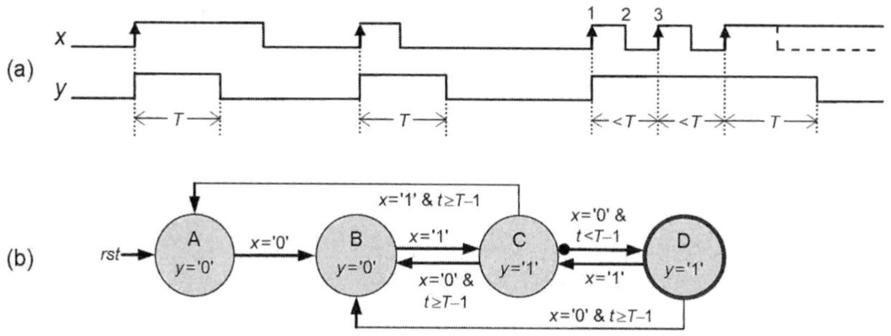
(a)
x
y
1 2 3
T
T
<T
<T
T
(b)
rst
A
y='0'
x='0'
B
y='0'
x='1'
x='1' & t≥T–1
C
y='1'
x='0' & t≥T–1
x='0' & t<T–1
D
y='1'
x='1'
x='0' & t≥T–1

Figure 8.24
(a) Signal produced by a triggered monostable circuit (note that the output always returns to zero). (b) A possible solution (the timer is not zeroed but it rather holds its value during the CD transition).

A solution for the case in figure 8.23b is presented in figure 8.23c. Note that in this machine, when reset is released, the output goes immediately to '1' if the input is '1', which is fine because this is a level-detecting machine (as opposed to edge-detecting machines, depicted in the next example).

Monostable (one-shot) circuits are generally more complex to design than bistable circuits. An example is shown in figure 8.24. Observe that now y always returns to '0' (initial value) after a certain time interval, regardless of x. As indicated by arrows in the figure, y is now edge-dependent rather than level-dependent. The output is triggered by a positive transition in x and detriggered T clock cycles later. Observe that retriggering during the time interval T is allowed (check the final part of the plot).

A solution for this problem is presented in figure 8.24b. Note that the timer must not be zeroed when the machine enters state D (the thick circle indicates that there is at least one transition into state D in which the timer should not be zeroed, while the different arrow, with a large dot at its origin, identifies that transition). Observe also that when reset is released the output does not go to '1' if the input is '1', but it rather waits for the next upward transition of x, which is proper of edge-detecting (as opposed to level-detecting) circuits.

Solutions for edge-triggered circuits (as in figures 8.24b) are among the few cases in which the timer control strategy #1 (section 8.5.2) cannot be applied completely because the timer cannot be zeroed in all state transitions (another example was seen in section 8.8). Anyhow, it will be shown in the designs with VHDL and SystemVerilog (sections 9.6 and 10.6, respectively) that preventing the timer from being zeroed during specific state transitions is very simple. Moreover, it will be shown in exercise 8.18 that this particular FSM can be broken into two FSMs, causing strategy #1 to be applicable without restrictions.

8.11.9 Pulse Shifter

This section presents a circuit that is a particular case of the triggered circuits family described above. It consists of a "pulse shifter," which, as the name says, shifts a pulse a certain number of time units. In other words, it makes a copy of a given pulse T_{shift} clock cycles later.

An example is presented in figure 8.25. The circuit ports are shown in figure 8.25a, where x is the input (original pulse) and y is the output (shifted pulse). An illustrative timing diagram is included if figure 8.25b, which shows that x can be synchronous or asynchronous. The time parameters are $T_{pulse} = 3\ T_{clk}$ and $T_{shift} = 8\ T_{clk}$. Note, however, that T_{pulse} is measured (inevitably) in *number of clock edges* rather than number of clock periods (these values coincide when x is synchronous). The last (positive) clock edge for which x = '0' was chosen as the reference for the shift; a different alternative would be the first (positive) clock edge for which x = '1'.

A solution for this problem is presented in figure 8.26a. Note the box above the state *shift*, which says that T_x is a (registered) copy of t, enabled when x is high. T_x is needed to keep track of the pulse's width, so the circuit can operate without any a-priori information on the value of T_{pulse}.

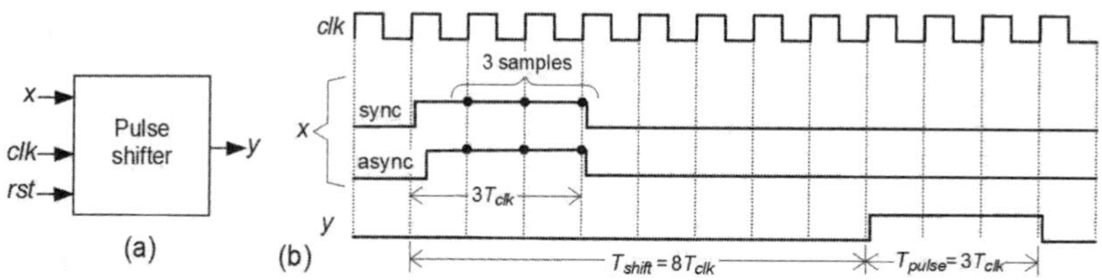
x
clk
rst
Pulse shifter
y
(a)
clk
3 samples
sync
async
3Tclk
y
(b)
Tshift = 8Tclk
Tpulse= 3Tclk

Figure 8.25
Pulse shifter. (a) Circuit ports. (b) Desired behavior for both synchronous and asynchronous input.

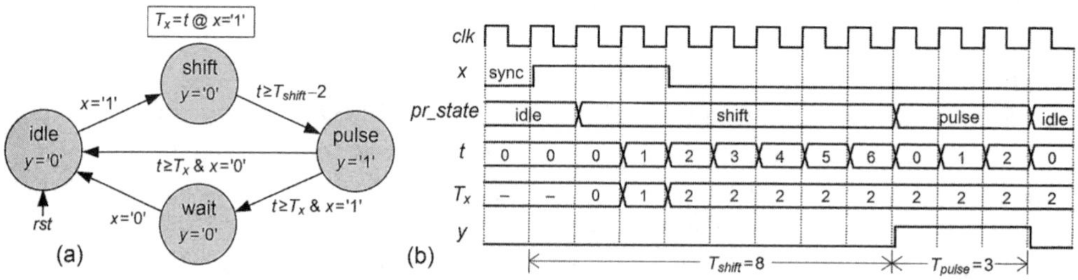
Tx=t @ x='1'
shift
y='0'
x='1'
t≥Tshift−2
idle
y='0'
pulse
y='1'
t≥Tx & x='0'
wait
y='0'
x='0'
t≥Tx & x='1'
rst
(a)
clk
x
sync
pr_state
idle
shift
pulse
t
Tx
y
(b)
Tshift=8
Tpulse=3

Figure 8.26
(a) FSM that implements the pulse shifter of figure 8.25. (b) Corresponding timing diagram, for $T_{pulse} = 3$ and $T_{shift} = 8$ clock periods.

An illustrative timing diagram for this FSM is shown in figure 8.26b. It is very important that the reader examine this diagram carefully and check the correctness of the circuit operation.

An application for pulse shifters is in the generation of enable signals (see section 3.11). In the case of figure 3.18, the input is synchronous and its width is just one clock period, being therefore simpler to generate than the generic case above (exercise 8.19).

8.11.10 Pulse Stretchers

This section presents another king of circuit that is a particular case of the triggered circuits family introduced in section 8.11.8. It consists of "pulse stretchers," which, as the name indicates, take a pulse of shorter duration (often one clock period) and stretch it to a longer length (in fact, one case was already seen in section 2.4 and exercises 2.4 and 2.5). In figure 8.29 an application for a pulse stretcher will be presented.

The circuit ports are shown in figure 8.27a, where *x* (short pulse) is the input and *y* (longer pulse) is the output. The desired behavior is depicted in figures 8.27b,c. In figure 8.27b the input is asynchronous and the output can be asynchronous (*y* goes to '1' as soon as *x* goes to '1') or synchronous (*y* changes only at clock edges). In figure 8.27c the input is synchronous and the output can again be asynchronous or synchronous. As usual, cases with synchronous output can be implemented with

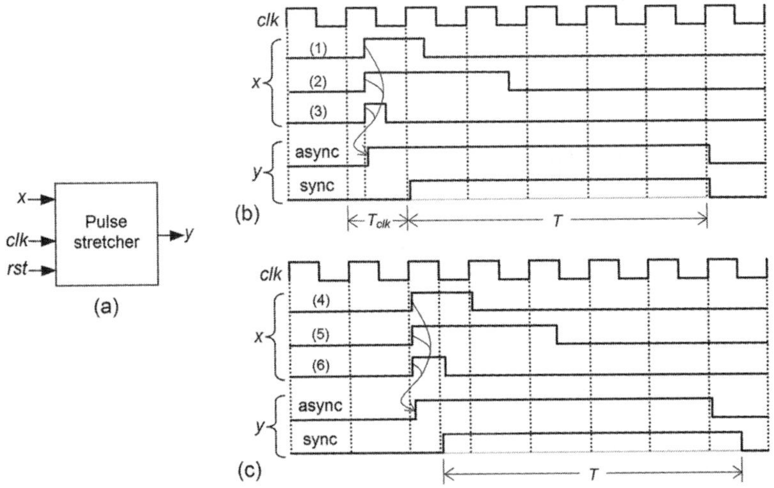
x
clk
rst
Pulse stretcher
y
(a)
clk
(1)
(2)
(3)
x
async
sync
y
(b)
T_{clk}
T
clk
(4)
(5)
(6)
x
async
sync
y
(c)
T

Figure 8.27
Pulse stretcher. (a) Circuit ports. Desired behavior for (b) asynchronous and (c) synchronous input, both with asynchronous or synchronous output.

Moore machines, whereas for asynchronous output a Mealy machine is the natural choice.

Figure 8.27b shows three options for the asynchronous input: in graph 1, x lasts one clock period; in graph 2, it lasts more than one clock period but not more than T; in graph 3, it lasts less than one (or even less than one-half of a) clock period. In all cases the output pulse (y) must have the same length T. Because stretchers can be synchronous or asynchronous, two options are shown for the output. In the asynchronous case (upper plot for y), the output goes up as soon as the input goes up (thus, the total length is obviously $>T$), whereas in the synchronous case (lower plot for y), the output changes only at clock edges. As usual, small propagation delays were included between clock transitions and the corresponding responses to portray a realistic situation.

Figure 8.27c shows three options for the synchronous input: in graph 4, x lasts one clock period; in graph 5, it lasts more than one but less than T clock periods; in graph 6, it lasts at least one-half of a clock period. Again, the output can be asynchronous (upper plot for y) or synchronous (lower plot for y). Note that the asynchronous output looks synchronous, but rigorously speaking it is not because its starting point is determined by x, not directly by the clock. Observe that in the truly synchronous case the negative clock edge was adopted for the FSM (so x and y are updated at opposite clock edges).

Solutions for two of the cases presented in figure 8.27 are shown in figure 8.28. The first solution is for the synchronous case of figure 8.27b, valid for inputs 1 and 2; because this circuit is synchronous, it was implemented with a Moore machine (figure

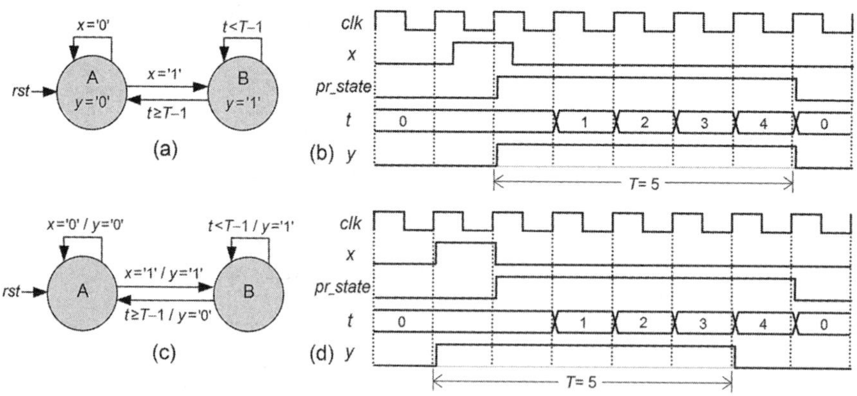
x='0'
t<T–1
rst
A
y='0'
x='1'
B
y='1'
t≥T–1
(a)
clk
x
pr_state
t
0
1
2
3
4
0
(b) y
T=5
x='0' / y='0'
t<T–1 / y='1'
x='1' / y='1'
rst
A
B
t≥T–1 / y='0'
(c)
clk
x
pr_state
t
0
1
2
3
4
0
(d) y
T=5

Figure 8.28

(a) Moore solution for the synchronous case of figure 8.27b, covering inputs 1–2, and (b) an illustrative timing diagram. (c) Mealy solution for the asynchronous case of figure 8.27c, covering inputs 4–5, and (d) an illustrative timing diagram.

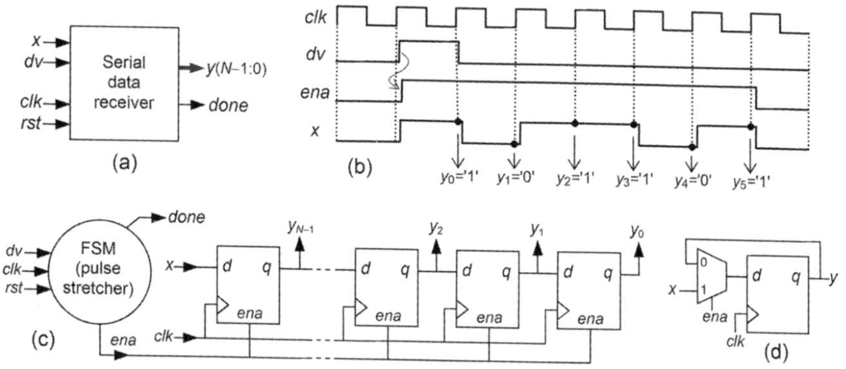
x
dv
clk
rst
Serial data receiver
y(N–1:0)
done
(a)
clk
dv
ena
x
(b)
y0='1' y1='0' y2='1' y3='1' y4='0' y5='1'
done
FSM (pulse stretcher)
dv
clk
rst
x
ena
clk
(c)
yN-1
y2
y1
y0
d q
ena
0
1
x
y
clk
(d)

Figure 8.29
Serial data receiver. (a) Circuit ports. (b) Desired behavior (*dv* is stretched to produce *ena*). (c) Solution with a shift register controlled by the pulse stretcher.

8.28a). The second solution is for the asynchronous case of figure 8.27c, valid for inputs 4 and 5; because this circuit is asynchronous, it was implemented with a Mealy machine (figure 8.28c). Solutions for other cases are treated in exercises 8.20 to 8.23.

Figures 8.28b and 8.28d present illustrative timing diagrams for the FSMs of figures 8.28a and 8.28c, respectively, for $T = 5$. It is very important that the reader examine these diagrams carefully and check the correctness of the circuit operation.

An application for a pulse stretcher is depicted in figure 8.29, which consists of a serial data receiver (a deserializer). The circuit ports are shown in figure 8.29a. The inputs are *x* (serial bit stream), *dv* (data valid bit, high during only one clock cycle, informing that data storage should start), plus the conventional clock and reset. The outputs are $y(N–1:0)$ (multibit one-dimensional register in which the received data must be stored) and *done* (high while the machine is idle). Some of these signals are shown in figure 8.29b, which also highlights the fact that the first bit of *x* is made available at the same time that *dv* is asserted, so data storage must start immediately.

A possible solution is presented in figure 8.29c. It consists of a shift register whose enable input is produced by an FSM (this is a simplified view; the enable port of a DFF, if not built-in, can be constructed using a multiplexer, as shown in figure 8.29d). When a *dv* = '1' pulse occurs, the FSM produces *ena* = '1' during *N* consecutive clock cycles, causing *N* bits of *x* to be shifted in, thus getting stored in the *N* flip-flops that comprise the shift register, producing $y(N–1:0)$.

Note that in this case the FSM is simply a pulse stretcher. Because the first bit of *x* is made available at the same time that *dv* is asserted, one must be careful not to skip that bit (see section 3.10). Consequently, we can employ either an asynchronous (Mealy) FSM, which would then produce the signal shown in the figure, or a synchronous (Moore) machine operating at the negative clock transition. The former option can be implemented with the FSM of figure 8.28c, with $T = N$. Note also that *done* can be computed as *ena'*. Another serial data receiver will be seen in section 11.7.7.

8.12 Exercises

Exercise 8.1: Machines Category

a) Why are the state machines in figures 8.12c, 8.14b, 8.20c, and 8.21c (among others) said to be of category 2?

b) What differentiates category 2 from category 1? (Compare figures 8.2 and 5.2.)

Exercise 8.2: Timer Interpretation #1

Consider the timed machine of figure 8.3, operating with $f_{clk} = 1$ MHz and $T = 13$ clock cycles.

a) Which states are timed (timer needed) and which are not?

b) Can any of the states last longer than T clock periods? Explain.

c) Can the timer control strategy #2 (section 8.5.3) be used to build the timer?

d) Since $T = 13$, we know that the range of interest is from 0 to 12. Assuming that strategy #1 (section 8.5.2) is adopted to build the timer, can we employ a timer that runs (when enabled, of course) up to 16 (a power of two)? What are the consequences of this?

e) Still assuming strategy #1 for the timer, is it necessary to specify a value for T (= 0, for example) in the untimed states? Is that the case also in strategy #2?

f) During how many microseconds will the machine stay in each state? Does your answer depend on x?

g) How many flip-flops are needed to build this FSM (with sequential encoding), including the timer? Does this answer depend on x?

Exercise 8.3: Timer Interpretation #2

Consider the timed machine of figure 8.3, operating with $T = 3$ clock cycles. Fill in the missing parts in the plots of figure 8.30. Note the intentional propagation delays left between the clock transitions and the respective responses to portray a realistic

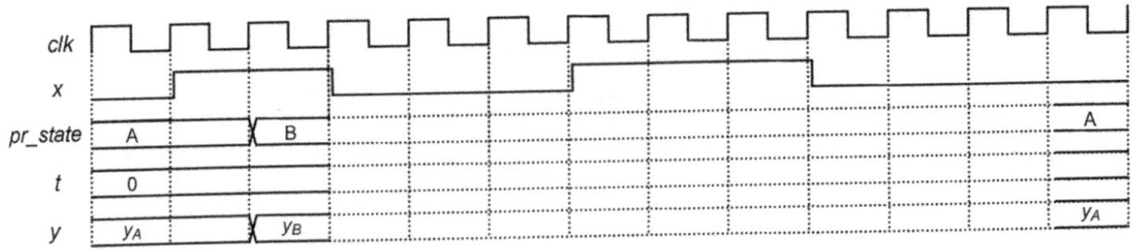
clk
x
pr_state
A
B
A
t
0
y
y_A
y_B
y_A

Figure 8.30

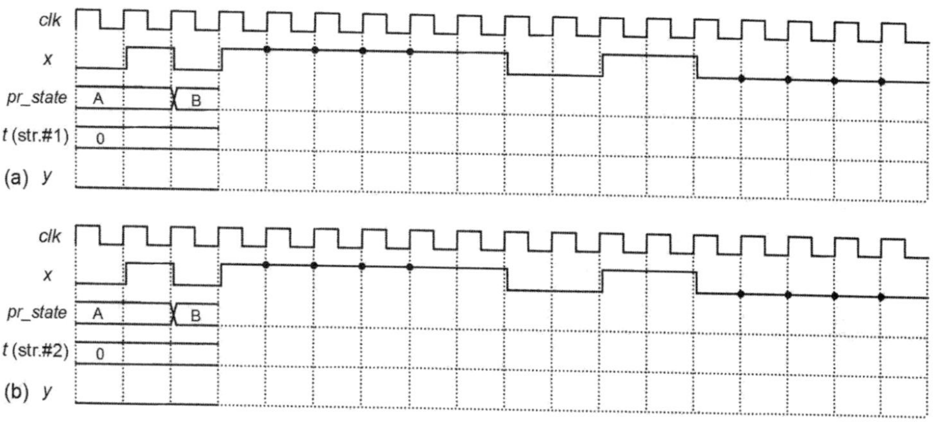
clk
x
pr_state
A
B
t (str.#1)
0
(a) y
clk
x
pr_state
A
B
t (str.#2)
0
(b) y

Figure 8.31

situation. Assume that strategy #1 (section 8.5.2) was adopted for the timer. (Regarding strategy #2, see the previous exercise.)

Exercise 8.4: Analysis of Timer Control Strategies #1 and #2

Assume that the switch debouncer of figure 8.16c was designed to operate with $T = 4$ clock cycles (more precisely, with 4 clock edges, because x is asynchronous).

a) Say that strategy #1 (section 8.5.2) was employed to design the timer. Complete the timing diagram of figure 8.31a for the given input x. As usual, a small propagation delay was included between clock transitions and corresponding responses to portray a more realistic situation. Call the states A, B, C, and D to simplify the notation.
b) Do the same for the timing diagram of figure 8.31b, assuming now that strategy #2 was used for the timer. Is the result (y) the same as for strategy #1?

Exercise 8.5: Blinking Light without Reset

It was said in section 8.11.1 that the light blinker of figure 8.12c might not require a reset signal, even if flip-flops with arbitrary initial states are used to implement it.

a) Prove that if sequential encoding is used and optimal (minimal) expressions are adopted for *nx_state* (i.e., d_1 and d_0), then this FSM can indeed operate without reset. (Suggestion: Review sections 3.8 and 3.9 and see exercise 3.11.)
b) Show that, on the other hand, if sequential encoding is used but all "don't care" bits are filled with '1's, then the machine is subject to deadlock, so a reset signal is needed.

Exercise 8.6: Blinking Light with Several Speeds

This exercise is an extension to the light blinker of figure 8.12c, which must now operate with a *programmable* speed, set by a pushbutton (called *spd*). The desired speeds are 1, 2, 4, and 8 Hz. The next speed must be selected every time the pushbutton is pressed, returning to 1 Hz after passing 8 Hz. One alternative (among others) is presented in figure 8.32, which consists of a cascade of FSMs. The first can be a debouncer + one-shot converter (similar to figure 8.18b), the second can be a reference-value definer (similar to figure 8.18c), and the third the light blinker proper (figure 8.12c). The purpose of the first pair of FSMs is to produce T_{ref}, which is then used as time parameter for the blinker. Assume a 1-ms debouncing interval, a 50-MHz clock, and sequential encoding for the FSMs.

a) Calculate the four values of T_{ref} (one for each speed); for example, for 1 Hz, $t_{ref} = 0.5$ s, so $T_{ref} = 25 \cdot 10^6$ clock cycles.

b) Draw a block diagram for your solution, splitting the big block of figure 8.32 into two blocks.

c) Show the state transition diagram for each FSM used in this circuit.

d) How many DFFs are needed to build the entire circuit?

Exercise 8.7: Pushbutton Debouncer plus Memory

Figure 8.33 shows a pushbutton that must be debounced and also "memorized," such that the stored value gets inverted every time the pushbutton is pressed (as in the stopwatch used by football referees, which alternately runs and stops every time the pushbutton is pressed). If a debouncer were not needed, the trivial solution of figure 8.33a could be used, in which x is connected directly to the clock input of a DFF (due to the inverted version of q connected back to d, it resembles a toggle-type flip-flop, so every time a positive clock edge occurs, the value of y gets inverted). Let us assume, however, the usual situation, in which the pushbutton must be debounced.

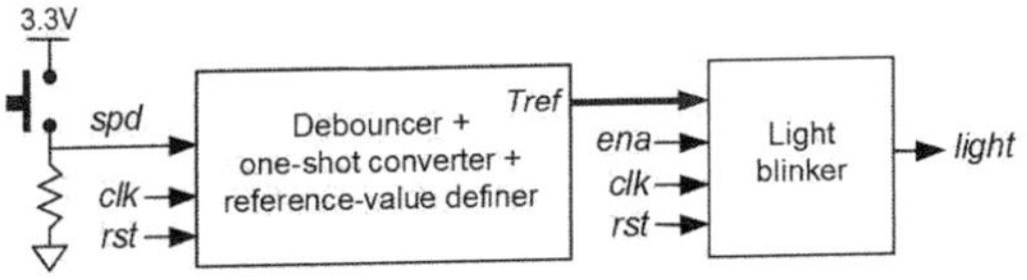
3.3V
spd
clk
rst
Debouncer +
one-shot converter +
reference-value definer
Tref
ena
clk
rst
Light
blinker
light

Figure 8.32

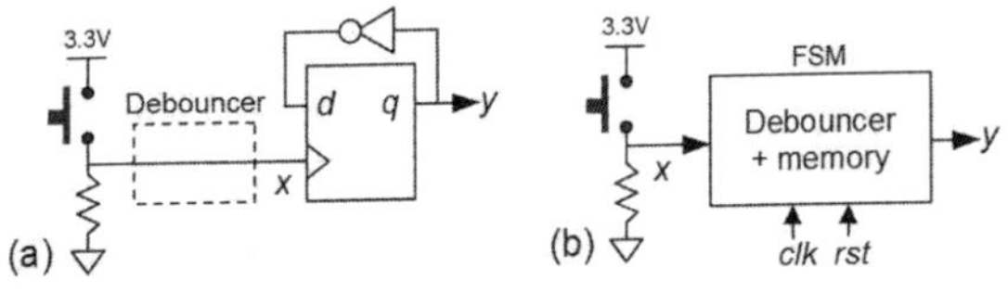
3.3V
Debouncer
d
q
y
x
(a)
3.3V
FSM
Debouncer
+ memory
x
y
clk
rst
(b)

Figure 8.33

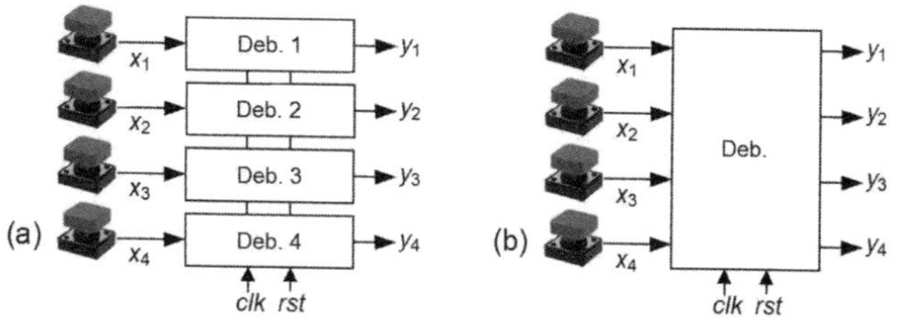
(a)
x_1
x_2
x_3
x_4
Deb. 1
Deb. 2
Deb. 3
Deb. 4
y_1
y_2
y_3
y_4
clk rst
(b)
x_1
x_2
x_3
x_4
Deb.
y_1
y_2
y_3
y_4
clk rst

Figure 8.34

a) Draw a flowchart for the combined solution of figure 8.33b (debouncer plus memory in one FSM).

b) Draw a Moore-type state diagram corresponding to the flowchart presented above.

c) Assuming that sequential encoding is used and that the debouncing time interval is 1 ms, with f_{clk} = 50 MHz, calculate the number of flip-flops needed to build this machine.

d) If the solution of figure 8.33a were employed, with the debouncer included, how many flip-flops would be required?

Exercise 8.8: Independent Multisignal Debouncer

Figure 8.34 shows four mechanical switches for which debouncers are needed. In figure 8.34a a complete debouncer for each signal is considered, whereas figure 8.34b considers a "combined" approach. Because the timer is the most expensive part, if a single timer could be used in the latter it would already represent a major gain. In this exercise the switches are independent of each other, so they might be activated simultaneously. Assume a 1-ms *minimum* debouncing interval, a 50-MHz clock, and sequential encoding for the FSMs.

a) How many flip-flops are needed to implement the option in figure 8.34a, employing the debouncer of figure 8.16c?

b) Draw a state transition diagram for an FSM capable of implementing the combined debouncer of figure 8.34b.

c) How many flip-flops are needed to implement your combined circuit?

Exercise 8.9: Dependent Multisignal Debouncer

Figure 8.35 shows a keypad (see details in exercise 5.14) for which debouncers are needed. Note that this exercise is an extension to that above, with the difference that now the signals are no longer independent of each other. Because only one key is supposed to be pressed at a time, the only valid codewords are "1111" (no key pressed), "0111" (key in row 1 pressed), "1011" (key in row 2 pressed), "1101" (key in row 3

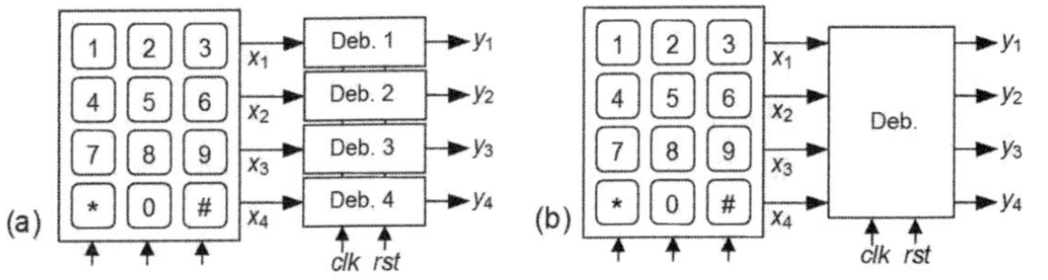
(a)
1
2
3
4
5
6
7
8
9
*
0
#
x_1
x_2
x_3
x_4
Deb. 1
Deb. 2
Deb. 3
Deb. 4
y_1
y_2
y_3
y_4
clk rst
(b)
1
2
3
4
5
6
7
8
9
*
0
#
x_1
x_2
x_3
x_4
Deb.
y_1
y_2
y_3
y_4
clk rst

Figure 8.35

pressed), and "1110" (key in row 4 pressed). Note that for the purposes of this exercise, as well as for testing the solution with VHDL or SystemVerilog, the arrangement in figure 8.34 can be used equivalently. Assume a 1-ms debouncing interval, a 50-MHz clock, and sequential encoding for the FSMs.

a) How many flip-flops are needed to implement the option in figure 8.35a (or 8.34a) with the debouncer of figure 8.16c? Is this option capable of filtering out invalid codewords?
b) Draw a state transition diagram for an FSM capable of implementing the combined debouncer of figure 8.35b (or 8.34b). It must also be able to filter out invalid codewords (hence, y can only be "1111," "0111," "1011," "1101," or "1110").
c) How many flip-flops are now required?

Exercise 8.10: Dependent Multisignal Debouncer with One-Shot Conversion

This exercise concerns the FSM1 block of figure 8.19b, which must implement a two-signal debouncer with one-shot output. Recall that *up* and *dn* are not supposed to be high at the same time, so the machine should be able to filter out invalid inputs (the only values allowed for x_1x_2 are "00", "10", and "01"). Draw a state transition diagram for this FSM.

Exercise 8.11: Arbitrary Reference-Value Definer with Up/Down Controls

Figure 8.19b shows an alternative for implementing a reference-value definer with up and down controls, which is advantageous when the reference values are few and irregular (arbitrary). The first block was already treated in the previous exercise. Draw a state transition diagram for an FSM capable of implementing the second block, with eight reference values (r_1, r_2, …, r_8). Recall that the inputs to this machine can only be x_1x_2 = "00" (*idle*), "10" (*up*), or "01" (*down*), with any nonzero input lasting only one clock period (as determined by the previous block).

Exercise 8.12: Regular Reference-Value Definer with Up/Down Controls

Figure 8.19a shows an alternative for implementing a reference-value definer with up and down controls, which is advantageous when the reference values are regularly distributed, so some type of counter/adder can be used to build it. Say that the circuit must generate values from 0 to 99, incremented (when *up* = '1') or decremented (when *dn* = '1') in steps of size 1.

a) Make a sketch for this circuit, detailing especially the connections of the two pushbuttons (feel free to add other contacts to the pushbuttons if you consider it necessary). How many debouncers are needed?
b) How many flip-flops are needed to build the entire circuit of figure 8.19a based on your sketch above? If any debouncer is needed, consider 1 ms for the debouncing interval and f_{clk} = 50 MHz.

Exercise 8.13: Traffic Light Controller
The questions below refer to the traffic light controller of figure 8.20c.

a) Explain why either strategy #1 or #2 (section 8.5) can be used to implement the timer in this machine.
b) Is a reset signal needed? Explain. (Suggestion: Review sections 3.8 and 3.9.)
c) Which inputs are asynchronous? (Suggestion: Review section 2.3.)
d) If debouncers are included in the asynchronous inputs, are synchronizers needed?
e) If debouncers are not used, are synchronizers indispensable in this application?
f) Redraw the state transition diagram including in it the following feature: Instead of having the yellow lights in both directions statically on while in standby mode, make them blink continuously (with a 0.5 Hz frequency) in that mode.
g) The number of DFFs needed to build the FSM after the feature above is included is still that determined in section 8.11.5? Explain.

Exercise 8.14: Car Alarm
The questions below refer to the car alarm of figure 8.21c. Assume that the chirps must last 0.3 s and f_{clk} = 50 MHz.

a) Explain why both strategies #1 and #2 (section 8.5) are appropriate to implement the timer in this machine. What are the advantages and disadvantages of each one?
b) How many DFFs are needed to build that circuit with sequential encoding for the machine states? And with one-hot encoding?
c) Is a reset signal needed? Explain. (Suggestion: Review sections 3.8 and 3.9.)
d) Which inputs are asynchronous? (Suggestion: Review section 2.3.)
e) If debouncers are included in the asynchronous inputs, are synchronizers needed?
f) If debouncers are not used, are synchronizers indispensable in this application?

Exercise 8.15: Password Detector
The questions below refer to the password detector of figure 8.22c.

a) Explain why both strategies #1 and #2 (section 8.5) are appropriate to implement the timer in this machine. What are the advantages and disadvantages of each one?
b) Is a reset signal needed? Explain. (Suggestion: Review sections 3.8 and 3.9.)
c) Is *key* an asynchronous input? (Suggestion: Review section 2.3.)
d) If *key* has already been processed by a debouncer (as in exercise 8.11), are synchronizers needed?
e) Why is the state *ready* needed in this FSM?
f) Why must the machine not go back to the *idle* state as soon as a wrong key is punched in?

Exercise 8.16: Triggered Circuits #1

The questions below concern the pulse generator of figure 8.24b, which produces the signal of figure 8.24a.

a) How many flip-flops are needed to build that circuit, for $T = 3$ clock cycles, sequential encoding, and not including the optional output register?
b) In which states is the timer not needed? How should the timer be operated in those states?
c) Complete the plots of figure 8.36 (for $T = 3$) and then comment on the results.

Exercise 8.17: Triggered Circuits #2

Two signals produced by triggered circuits are exhibited in figure 8.37.

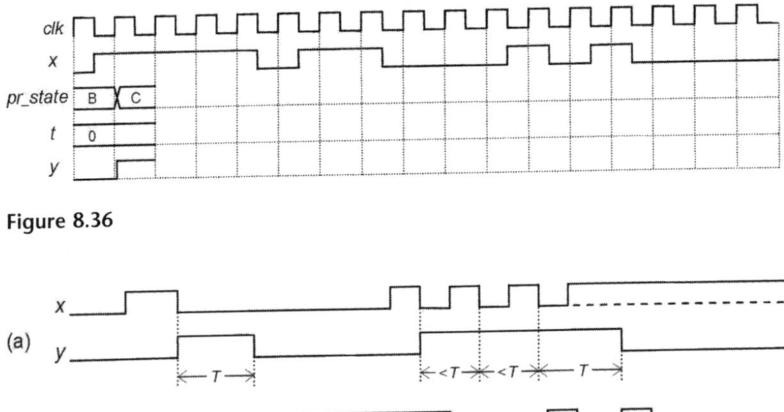
clk
x
pr_state
B
C
t
0
y

Figure 8.36

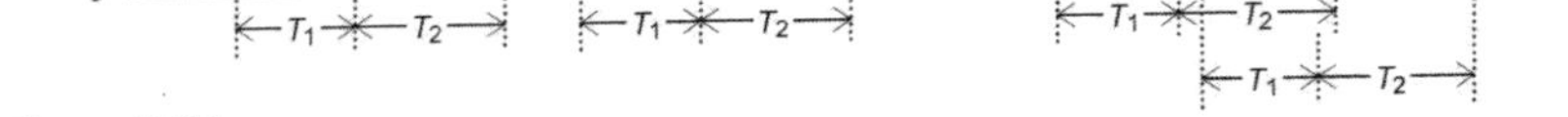
(a)
x
y
T
<T
<T
T
(b)
x
y
T_1
T_2

Figure 8.37

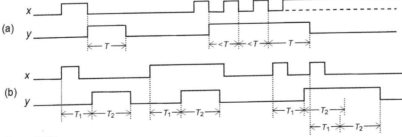

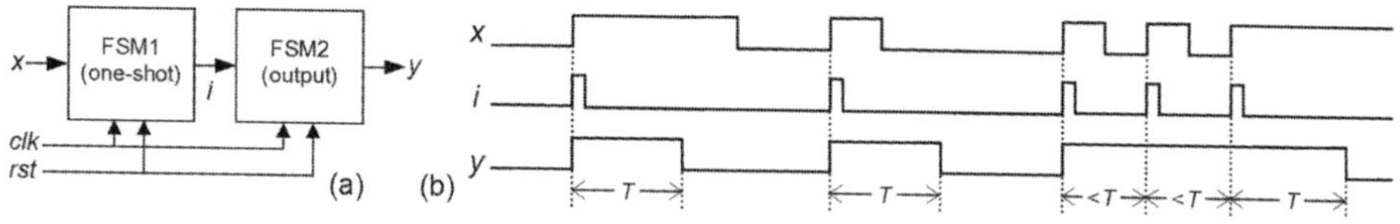
FSM1
(one-shot)
FSM2
(output)
x
i
y
clk
rst
(a)
(b)
T
<T

Figure 8.38

a) Present a state transition diagram for an FSM capable of producing the signal of figure 8.37a. Estimate the number of DFFs needed to build your complete circuit for $T = 3$ clock cycles and sequential encoding.
b) Do the same for the signal of figure 8.37b. Assume sequential encoding and $T_1 = 3$ and $T_2 = 5$ clock cycles.

Exercise 8.18: Triggered Circuits #3

Figure 8.38a shows a two-machine arrangement for the implementation of an edge-detecting triggered circuit. We want to use this arrangement to implement a circuit that generates the signal of figure 8.24a. A solution for that case was already seen in figure 8.24b, using a single machine. The advantage of the approach discussed here is that the timer can be zeroed every time the machine changes its state; hence, contrary to figure 8.24b, strategy #1 (section 8.5.2) can be applied without exceptions. In figure 8.38a, FSM1 is a one-shot circuit (discussed in section 5.4.3) that converts x into a short pulse (internal signal i), from which FSM2 must produce the actual output signal, y, as illustrated in figure 8.38b. Present two state transitions diagrams (one for each machine) to solve this problem with the timer allowed to be zeroed at every state transition. Can the timer control strategy #2 also be employed in your FSMs?

Exercise 8.19: Pulse Shifter

This exercise concerns the pulse shifter introduced in section 8.11.9.

a) Draw a state transition diagram and present an illustrative timing diagram for an FSM capable of producing a signal similar to that in figure 8.25b, but knowing that x (still synchronous) now lasts exactly one clock period.
b) How many flip-flops are needed to build your circuit, for $T_{shift} = 16\ T_{clk}$, using sequential encoding for the FSM?

Exercise 8.20: Synchronous Pulse Stretcher #1

This exercise concerns the synchronous version of a pulse stretcher whose behavior was depicted in figure 8.27b.

a) If the falling edge of signal 2 is beyond the falling edge of (sync) y, will the solution presented in figure 8.28a still produce the same result? If not, modify it to accommodate this situation as well.

b) Develop a circuit capable of processing signal 3. (Suggestion: See section 2.4.)
c) Draw an illustrative timing diagram for your circuit (as in figure 8.28b), demonstrating that it indeed covers case 3.

Exercise 8.21: Synchronous Pulse Stretcher #2

This exercise concerns the synchronous version of a pulse stretcher whose behavior was depicted in figure 8.27c.

a) Present a state transition diagram for a Moore FSM capable of processing all three signals (4 to 6). Should it operate at the positive or negative clock transition?
b) Draw an illustrative timing diagram for your FSM (as in figure 8.28b), demonstrating that it indeed works as expected.

Exercise 8.22: Asynchronous Pulse Stretcher #1

This exercise concerns the asynchronous version of a pulse stretcher whose behavior was depicted in figure 8.27b.

a) Present a state transition diagram for a Mealy FSM capable of processing signals 1–2.
b) Draw an illustrative timing diagram for your FSM (as in figure 8.28d), demonstrating that it indeed works as expected.
c) Develop a circuit capable of processing signal 3. (Suggestion: See section 2.4.)

Exercise 8.23: Asynchronous Pulse Stretcher #2

This exercise concerns the asynchronous version of a pulse stretcher whose behavior was depicted in figure 8.27c.

a) If the falling edge of signal 5 is beyond the falling edge of (async) y, will the solution presented in figure 8.28c still produce the same result? If not, modify it to accommodate this situation as well.
b) Develop a circuit capable of processing signal 6. (Suggestion: See section 2.4.)
c) Draw an illustrative timing diagram for your circuit (as in figure 8.28d), demonstrating that it indeed covers case 6.

Exercise 8.24: Eliminating Conditional-Timed Transitions

Because the conditional-timed transition (CD) in figure 8.3 is the only transition that departs from state C, it can be eliminated by splitting it into a simple timed transition followed by a simple conditional transition.

a) Apply the principle described above to the FSM of figure 8.3.
b) Can strategy #2 be now used to build the timer? Why couldn't it be used in the original machine of figure 8.3?

9 VHDL Design of Timed (Category 2) State Machines

9.1 Introduction

This chapter presents several VHDL designs of category 2 state machines. It starts by presenting two VHDL templates, for Moore- and Mealy-based implementations, which are used subsequently to develop a series of designs related to the examples introduced in chapter 8.

The codes are always complete (not only partial sketches) and are accompanied by comments and simulation results, illustrating the design's main features. All circuits were synthesized using Quartus II (from Altera) or ISE (from Xilinx). The simulations were performed with Quartus II or ModelSim (from Mentor Graphics). The default encoding scheme for the states of the FSMs was regular sequential encoding (see encoding options in section 3.7; see ways of selecting the encoding scheme at the end of section 6.3).

The same designs are presented in chapter 10 using SystemVerilog, so the reader can make a direct comparison between the codes.

Note: See suggestions of VHDL books in the bibliography.

9.2 VHDL Template for Timed (Category 2) Moore Machines

The template is presented below. Because it is an extension to the Moore template for category 1, described in section 6.3, a review of that template is suggested before this one is examined because only the differences are described.

The only differences are those needed for the inclusion of a timer (external to the FSM—see figure 8.2a). Recall, however, that the FSM itself is responsible for controlling the timer. For that purpose, two strategies were developed in chapter 8, being the first generic (section 8.5.2), and the second (section 8.5.3), non-generic. It is very important that the reader review those two sections before proceeding.

The first of the two templates that follow is for timed Moore machines with the timer implemented using strategy #1. The timer-related constants (T_1, T_2, . . .) can be declared either as *generic* constants (lines 6–7; see details in the template for category 1 in section 6.3), therefore in the entity, or in the declarative part of the architecture, as shown in lines 18–20. The signal *t* must obviously stay where it is (line 21). As seen in section 8.5.2, the timer must obey $t_{max} \geq \max(T_1, T_2, \ldots) - 1$.

The first process (lines 26–37) implements the timer. Note that it is a straight copy of the code presented in section 8.5.2.

The second process (line 40) implements the machine's state register, exactly as for category 1 Moore (section 6.3).

The third process (lines 43–71) implements the machine's combinational logic. It is the same as for category 1 except for the fact that *t* might appear in the conditions for *nx_state* (lines 50, 52, . . .). The use of $t \geq T - 1$ instead of $t = T - 1$ is required in the conditional-timed transitions with $T - 1 < t_{max}$. Note that t_{max} does not need to be defined in all states, which is not true for strategy #2.

The fourth and final process (line 74) implements the optional output register, exactly as for category 1.

Note: See also the comments in section 6.4 on template variations.

```
1     --Timed Moore machine with timer control strategy #1---------
2     library ieee;
3     use ieee.std_logic_1164.all;
4     -------------------------------------------------------------
5     entity circuit is
6        generic (
7           (timer-related constants of lines 18-20 can go here)
8        port (
9           (same as for category 1 Moore, section 6.3)
10    end entity;
11    -------------------------------------------------------------
12    architecture moore_fsm of circuit is
13
14       --FSM-related declarations:
15       (same as for category 1 Moore, section 6.3)
16
17       --Timer-related declarations:
18       constant T1: natural := <value>;
19       constant T2: natural := <value>;   ...
20       constant tmax: natural := <value>; --tmax≥max(T1,T2,...)-1
21       signal t: natural range 0 to tmax;
22
23    begin
24
25       --Timer (strategy #1, section 8.5.2):
26       process (clk, rst)
27       begin
28          if (rst='1') then
29             t <= 0;
```

```
      elsif rising_edge(clk) then
         if pr_state /= nx_state then
            t <= 0;
         elsif t /= tmax then
            t <= t + 1;
         end if;
      end if;
   end process;

   --FSM state register:
   (same as for category 1 Moore, section 6.3)

   --FSM combinational logic:
   process (all) --list proc. inputs if "all" not supported
   begin
      case pr_state is
         when A =>
            output1 <= <value>;
            output2 <= <value>;
            ...
            if ... and t>=T1-1 then
               nx_state <= B;
            elsif ... and t>=T2-1 then
               nx_state <= ...;
            else
               nx_state <= A;
            end if;
         when B =>
            output1 <= <value>;
            output2 <= <value>;
            ...
            if ... and t>=T3-1 then
               nx_state <= C;
            elsif ... then
               nx_state <= ...;
            else
               nx_state <= B;
            end if;
         when C =>
            ...
      end case;
   end process;

   --Optional output register:
   (same as for category 1 Moore, section 6.3)

end architecture;
--------------------------------------------------------------
```

The next template is for timed Moore machines employing strategy #2 to implement the timer.

The first difference is in line 18, which now includes also t_{max}.

The second difference is in the process for the timer (lines 23–34), which is a copy of the code presented in section 8.5.3.

The third and final difference is in the process for the combinational logic block (lines 40–70), which requires now the value of t_{max} to be specified in each state (lines 47, 59, . . .), even if the state is untimed ($t_{max} = 0$). This code can obviously be simplified in several ways when there are no conditional-timed transitions and/or t_{max} is the same in all or most states.

```
1    --Timed Moore machine with timer control strategy #2-----------
2    library ieee;
3    use ieee.std_logic_1164.all;
4    ---------------------------------------------------------------
5    entity circuit is
6          (same as template above)
7    end entity;
8    ---------------------------------------------------------------
9    architecture moore_fsm of circuit is
10
11      --FSM-related declarations:
12      (same as for category 1 Moore, section 6.3)
13
14      --Timer-related declarations:
15      constant T1: natural := <value>;
16      constant T2: natural := <value>; ...
17      constant tmax_timer: natural := <value>; -- ≥max(T1,T2,...)-1
18      signal t, tmax: natural range 0 to tmax_timer;
19
20   begin
21
22      --Timer (strategy #2, section 8.5.3):
23      process (clk, rst)
24      begin
25         if (rst='1') then
26               t <= 0;
27         elsif rising_edge(clk) then
28               if t < tmax then
29                  t <= t + 1;
30               else
31                  t <= 0;
32               end if;
33         end if;
34      end process;
35
36      --FSM state register:
37      (same as for category 1 Moore, section 6.3)
38
39      --FSM combinational logic:
40      process (all) --list proc. inputs if "all" not supported
41      begin
42         case pr_state is
43            when A =>
44               output1 <= <value>;
45               output2 <= <value>;
46               ...
47               tmax <= T1-1;
48               if ... and t=tmax then
49                  nx_state <= B;
```

```
            elsif ... then
               nx_state <= ...;
            else
               nx_state <= A;
            end if;
         when B =>
            output1 <= <value>;
            output2 <= <value>;
            ...
            tmax <= T2-1;
            if ... and t=tmax then
               nx_state <= C;
            elsif ... then
               nx_state <= ...;
            else
               nx_state <= B;
            end if;
         when C =>
            ...
      end case;
   end process;

   --Optional output register:
   (same as for category 1 Moore, section 6.3)

end architecture;
-------------------------------------------------------------
```

9.3 VHDL Template for Timed (Category 2) Mealy Machines

The template is presented below, using strategy #1 to implement the timer. The only difference with respect to the Moore template (section 9.2) is in the process for the combinational logic block (lines 20–60) because the outputs are specified differently here (see the template for category 1 Mealy machines in section 6.5). Recall that in a Mealy machine the output depends not only on the FSM's state, but also on the input, so **if** statements are expected for the output in one or more states because the output values might not be unique.

Please review the following comments, which can be easily adapted from the Moore case to the Mealy case:

—On the Moore template for category 1, in section 6.3, especially comment 10.
—On the *enum_encoding* and *fsm_encoding* attributes, also in section 6.3.
—On possible code variations, in section 6.4.
—On the Mealy template for category 1, in section 6.5.
—On the Moore templates for category 2, in section 9.2.

```
1    --Timed Mealy machine with timer control strategy #1----
2    library ieee;
3    use ieee.std_logic_1164.all;
4    ------------------------------------------------------
```

```
entity circuit is
   (same as for Moore, Section 9.2));
end entity;
------------------------------------------------------------
architecture mealy_fsm of circuit is
   (same as for Moore, section 9.2)
begin

   --Timer (using timer control strategy #1):
   (same as for Moore, section 9.2)

   --FSM state register:
   (same as for Moore, section 9.2)

   --FSM combinational logic:
   process (all)
   begin
      case pr_state is
         when A =>
            if ... and t>=T1-1 then
               output1 <= <value>;
               output2 <= <value>;
               ...
               nx_state <= B;
            elsif ... and t>=T2-1 then
               output1 <= <value>;
               output2 <= <value>;
               ...
               nx_state <= ...;
            else
               output1 <= <value>;
               output2 <= <value>;
               ...
               nx_state <= A;
            end if;
         when B =>
            if ... and t>=T3-1 then
               output1 <= <value>;
               output2 <= <value>;
               ...
               nx_state <= C;
            elsif ... then
               output1 <= <value>;
               output2 <= <value>;
               ...
               nx_state <= ...;
            else
               output1 <= <value>;
               output2 <= <value>;
               ...
               nx_state <= B;
            end if;
         when C =>
            ...
      end case;
   end process;
```

```

     --Optional output register:
     (same as for Moore, section 9.2)

end architecture;
-----------------------------------------------------
```

9.4 Design of a Light Rotator

This section presents a VHDL-based design for the light rotator introduced in section 8.11.2. The Moore template of section 9.2 is used to implement the FSM of figure 8.14b. Either strategy #1 (section 8.5.2) or #2 (section 8.5.3) can be used to build the timer (both templates were presented in section 9.2); the former is employed in the code below, while the latter is explored in exercise 9.1.

The entity, called *light_rotator*, is in lines 5–9. All ports are of type *std_logic* or *std_logic_vector* (industry standard).

The architecture, called *moore_fsm*, is in lines 11–105. As usual, it contains a declarative part and a statements part, with three processes in the latter.

The declarative part of the architecture (lines 13–21) contains FSM- and timer-related declarations. In the former, the enumerated type *state* is created to represent the machine's present and next states. In the latter, the values chosen for T_1 and T_2 are such that 120 ms and 35 ms result, respectively, assuming f_{clk} = 50 MHz.

The first process (lines 26–37) implements the timer (using strategy #1). Except for the inclusion of *stp* (lines 26 and 30), this code is exactly as in the template.

The second process (lines 40–47) implements the FSM's state register, exactly as in the template.

The third and final process (lines 50–103) implements the entire combinational logic section. It is just a list of all states (indeed, because this code is repetitive, some of the states were not detailed in order to save some space), each containing the output (*ssd*) value and the next state. Note that in each state the output value is unique because in a Moore machine the output depends only on the state in which the machine is.

In this kind of application the "–1" term present in the determination of the total time (lines 20, 55, 62, . . .) does not make any difference, but it was maintained as a reminder of the accurate value.

Observe the correct use of registers and the completeness of the code, as described in comment 10 of section 6.3. Note in particular the following:

1) Regarding the use of registers: The circuit is not overregistered. This can be observed in the **elsif rising_edge(clk)** statement of line 44 (responsible for the inference of flip-flops), which is closed in line 46, guaranteeing that only the machine state (line

45) gets stored (the timer is in a separate circuit—see figure 8.2a). The output (*ssd*) is in the next process, which is purely combinational (thus not registered).

2) Regarding the outputs: The list of outputs (just *ssd* in this example) is exactly the same in all states (see lines 54, 61, 68, . . .), and the corresponding values are always properly declared.

3) Regarding the next state: Again, the coverage is complete because all states are included (see lines 53, 60, 67, . . .), and in each state the next state is always properly declared (lines 55-59, 62-66, 69-73, . . .).

The total number of flip-flops inferred by the compiler using the code below was 27 for sequential or Gray encoding, 29 for Johnson, and 35 for one-hot, which agree with the predictions made in section 8.11.2.

Because this particular machine has only simple timed transitions, a few simplifications could be made in the code below, but with no impact on the resulting circuit (thus with no reason to depart from the proposed template).

```
1    ----------------------------------------------------------
2    library ieee;
3    use ieee.std_logic_1164.all;
4    ----------------------------------------------------------
5    entity light_rotator is
6       port (
7          stp, clk, rst: in std_logic;
8          ssd: out std_logic_vector(6 downto 0));
9    end entity;
10   ----------------------------------------------------------
11   architecture moore_fsm of light_rotator is
12
13      --FSM-related declarations:
14      type state is (A, AB, B, BC, C, CD, D, DE, E, EF, F, FA);
15      signal pr_state, nx_state: state;
16
17      --Timer-related declarations:
18      constant T1: natural := 6_000_000;   --120ms @ fclk=50MHz
19      constant T2: natural := 1_750_000;   --35ms @ fclk=50MHz
20      constant tmax: natural := T1-1;      --tmax≥max(T1,T2)-1
21      signal t: natural range 0 to tmax;
22
23   begin
24
25      --Timer (using strategy #1):
26      process (clk, rst, stp)
27      begin
28         if rst='1' then
29            t <= 0;
30         elsif rising_edge(clk) and stp='0' then
31            if pr_state /= nx_state then
32               t <= 0;
33            elsif t /= tmax then
34               t <= t + 1;
```

```
         end if;
      end if;
   end process;

   --FSM state register:
   process (clk, rst)
   begin
      if rst='1' then
         pr_state <= A;
      elsif rising_edge(clk) then
         pr_state <= nx_state;
      end if;
   end process;

   --FSM combinational logic:
   process (all)
   begin
      case pr_state is
         when A =>
            ssd <= "0111111";
            if t>=T1-1 then -- or t=T1-1
               nx_state <= AB;
            else
               nx_state <= A;
            end if;
         when AB =>
            ssd <= "0011111";
            if t>=T2-1 then -- or t=T2-1
               nx_state <= B;
            else
               nx_state <= AB;
            end if;
         when B =>
            ssd <= "1011111";
            if t>=T1-1 then
               nx_state <= BC;
            else
               nx_state <= B;
            end if;
         when BC =>
            ssd <= "1001111";
            if t>=T2-1 then
               nx_state <= C;
            else
               nx_state <= BC;
            end if;
         when C =>
            ...
         when CD =>
            ...
         when D =>
            ...
         when DE =>
            ...
         when E =>
            ...
         when EF =>
```

```
92              ...
93           when F =>
94              ...
95           when FA =>
96              ssd <= "0111101";
97              if t=T2-1 then
98                 nx_state <= A;
99              else
100                nx_state <= FA;
101             end if;
102       end case;
103    end process;
104
105 end architecture;
106 -------------------------------------------------------------
```

9.5 Design of a Car Alarm (with Chirps)

This section presents a VHDL-based design for the car alarm with chirps introduced in section 8.11.6. The Moore template of section 9.2 is employed to implement the FSM of figure 8.21c. Again, either strategy #1 or #2 can be used to build the timer; the latter was adopted in the code below.

The entity, called *car_alarm_with_chirps*, is in lines 5–9. All ports are of type *std_logic* (industry standard).

The architecture, called *moore_fsm*, is in lines 11–138. As usual, it contains a declarative part and a statements part, with three processes in the latter.

The declarative part of the architecture (lines 13–21) contains FSM- and timer-related declarations. In the former the enumerated type *state* is created to represent the machine's present and next states. In the latter the value chosen for *chirpON* and *chirpOFF* is such that the chirp and the time interval between chirps last 0.3 s, assuming f_{clk} = 50 MHz.

The first process (lines 26–37) implements the timer, using strategy #2, exactly as in the template.

The second process (lines 40–47) implements the state register, again exactly as in the template.

The third and final process (lines 50–136) implements the entire combinational logic section. It is just a list of all states, each containing the output (*siren*) value, the value of t_{max}, and the next state. Note that in each state the output value is unique because in a Moore machine the output depends only on the state in which the machine is.

In this kind of application the "–1" term present in the determination of t_{max} (lines 21, 63, 97, . . .) does not make any difference, but it was maintained as a reminder of the accurate value.

Observe the correct use of registers and the completeness of the code, as described in comment 10 of section 6.3. Note in particular the following:

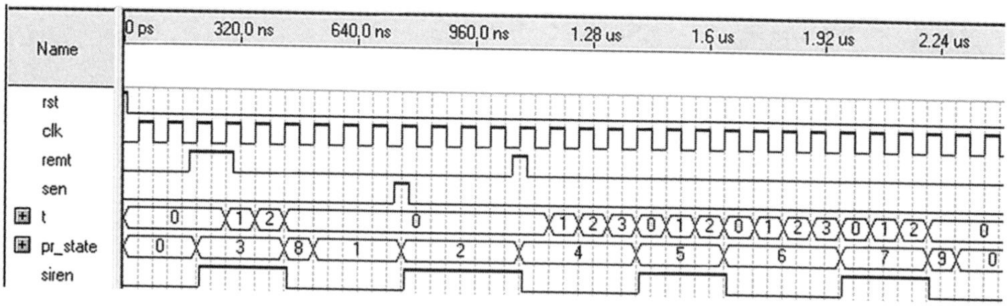
Name
0 ps
320.0 ns
640.0 ns
960.0 ns
1.28 us
1.6 us
1.92 us
2.24 us
rst
clk
remt
sen
t
pr_state
siren

Figure 9.1
Simulation results from the VHDL code for the car alarm of figure 8.21c.

1) Regarding the use of registers: The circuit is not overregistered. This can be observed in the **elsif rising_edge(clk)** statement of line 44 (responsible for the inference of flip-flops), which is closed in line 46, guaranteeing that only the machine state (line 45) gets stored (besides the timer, of course, designed in the previous process). The output (*siren*) is in the next process, which is purely combinational (thus not registered).
2) Regarding the outputs: The list of outputs (just *siren* in this example) and time parameters (t_{max}) is exactly the same in all states (see lines 54–55, 62–63, 70–71, . . .), and the corresponding values are always properly declared.
3) Regarding the next state: Again, the coverage is complete because all states are included (see lines 53, 61, 69, . . .), and in each state the conditional declarations for the next state are always finalized with an **else** statement (lines 58, 66, 74, . . .), guaranteeing that no condition is left unchecked.

The total number of flip-flops inferred by the compiler on synthesizing this code was 28 for sequential or Gray encoding, 29 for Johnson, and 34 for one-hot. Compare these results against your predictions made in exercise 8.14.

Simulation results are shown in figure 9.1.

```
1  ---------------------------------------------------------------
2  library ieee;
3  use ieee.std_logic_1164.all;
4  ---------------------------------------------------------------
5  entity car_alarm_with_chirps is
6     port (
7        remt, sen, clk, rst: in std_logic;
8        siren: out std_logic);
9  end entity;
10 ---------------------------------------------------------------
11 architecture moore_fsm of car_alarm_with_chirps is
12
13    --FSM-related declarations:
14    type state is (disarmed, armed, alarm, chirp1, chirp2, chirp3,
15       chirp4, chirp5, wait1, wait2);
```

```
   signal pr_state, nx_state: state;

   --Timer-related declarations:
   constant chirpON: natural := 15_000_000; --0.3s @fclk=50MHz
   constant chirpOFF: natural := 15_000_000;
   signal t, tmax: natural range 0 to chirpOFF-1; --range ≥ max time

begin

   --Timer (using strategy #2):
   process (clk, rst)
   begin
      if rst='1' then
         t <= 0;
      elsif rising_edge(clk) then
         if t < tmax then
            t <= t + 1;
         else
            t <= 0;
         end if;
      end if;
   end process;

   --FSM state register:
   process (clk, rst)
   begin
      if rst='1' then
         pr_state <= disarmed;
      elsif rising_edge(clk) then
         pr_state <= nx_state;
      end if;
   end process;

   --FSM combinational logic:
   process (all)
   begin
      case pr_state is
         when disarmed =>
            siren <= '0';
            tmax <= 0;
            if remt='1' then
               nx_state <= chirp1;
            else
               nx_state <= disarmed;
            end if;
         when chirp1 =>
            siren <= '1';
            tmax <= chirpON-1;
            if t=tmax then
               nx_state <= wait1;
            else
               nx_state <= chirp1;
            end if;
         when wait1 =>
            siren <= '0';
            tmax <= 0;
```

```
         if remt='0' then
            nx_state <= armed;
         else
            nx_state <= wait1;
         end if;
      when armed =>
         siren <= '0';
         tmax <= 0;
         if sen='1' then
            nx_state <= alarm;
         elsif remt='1' then
            nx_state <= chirp3;
         else
            nx_state <= armed;
         end if;
      when alarm =>
         siren <= '1';
         tmax <= 0;
         if remt='1' then
            nx_state <= chirp2;
         else
            nx_state <= alarm;
         end if;
      when chirp2 =>
         siren <= '0';
         tmax <= chirpOFF-1;
         if t=tmax then
            nx_state <= chirp3;
         else
            nx_state <= chirp2;
         end if;
      when chirp3 =>
         siren <= '1';
         tmax <= chirpON-1;
         if t=tmax then
            nx_state <= chirp4;
         else
            nx_state <= chirp3;
         end if;
      when chirp4 =>
         siren <= '0';
         tmax <= chirpOFF-1;
         if t=tmax then
            nx_state <= chirp5;
         else
            nx_state <= chirp4;
         end if;
      when chirp5 =>
         siren <= '1';
         tmax <= chirpON-1;
         if t=tmax then
            nx_state <= wait2;
         else
            nx_state <= chirp5;
         end if;
      when wait2 =>
         siren <= '0';
```

```
            tmax <= 0;
            if remt='0' then
               nx_state <= disarmed;
            else
               nx_state <= wait2;
            end if;
      end case;
   end process;

end architecture;
-------------------------------------------------------------------
```

9.6 Design of a Triggered Monostable Circuit

This section presents a VHDL-based design for the triggered monostable circuit of figure 8.24b, which is capable of generating the signal of figure 8.24a. Again, the code that follows is a straightforward application of the VHDL template for category 2 Moore machines introduced in section 9.2. Note, however, that in this FSM the timer control strategy #2 (section 8.5.3) cannot be used. Indeed, even strategy #1 (section 8.5.2) cannot be applied completely because in one of the state transitions the timer must not be zeroed.

The entity, called *triggered_mono*, is in lines 5–10. All ports are of type *std_logic* (industry standard).

The architecture, called *moore_fsm*, is in lines 12–99. As usual, it contains a declarative part and a statements part, with four processes in the latter.

The declarative part of the architecture (lines 14–20) contains FSM- and timer-related declarations. In the former the enumerated type *state* is created to represent the machine's present and next states. In the latter a small value was used for T (called *delay* in the code; note *delay* = 3 in line 19) in order to ease the inspection of the simulation results (shown later).

The first process (lines 25–38) implements the timer (with strategy #1, adapted). Observe how the timer is prevented from being zeroed when the machine enters state D, done with just the introduction of lines 31 and 33.

The second process (lines 41–48) implements the state register, exactly as in the template.

The third process (lines 51–89) implements the entire combinational logic section. It is just a list of all states, each containing the output (y) value and the next state. Note that in each state the output value is unique because in a Moore machine the output depends only on the state in which the machine is.

The fourth and final process (lines 92–97) implements the optional output register, exactly as in the template. The output register was included because in this kind of application glitches are generally not acceptable. Even though y could come directly from a DFF (hence glitch-free), that is not guaranteed because it depends on the encoding scheme used in the machine.

Observe the correct use of registers and the completeness of the code, as described in comment 10 of section 6.3. Note in particular the following:

1) Regarding the use of registers: The circuit is not overregistered. This can be observed in the **elsif rising_edge(clk)** statement of line 45 (responsible for the inference of flip-flops), which is closed in line 47, guaranteeing that only the machine state (line 46) gets stored (besides the timer and the output register, of course, designed in other processes). The output (*y*) is in the next process, which is purely combinational (thus not registered).
2) Regarding the outputs: The list of outputs (just *y* in this example) is exactly the same in all states (see lines 55, 62, 69, . . .), and the corresponding values are always properly declared.
3) Regarding the next state: Again, the coverage is complete because all states are included (see lines 54, 61, 68, . . .), and in each state the conditional declarations for the next state are always finalized with an **else** statement (lines 58, 65, 76, . . .), guaranteeing that no condition is left unchecked.

The total number of flip-flops inferred by the compiler on synthesizing the code below, with regular sequential encoding for the machine states, was 5 for $T = 3$ and 15 for $T = 3000$.

Simulation results, for $T = 3$ clock cycles, are depicted in figure 9.2. Analyze the plots to confirm the correctness of the circuit operation.

```
1    ------------------------------------------------
2    library ieee;
3    use ieee.std_logic_1164.all;
4    ------------------------------------------------
```

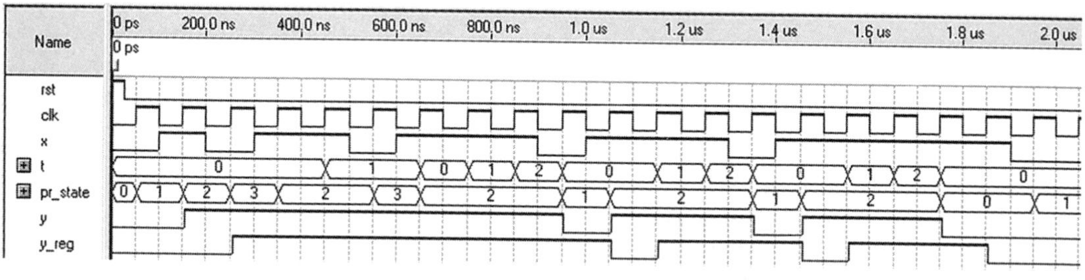
Name
0 ps
200.0 ns
400.0 ns
600.0 ns
800.0 ns
1.0 us
1.2 us
1.4 us
1.6 us
1.8 us
2.0 us
rst
clk
x
t
pr_state
y
y_reg

Figure 9.2
Simulation results from the VHDL code for the triggered monostable circuit of figure 8.24b for $T = 3$ clock periods.

```
entity triggered_mono is
   port (
      x, clk, rst: in std_logic;
      y: buffer std_logic;
      y_reg: out std_logic);
end entity;
------------------------------------------------------------
architecture moore_fsm of triggered_mono is

   --FSM-related declarations:
   type state is (A, B, C, D);
   signal pr_state, nx_state: state;

   --Timer-related declarations:
   constant delay: natural := 3; --any value >=2
   signal t: natural range 0 to delay-1; --tmax≥delay-1

begin

   --Timer (strategy #1, adapted):
   process (clk, rst)
   begin
      if rst='1' then
         t <= 0;
      elsif rising_edge(clk) then
         if pr_state /= nx_state then
            if nx_state/=D then
               t <= 0;
            end if;
         elsif t/=delay-1 then
            t <= t + 1;
         end if;
      end if;
   end process;

   --FSM state register:
   process (clk, rst)
   begin
      if rst='1' then
         pr_state <= A;
      elsif rising_edge(clk) then
         pr_state <= nx_state;
      end if;
   end process;

   --FSM combinational logic:
   process (all)
   begin
      case pr_state is
         when A =>
            y <= '0';
            if x='0' then
               nx_state <= B;
            else
               nx_state <= A;
            end if;
```

```
         when B =>
            y <= '0';
            if x='1' then
               nx_state <= C;
            else
               nx_state <= B;
            end if;
         when C =>
            y <= '1';
            if x='0' and t<delay-1 then
               nx_state <= D;
            elsif x='0' and t>=delay-1 then
               nx_state <= B;
            elsif x='1' and t>=delay-1 then
               nx_state <= A;
            else
               nx_state <= C;
            end if;
         when D =>
            y <= '1';
            if x='1' then
               nx_state <= C;
            elsif x='0' and t>=delay-2 then
               nx_state <= B;
            else
               nx_state <= D;
            end if;
      end case;
   end process;

   --Optional output register:
   process (clk)
   begin
      if rising_edge(clk) then
         y_reg <= y;
      end if;
   end process;

end architecture;
-------------------------------------------------------
```

9.7 Exercises

Exercise 9.1: Timer Control Strategies Analysis (Light Rotator)

This exercise concerns the light rotator of figure 8.14b, implemented with VHDL in section 9.4.

a) Compile the code of section 9.4 for the following options and write down the number of logic elements and registers inferred by the compiler in each case: 1) Using strategy #1 for the timer and sequential encoding for the machine; 2) With strategy #1 and one-hot encoding; 3) With strategy #2 and sequential encoding; 4) With strategy #2 and one-hot encoding.

b) Compare the results above. Was the difference between the two strategies more relevant for sequential or one-hot encoding? Explain.

Exercise 9.2: Blinking Light

This exercise concerns the blinking light FSM of figure 8.12c.

a) Which of the two timer control strategies (#1, section 8.5.2, or #2, section 8.5.3), if any, can be adopted in the implementation of this FSM?
b) Implement it using VHDL. Check whether the number of DFFs inferred by the compiler matches the prediction made in section 8.11.1 for each encoding option (sequential, Gray, Johnson, and one-hot). Recall that the predictions must be adjusted in case the clock frequency is different from 50 MHz.
c) Physically test your design in the FPGA development board. Use two switches to produce *rst* and *ena* and use an LED to display the output.

Exercise 9.3: Switch Debouncer

This exercise concerns the switch debouncer of figure 8.16c, which was inserted into the circuit of figure 9.3. The figure also shows two counters; the signal produced by the switch (*sw*) acts as clock to counter1, and its debounced version (*sw_deb*) acts as clock to counter2. Every time the pushbutton is pressed (or a toggle switch is flipped), both counters will be incremented, but counter1 might occasionally be incremented by more than one unit (the more the switch bounces, the bigger the difference between the values on the displays). Assume a 2-ms debouncing interval (check the clock frequency in your development board) and sequential encoding for the FSM, with the counters able to count from 0h to Fh.

a) Which of the two timer control strategies (#1, section 8.5.2, or #2, section 8.5.3), if any, can be adopted in the implementation of this FSM?
b) Estimate the number of DFFs needed to build the complete circuit. Does this number depend on the answer to part a above?
c) Design the circuit using VHDL. Check whether the number of DFFs inferred by the compiler matches your prediction.

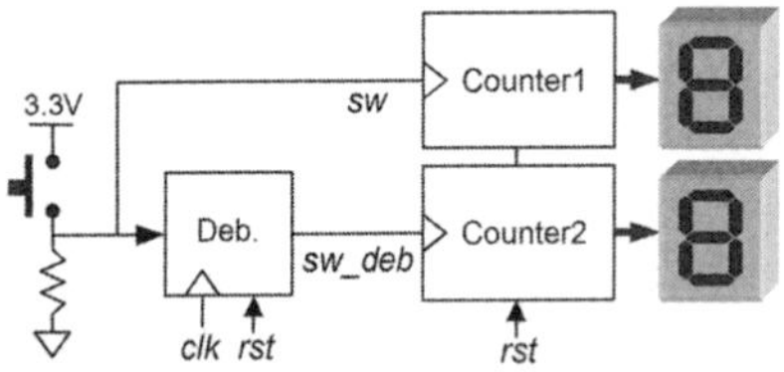
3.3V
sw
Counter1
Deb.
sw_deb
Counter2
clk rst
rst

Figure 9.3

d) Physically test your design in the FPGA development board for several switches (both toggle and pushbutton types).

Exercise 9.4: Reference-Value Definer

This exercise concerns the reference-value definer of figure 8.17b, which must produce the following consecutive values (the value must change every time the pushbutton is pressed): 250, 180, 130, 100, 70, and 40 (thus *ref* is an eight-bit signal). These values must be displayed on your development board using either three SSDs or eight LEDs (if the former is chosen, an SSD driver must be included in the design). In this exercise it is requested that the clock frequency be divided down to 1 kHz; this 1-kHz signal (*clk1k*) is the clock to be employed in the circuit.

a) Assume a 3-ms debouncing interval. Consequently, only four consecutive equal readings are needed for the pushbutton value to be considered valid. Is an FSM still desired for the debouncer (plus one-shot conversion)? If so, does it need to be a timed machine, as in figure 8.18b?
b) Draw a block diagram for your circuit, including in it the clock divider and the output display.
c) Draw the state transition diagram for each FSM used in the design.
d) Estimate the number of DFFs that will be needed to build the entire circuit (including the clock divider). Assume sequential encoding for the FSM(s) and check the clock frequency in your development board.
e) Implement the circuit using VHDL. Check whether the number of DFFs inferred by the compiler matches your prediction.
f) Physically demonstrate your design in the FPGA development board.

Exercise 9.5: Blinking Light with Several Speeds

This exercise is an extension to the light blinker of figure 8.12c, which must now operate with a *programmable* speed, set by a pushbutton, called *spd* (see the general diagram of figure 9.4). The next speed must be selected every time the pushbutton is pressed. The speed is determined by the on–off time interval (T_{ref}), which must be one of the following: 250, 180, 130, 100, 70, or 40 ms. As in exercise 9.4, the frequency of the system clock should be divided down, producing a 1-kHz

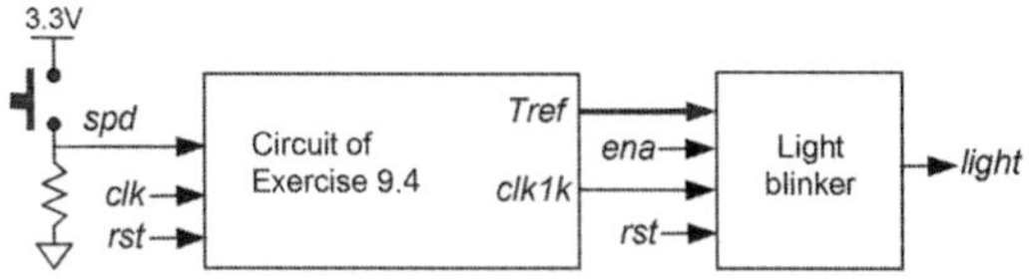
3.3V
spd
clk
rst
Circuit of Exercise 9.4
Tref
clk1k
ena
rst
Light blinker
light

Figure 9.4

clock for the present circuit. Note that a debouncer is indispensable for the speed pushbutton.

a) Draw a block diagram for your circuit, including the pushbutton, clock divider, and FSMs.
b) Draw the state transition diagram for each FSM to be used in the design.
c) Estimate the number of DFFs that will be needed to build the entire circuit.
d) Implement the circuit using VHDL. Check whether the number of DFFs inferred by the compiler matches your prediction.
e) Physically demonstrate your design in the FPGA development board.

Suggestion: Before solving this problem, solve exercises 9.3 and 9.4 if not done yet.

Exercise 9.6: Light Rotator with Additional Features

This exercise concerns the light rotator of figure 8.14, to which the following features must be added:

i) An input called *dir* (produced by a switch) that selects the rotating direction (clockwise when *dir* = '1', counterclockwise otherwise).
ii) An input called *spd* (produced by a pushbutton) that selects the rotating speed, as in exercises 9.4 and 9.5. Every time the pushbutton is pressed, the next speed must be selected. The speed is determined by the time interval during which the machine stays in states A, B, C, . . . , which must be one of the following: 250, 180, 130, 100, 70, or 40 ms. The duration of states AB, BC, CD, . . . must be always 20 ms. Note that a debouncer is necessary for the speed pushbutton.

As in exercises 9.4 and 9.5, the system clock should be divided down, producing a 1-kHz clock for the present circuit. (Suggestion: Before solving this problem, solve exercises 9.4 and 9.5 if not done yet.)

a) Draw a block diagram for your circuit.
b) Draw the state transition diagram for each FSM to be used in the design.
c) Estimate the number of DFFs that will be needed to build the entire circuit. Assume sequential encoding for the FSM(s).
d) Design the circuit using VHDL. Check whether the number of DFFs inferred by the compiler matches your prediction.
e) Physically demonstrate your design in the FPGA development board.

Exercise 9.7: Garage Door Controller

This exercise concerns the garage door controller seen in section 5.4.5, designed with VHDL and SystemVerilog in sections 6.7 and 7.6, respectively. Make the modifications needed in the VHDL code to incorporate the following feature: the door must close automatically 30 s after arriving at the completely open position. This feature should be optional, so an input must be added to the circuit to allow the user to choose between enabling it or not. How many DFFs will be needed to build the entire circuit now?

Exercise 9.8: Traffic Light Controller

This exercise concerns the traffic light controller of figure 8.20c.

a) Which of the two timer control strategies (#1, section 8.5.2, or #2, section 8.5.3), if any, can be adopted to implement this FSM?
b) Implement it using VHDL. Check whether the number of DFFs inferred by the compiler matches the prediction made in section 8.11.5 for each encoding option (sequential, Gray, Johnson, and one-hot). Recall that the predictions must be adjusted in case the clock frequency is different from 50 MHz.
c) Physically test your design in the FPGA development board. Use three switches to produce *stby*, *test*, and *rst*, and six LEDs to display the outputs.
d) At this point add the following feature (modify the design and download it to the FPGA board): the yellow lights should blink (at 0.5 Hz) while the circuit is in the standby mode.

Exercise 9.9: Password Detector

This exercise concerns the password detector of figure 8.22. A general block diagram for the present design is shown in figure 9.5, where, to ease the experiment, four pushbuttons (from the FPGA board itself) replace the keypad. A multisignal debouncer (treated in exercise 8.9) is also included. Only the following values are valid inputs (x) to the password detector: "1111" (no pushbutton pressed), "0111" (top pushbutton pressed), "1011", "1101", and "1110" (bottom pushbutton pressed).

a) Solve exercise 8.9 if not done yet.
b) Present a state transition diagram for each FSM to be used in this design.
c) How many DFFs are needed to build the entire circuit? Adopt a 1-ms debouncing interval and sequential encoding for the FSMs. Check the clock frequency in your FPGA development board.

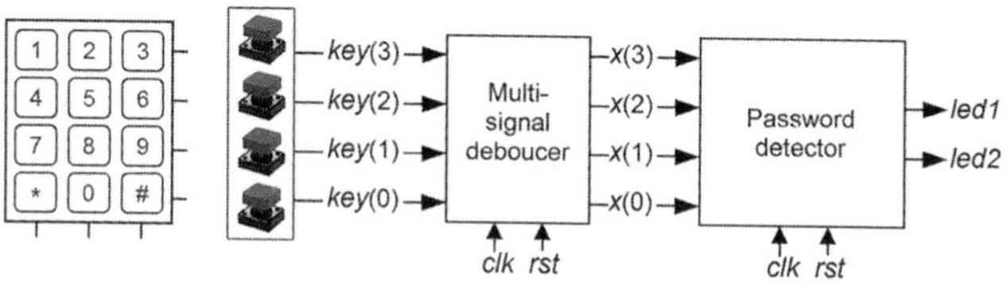
1 2 3
4 5 6
7 8 9
* 0 #
key(3)
key(2)
key(1)
key(0)
Multi-
signal
deboucer
clk rst
x(3)
x(2)
x(1)
x(0)
Password
detector
clk rst
led1
led2

Figure 9.5

d) Design the circuit using VHDL. Enter the multisignal debouncer as a *component* in the main code. After compilation, check whether the inferred number of DFFs matches your prediction.
e) Physically test your design in the FPGA development board. Test it for the following passwords: *abc*, *aad*, and *aaa* (where a = "0111", b = "1011", c = "1101", and d = "1110").

Exercise 9.10: Triggered Circuits

This exercise concerns the triggered FSMs treated in exercise 8.17.

a) Solve exercise 8.17 if not done yet.
b) Using VHDL, implement the FSM devised in part a of that exercise. Check whether the number of DFFs inferred by the compiler matches your prediction. Show corresponding simulation results.
c) Do the same for the FSM of part b in that exercise.

Exercise 9.11: Pulse Shifter

This exercise concerns the pulse shifter of figure 8.26a. Implement it using VHDL. Simulate it using the same stimuli of figure 8.26b and check whether the same waveforms result.

Exercise 9.12: Synchronous Pulse Stretcher

This exercise concerns the synchronous pulse stretcher introduced in figure 8.28a.

a) How many DFFs are needed to build it for T = 64 clock cycles and sequential encoding?
b) Implement it using VHDL. Check whether the number of DFFs inferred by the compiler matches your estimate.
c) Recompile it for T = 5 and simulate it with the same stimuli of figure 8.28b, checking if the same waveforms result.

Exercise 9.13: Asynchronous Pulse Stretcher

This exercise concerns the asynchronous pulse stretcher introduced in figure 8.28c.

a) How many DFFs are needed to build it for T = 64 clock cycles and sequential encoding?
b) Implement it using VHDL. Check whether the number of DFFs inferred by the compiler matches your estimate.
c) Recompile it for T = 5 and simulate it with the same stimuli of figure 8.28d, checking if the same waveforms result.

10 SystemVerilog Design of Timed (Category 2) State Machines

10.1 Introduction

This chapter presents several SystemVerilog designs of category 2 state machines. It starts by presenting two SystemVerilog templates, for Moore- and Mealy-based implementations, which are used subsequently to develop a series of designs related to the examples introduced in chapter 8.

The codes are all complete (not only partial sketches) and are accompanied by comments and simulation results, illustrating the design's main features. All circuits were synthesized using Quartus II (from Altera) or ISE (from Xilinx). The simulations were performed with Quartus II or ModelSim (from Mentor Graphics). The default encoding scheme for the states of the FSMs was regular sequential encoding (see encoding options in section 3.7).

The same designs were developed in chapter 9 using VHDL, so the reader can make a direct comparison between the codes.

Note: See suggestions of SystemVerilog books in the bibliography.

10.2 SystemVerilog Template for Timed (Category 2) Moore Machines

The template is presented below. Because it is an extension to the Moore template for category 1, described in section 7.3, a review of that template is suggested before this one is examined because only the differences are described.

The only differences are those needed for the inclusion of a timer (external to the FSM—see figure 8.2a). Recall, however, that the FSM itself is responsible for controlling the timer. For that purpose, two strategies were developed in chapter 8, being the first generic (section 8.5.2), and the second (section 8.5.3), non-generic. It is very important that the reader review those two sections before proceeding.

The first of the two templates that follow is for timed Moore machines with the timer implemented using strategy #1. The timer-related constants (T_1, T_2, . . .) can be declared either as global parameters (in the module header—see lines 3–5 in the template of section 7.3) or as local parameters, as shown in lines 11–13 of the template below. The variable *t* (line 14) must obviously stay where it is. As seen in section 8.5.2, the timer must obey $t_{max} \geq \max(T_1, T_2, \ldots) - 1$.

In the statements part of the code (lines 16–55), there are two differences.

The first difference is an additional **always_ff** block (lines 19–22), which implements the timer, according to the strategy described in section 8.5.2.

The second difference is in the **always_comb** block that implements the FSM's combinational logic section (lines 28–50), because *t* might now appear in the conditions for *nx_state* (lines 34, 35, 42, . . .). The use of $t \geq T - 1$ instead of $t = T - 1$ is required in the conditional-timed transitions with $T - 1 < t_{max}$. Note that t_{max} does not need to be defined in all states, which is not true for strategy #2.

```
1     //Timed Moore machine with timer control strategy #1
2     //Part 1: Module header:-----------------------------------
3        (same as for category 1 Moore, section 7.3)
4
5     //Part 2: Declarations:-----------------------------------
6
7        //FSM-related declarations:
8        (same as for category 1 Moore, section 7.3)
9
10       //Timer-related declarations:
11       const logic [7:0] T1 = <value>;
12       const logic [7:0] T2 = <value>;
13       const logic [7:0] tmax = <value>;//tmax≥max(T1,T2,...)-1
14       logic [7:0] t;
15
16    //Part 3: Statements:-------------------------------------
17
18       //Timer (strategy #1, section 8.5.2):
19       always_ff @(posedge clk, posedge rst)
20          if (rst) t <= 0;
21          else if (pr_state != nx_state) t <= 0;
22          else if (t != tmax) t <= t + 1;
23
24       //FSM state register:
25       (same as for category 1 Moore, Section 7.3)
26
27       //FSM combinational logic:
28       always_comb
29          case (pr_state)
30             A: begin
31                outp1 <= <value>;
32                outp2 <= <value>;
33                ...
34                if (... and t>=T1-1) nx_state <= B;
35                else if (... and t>=T2-1) nx_state <= ...;
36                else nx_state <= A;
```

```
         end
         B: begin
            outp1 <= <value>;
            outp2 <= <value>;
            ...
            if (... and t>=T3-1) nx_state <= C;
            else if (...) nx_state <= ...;
            else nx_state <= B;
         end
         C: begin
            ...
         end
         ...
      endcase

   //Optional output register:
   (same as for category 1 Moore, section 7.3)

endmodule
//------------------------------------------------------------
```

The next template is for timed Moore machines employing strategy #2 to implement the timer.

The first difference is in line 14, which now includes also t_{max}.

The second difference is in the **always_ff** block for the timer (lines 19–22), which is now based on the strategy described in section 8.5.3.

The third and final difference is in **always_comb** block that implements the FSM's combinational logic section (lines 28–52), which requires now the value of t_{max} to be specified in each state (lines 34, 43, . . .), even if the state is untimed (t_{max} = 0). This code can obviously be simplified in several ways when there are no conditional-timed transitions and/or t_{max} is the same in all or most states.

```
1    //Timed Moore machine with timer control strategy #2
2    //Part 1: Module header:------------------------------
3       (same as template above)
4
5    //Part 2: Declarations:-------------------------------
6
7       //FSM-related declarations:
8       (same as for category 1 Moore, section 7.3)
9
10      //Timer-related declarations:
11      const logic [7:0] T1 = <value>;
12      const logic [7:0] T2 = <value>;
13      ...
14      logic [7:0] t, tmax;
15
16   //Part 3: Statements:--------------------------------
17
18      //Timer (strategy #2, section 8.5.3):
```

```
   always_ff @(posedge clk, posedge rst)
      if (rst) t <= 0;
      else if (t < tmax) t <= t + 1;
      else t <= 0;

   //FSM state register:
   (same as for category 1 Moore, Section 7.3)

   //FSM combinational logic:
   always_comb
      case (pr_state)
         A: begin
            outp1 <= <value>;
            outp2 <= <value>;
            ...
            tmax <= T1-1;
            if (... and t=tmax) nx_state <= B;
            else if (...) nx_state <= ...;
            else nx_state <= A;
         end
         B: begin
            outp1 <= <value>;
            outp2 <= <value>;
            ...
            tmax <= T2-1;
            if (... and t=tmax) nx_state <= C;
            else if (...) nx_state <= ...;
            else nx_state <= B;
         end
         C: begin
            ...
         end
         ...
      endcase

   //Optional output register:
   (same as for category 1 Moore, section 7.3)

endmodule
//-----------------------------------------------
```

10.3 SystemVerilog Template for Timed (Category 2) Mealy Machines

The template is presented below, using strategy #1 to implement the timer. The only difference with respect to the Moore template just described is in the **always_comb** block for the combinational logic (lines 22–64) because the output is specified differently now. Recall that in a Mealy machine the output depends not only on the FSM's state but also on its input, so **if** statements are expected for the output in one or more states because the output values might not be unique. This is achieved by including such values within the conditional statements for *nx_state*. For example, observe in lines 24–42, relative to state A, that the output values are now conditional. Compare these lines against lines 30–37 in the previous template.

```
1   //Timed Mealy machine with timer control strategy #1
2   //Part 1: Module header:---------------------------
3      (same as for category 2 Moore, section 10.2)
4
5   //Part 2: Declarations:----------------------------
6
7      //FSM-related declarations:
8      (same as for category 2 Moore, section 10.2)
9
10     //Timer-related declarations:
11     (same as for category 2 Moore, section 10.2)
12
13  //Part 3: Statements:------------------------------
14
15     //Timer (using timer control strategy #1):
16     (same as for category 2 Moore, section 10.2)
17
18     //FSM state register:
19     (same as for category 2 Moore, section 10.2)
20
21     //FSM combinational logic:
22     always_comb
23        case (pr_state)
24           A:
25              if (... and t>=T1-1) begin
26                 outp1 <= <value>;
27                 outp2 <= <value>;
28                 ...
29                 nx_state <= B;
30              end
31              else if (... and t>=T2-1) begin
32                 outp1 <= <value>;
33                 outp2 <= <value>;
34                 ...
35                 nx_state <= ...;
36              end
37              else begin
38                 outp1 <= <value>;
39                 outp2 <= <value>;
40                 ...
41                 nx_state <= A;
42              end
43           B:
44              if (... and t>=T3-1) begin
45                 outp1 <= <value>;
46                 outp2 <= <value>;
47                 ...
48                 nx_state <= C;
49              end
50              else if (condition) begin
51                 outp1 <= <value>;
52                 outp2 <= <value>;
```

```
                    ...
                    nx_state <= ...;
                 end
                 else begin
                    outp1 <= <value>;
                    outp2 <= <value>;
                    ...
                    nx_state <= B;
                 end
              C: ...
              ...
        endcase

        //Optional output register:
        (same as for category 2 Moore, section 10.2)

    endmodule
    //---------------------------------------------------
```

10.4 Design of a Light Rotator

This section presents a SystemVerilog-based design for the light rotator introduced in section 8.11.2. The Moore template of section 10.2 is used to implement the FSM of figure 8.14b. Either strategy #1 (section 8.5.2) or #2 (section 8.5.3) can be used to build the timer (both templates are shown in section 10.2); the former is employed in the code below, while the latter is explored in exercise 10.1.

The first part of the code (*module header*) is in lines 1–4. The module's name is *light_rotator*. Note that all ports are of type **logic**.

The second part of the code (*declarations*) is in lines 6–17. In the FSM-related declarations (lines 9–11), the enumerated type *state* is created to represent the machine's present and next states. In the timer-related declarations (lines 14–17), the values chosen for T_1 and T_2 are such that 120 ms and 35 ms result, respectively, assuming f_{clk} = 50 MHz.

The third and final part of the code (*statements*) is in lines 19–85. It contains three **always** blocks, described next.

The first **always** block (lines 22–27) is an **always_ff** that implements the timer, using strategy #1. Except for the presence of *stp*, it is exactly as in the template.

The second **always** block (lines 30–32) is an **always_ff** that implements the FSM's state register, exactly as in the template.

The third and final **always** block (lines 35–83) is an **always_comb**, which implements the entire combinational logic section. It is just a list of all states (indeed, because this code is repetitive, some of the states were not detailed in order to save some space), each containing the output (*ssd*) value and the next state. Note that in each state the output value is unique because in a Moore machine the output depends only on the state in which the machine is.

In this kind of application, the "–1" term present in the definition of the total time (lines 16, 39, 44, 49, . . .) does not make any difference, but it was maintained as a reminder of the precise value. Also, in this application possible glitches during (positive) clock transitions are not a problem, so the optional output register shown in the last part of the template was not employed.

The reader is invited to compile this code and play with the circuit in the FPGA development board. Also, check whether the number of DFFs inferred by the compiler matches the prediction made in section 8.11.2 for each encoding style.

```
1    //Module header:-----------------------------------
2    module light_rotator (
3       input logic stp, clk, rst,
4       output logic [6:0] ssd);
5
6    //Declarations:------------------------------------
7
8       //FSM-related declarations:
9       typedef enum logic [3:0] {A, AB, B, BC, C, CD, D, DE, E, EF,
10         F, FA} state;
11      state pr_state, nx_state;
12
13      //Timer-related declarations:
14      const logic [22:0] T1 = 6_000_000; //120ms @fclk=50MHz
15      const logic [22:0] T2 = 1_750_000; //35ms @fclk=50MHz
16      const logic [22:0] tmax = T1-1; //tmax≥max(T1,T2)-1
17      logic [22:0] t;
18
19   //Statements:--------------------------------------
20
21      //Timer (using strategy #1):
22      always_ff @(posedge clk, posedge rst)
23         if (rst) t <= 0;
24         else if (~stp) begin
25            if (pr_state != nx_state) t <= 0;
26            else if (t != tmax) t <= t + 1;
27         end
28
29      //FSM state register:
30      always_ff @(posedge clk, posedge rst)
31         if (rst) pr_state <= A;
32         else pr_state <= nx_state;
33
34      //FSM combinational logic:
35      always_comb
36         case (pr_state)
37            A: begin
38               ssd <= 7'b0111111;
39               if (t>=T1-1) nx_state <= AB; //or t==T1-1
40               else nx_state <= A;
41            end
42            AB: begin
43               ssd <= 7'b0011111;
44               if (t>=T2-1) nx_state <= B; //or t==T2-1
45               else nx_state <= AB;
```

```
               end
            B: begin
               ssd <= 7'b1011111;
               if (t>=T1-1) nx_state <= BC;
               else nx_state <= B;
            end
            BC: begin
               ssd <= 7'b1001111;
               if (t>=T2-1) nx_state <= C;
               else nx_state <= BC;
            end
            C: begin
               ...
            end
            CD: begin
               ...
            end
            D: begin
               ...
            end
            DE: begin
               ...
            end
            E: begin
               ...
            end
            EF: begin
               ...
            end
            F: begin
               ...
            end
            FA: begin
               ssd <= 7'b0111101;
               if (t==T2-1) nx_state <= A;
               else nx_state <= FA;
            end
         endcase

endmodule
//----------------------------------------------------
```

10.5 Design of a Car Alarm (with Chirps)

This section presents a SystemVerilog-based design for the car alarm with chirps introduced in section 8.11.6. The Moore template of section 10.2 is employed to implement the FSM of figure 8.21c. Again, either strategy #1 or #2 can be used to build the timer; the latter was adopted in the code below.

The first part of the code (*module header*) is in lines 1–4. The module's name is *car_alarm_with_chirps*. Note that all ports are of type **logic**.

The second part of the code (*declarations*) is in lines 6–16. In the FSM-related declarations (lines 9–11), the enumerated type *state* is created to represent the machine's present and next states. In the timer-related declarations (lines 14–16), the value chosen for *chirpON* and *chirpOFF* is such that the chirp and the time interval between chirps last 0.3 s, assuming f_{clk} = 50 MHz.

The third and final part of the code (*statements*) is in lines 18–97. It contains three **always** blocks, described next.

The first **always** block (lines 21–24) is an **always_ff** that implements the timer, using strategy #2, exactly as in the template.

The second **always** block (lines 27–29) is another **always_ff**, implementing the machine's state register, also as in the template.

The third and final **always** block (lines 32–95) is an **always_comb**, which implements the entire combinational logic section. It is just a list of all states, each containing the output (*siren*) value, the value of t_{max}, and the next state. Note that in each state the output value is unique because in a Moore machine the output depends only on the state in which the machine is.

In this kind of application the "–1" term present in the determination of t_{max} (lines 42, 67, 73, . . .) does not make any difference, but it was maintained as a reminder of the precise value. Also, in this kind of application possible glitches during (positive) clock transitions are generally not a problem, so the optional output register shown in the final portion of the template was not employed.

Finally, and very importantly, observe the correct use of registers and the completeness of the code, as described in comment 8 of section 7.3. Observe in particular the following: 1) all states are included; 2) the list of outputs is exactly the same in all states, and the corresponding values are properly declared; 3) the specifications for *nx_state* are always finalized with an **else** statement, so no condition is left unchecked.

The total number of flip-flops inferred by the compiler on synthesizing this code was 28 for sequential or Gray encoding, 29 for Johnson, and 34 for one-hot. Compare these results against your predictions made in exercise 8.14.

Simulation results are shown in figure 10.1.

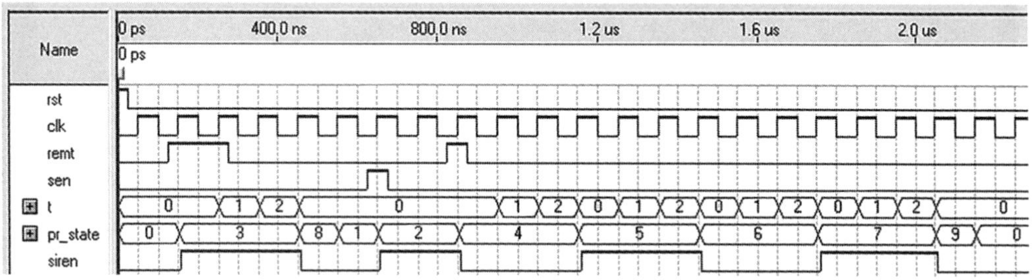
Name
0 ps
400,0 ns
800,0 ns
1.2 us
1.6 us
2.0 us
0 ps
rst
clk
remt
sen
t
pr_state
siren

Figure 10.1
Simulation results from the SystemVerilog code for the car alarm of figure 8.21c for *chirpON* = *chirpOFF* = 3 clock cycles.

```
//Module header:-----------------------------------------------
module car_alarm_with_chirps (
   input logic remt, sen, clk, rst,
   output logic siren);

//Declarations:------------------------------------------------

   //FSM-related declarations:
   typedef enum logic [3:0] {disarmed, armed, alarm, chirp1,
      chirp2, chirp3, chirp4, chirp5, wait1, wait2} state;
   state pr_state, nx_state;

   //Timer-related declarations:
   const logic [23:0] chirpON = 15_000_000; //0.3s @fclk=50MHz
   const logic [23:0] chirpOFF = 15_000_000;
   logic [23:0] t, tmax; --range≥max(chirpON,chirpOFF)

//Statements:--------------------------------------------------

   //Timer (using strategy #2):
   always_ff @(posedge clk, posedge rst)
      if (rst) t <= 0;
      else if (t < tmax) t <= t + 1;
      else t <= 0;

   //FSM state register:
   always_ff @(posedge clk, posedge rst)
      if (rst) pr_state <= disarmed;
      else pr_state <= nx_state;

   //FSM combinational logic:
   always_comb
      case (pr_state)
         disarmed: begin
            siren <= 1'b0;
            tmax <= 0;
            if (remt) nx_state <= chirp1;
            else nx_state <= disarmed;
         end
         chirp1: begin
            siren <= 1'b1;
            tmax <= chirpON-1;
            if (t==tmax) nx_state <= wait1;
            else nx_state <= chirp1;
         end
         wait1: begin
            siren <= 1'b0;
            tmax <= 0;
            if (~remt) nx_state <= armed;
            else nx_state <= wait1;
         end
         armed: begin
            siren <= 1'b0;
            tmax <= 0;
            if (sen) nx_state <= alarm;
```

```
            else if (remt) nx_state <= chirp3;
            else nx_state <= armed;
         end
         alarm: begin
            siren <= 1'b1;
            tmax <= 0;
            if (remt) nx_state <= chirp2;
            else nx_state <= alarm;
         end
         chirp2: begin
            siren <= 1'b0;
            tmax <= chirpOFF-1;
            if (t==tmax) nx_state <= chirp3;
            else nx_state <= chirp2;
         end
         chirp3: begin
            siren <= 1'b1;
            tmax <= chirpON-1;
            if (t==tmax) nx_state <= chirp4;
            else nx_state <= chirp3;
         end
         chirp4: begin
            siren <= 1'b0;
            tmax <= chirpOFF-1;
            if (t==tmax) nx_state <= chirp5;
            else nx_state <= chirp4;
         end
         chirp5: begin
            siren <= 1'b1;
            tmax <= chirpON-1;
            if (t==tmax) nx_state <= wait2;
            else nx_state <= chirp5;
         end
         wait2: begin
            siren <= 1'b0;
            tmax <= 0;
            if (~remt) nx_state <= disarmed;
            else nx_state <= wait2;
         end
      endcase

endmodule
//----------------------------------------------------------------
```

10.6 Design of a Triggered Monostable Circuit

This section presents a SystemVerilog-based design for the triggered monostable circuit of figure 8.24b, which is capable of generating the signal of figure 8.24a. Again, the code that follows is a straightforward application of the SystemVerilog template for category 2 Moore machines introduced in section 10.2. Note, however, that in this FSM the timer control strategy #2 (section 8.5.3) cannot be used. Indeed, even strategy #1 (section 8.5.2) cannot be applied completely because in one of the state transitions the timer must not be zeroed.

The first part of the code (*module header*) is in lines 1–4. The module's name is *triggered_mono*. Note that all ports are of type **logic**.

The second part of the code (*declarations*) is in lines 6–15. In the FSM-related declarations (lines 9–11), the enumerated type *state* is created to represent the machine's present and next states; also, a variable called *y* is defined because the optional output register (which will produce a registered version of *y*, called *y_reg*) is needed here to remove possible glitches. In the timer-related declarations (lines 14–15), a small value was used for *T* (called *delay* in the code; note *delay* = 3 in line 14) in order to ease the inspection of the simulation results.

The third and final part of the code (*statements*) is in lines 17–62. It contains four **always** blocks, described next.

The first **always** block (lines 20–23) is an **always_ff** that implements the timer. Note that the timer is not zeroed when the machine enters state D.

The second **always** block (lines 26–28) is another **always_ff**, implementing the machine's state register, exactly as in the template.

The third **always** block (lines 31–56) is an **always_comb**, which implements the entire combinational logic section. It is just a list of all states, each containing the output (*y*) value and the next state. Note that in each state the output value is unique because in a Moore machine the output depends only on the state in which the machine is.

The fourth and final **always** block (lines 59–60) implements the optional output register, exactly as in the template. Even though *y* could come directly from a DFF (hence glitch-free), that is not guaranteed because it depends on the encoding scheme used in the machine.

Finally, and very importantly, observe the correct use of registers and the completeness of the code, as described in comment 8 of section 7.3. Observe in particular the following: 1) all states are included; 2) the list of outputs is exactly the same in all states, and the corresponding values are properly declared; 3) the specifications for *nx_state* are always finalized with an **else** statement, so no condition is left unchecked.

The total number of flip-flops inferred by the compiler on synthesizing the code below, with regular sequential encoding for the machine states, was 5 for $T = 3$ and 15 for $T = 3000$.

Simulation results are similar to those in figure 9.2, where the same circuit was implemented using VHDL.

```
1    //Module header:-------------------------------------
2    module triggered_mono (
3       input logic x, clk, rst,
4       output logic y_reg);
5
6    //Declarations:--------------------------------------
7
```

```
    //FSM-related declarations:
    typedef enum logic [1:0] {A, B, C, D} state;
    state pr_state, nx_state;
    logic y;

    //Timer-related declarations:
    const logic [1:0] delay = 3; //any value >1
    logic [1:0] t; //tmax≥delay-1

 //Statements:---------------------------------------

    //Timer (strategy #1, adapted):
    always_ff @(posedge clk, posedge rst)
       if (rst) t <= 0;
       else if (pr_state!=nx_state & nx_state!=D) t <= 0;
       else if (pr_state==nx_state & t!=delay-1) t <= t + 1;

    //FSM state register:
    always_ff @(posedge clk, posedge rst)
       if (rst) pr_state <= A;
       else pr_state <= nx_state;

    //FSM combinational logic:
    always_comb
       case (pr_state)
          A: begin
             y <= 1'b0;
             if (~x) nx_state <= B;
             else nx_state <= A;
          end
          B: begin
             y <= 1'b0;
             if (x) nx_state <= C;
             else nx_state <= B;
          end
          C: begin
             y <= 1'b1;
             if (~x & t<delay-1) nx_state <= D;
             else if (~x & t>=delay-1) nx_state <= B;
             else if (x & t==delay-1) nx_state <= A;
             else nx_state <= C;
          end
          D: begin
             y <= 1'b1;
             if (x) nx_state <= C;
             else if (~x & t>=delay-2) nx_state <= B;
             else nx_state <= D;
          end
       endcase

    //Optional output register:-------
    always_ff @(posedge clk)
       y_reg <= y;

 endmodule
 //-----------------------------------------------
```

10.7 Exercises

Exercise 10.1: Timer Control Strategies Analysis (Light Rotator)
Solve exercise 9.1 using SystemVerilog instead of VHDL.

Exercise 10.2: Blinking Light
Solve exercise 9.2 using SystemVerilog instead of VHDL.

Exercise 10.3: Switch Debouncer
Solve exercise 9.3 using SystemVerilog instead of VHDL.

Exercise 10.4: Reference-Value Definer
Solve exercise 9.4 using SystemVerilog instead of VHDL.

Exercise 10.5: Blinking Light with Several Speeds
Solve exercise 9.5 using SystemVerilog instead of VHDL.

Exercise 10.6: Light Rotator with Additional Features
Solve exercise 9.6 using SystemVerilog instead of VHDL.

Exercise 10.7: Garage Door Controller
Solve exercise 9.7 using SystemVerilog instead of VHDL.

Exercise 10.8: Traffic Light Controller
Solve exercise 9.8 using SystemVerilog instead of VHDL.

Exercise 10.9: Password Detector
Solve exercise 9.9 using SystemVerilog instead of VHDL.

Exercise 10.10: Triggered Circuits
Solve exercise 9.10 using SystemVerilog instead of VHDL.

Exercise 10.11: Pulse Shifter
Solve exercise 9.11 using SystemVerilog instead of VHDL.

Exercise 10.12: Synchronous Pulse Stretcher
Solve exercise 9.12 using SystemVerilog instead of VHDL.

Exercise 10.13: Asynchronous Pulse Stretcher
Solve exercise 9.13 using SystemVerilog instead of VHDL.

11 Recursive (Category 3) State Machines

11.1 Introduction

We know that, from a hardware perspective, state machines can be classified into two types, based on their *input connections*, as follows.

1) *Moore machines*: The input, if it exists, is connected only to the logic block that computes the next state.
2) *Mealy machines*: The input is connected to both logic blocks, that is, for the next state and for the actual output.

In section 3.6 we introduced a new classification, also from a hardware point of view, based on the *transition types* and *nature of the outputs*, as follows (see figure 11.1).

1) *Regular (category 1) state machines*: This category, illustrated in figure 11.1a and studied in chapters 5 to 7, consists of machines with only untimed transitions and outputs that do not depend on previous (past) output values so none of the outputs need to be registered for the machine to function.
2) *Timed (category 2) state machines*: This category, illustrated in figure 11.1b and studied in chapters 8 to 10, is similar to category 1, except for the fact that one or more of its transitions depend on time (so these FSMs can have all four transition types: conditional, timed, conditional-timed, and unconditional).
3) *Recursive (category 3) state machines*: This category is illustrated in figure 11.1c and studied in chapters 11 to 13. It can have all four types of transitions, but one or more outputs depend on previous (past) output values so such outputs must be registered. Recall that in the standard architecture the outputs are produced by the FSM's *combinational* logic block, so the current output values are "forgotten" after the machine leaves that state; consequently, to implement a recursive (recurrent) machine, some sort of extra memory is needed.

The name "recursive" for category 3 is due to the fact that when an output depends on a previous output value that value is generally from that same output, so a recursive

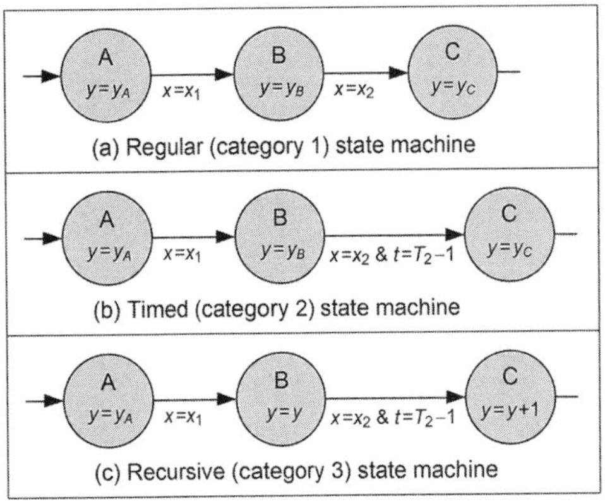
A
y=yA
x=x1
B
y=yB
x=x2
C
y=yC
(a) Regular (category 1) state machine
A
y=yA
x=x1
B
y=yB
x=x2 & t=T2−1
C
y=yC
(b) Timed (category 2) state machine
A
y=yA
x=x1
B
y=y
x=x2 & t=T2−1
C
y=y+1
(c) Recursive (category 3) state machine

Figure 11.1
State machine categories (from a hardware perspective).

equation results (i.e., the output is a function of itself). For example, $y = y$, $y = y'$, and $y = y + 1$ mean that y (which is an output) should keep in the present state the same value that it had in the previous state, or the complement of that value, or the incremented version of that value, respectively. Equivalently, one could write $y_{new} = y_{old}$, $y_{new} = y_{old}'$, and $y_{new} = y_{old} + 1$. Occasionally, an output might be a function of a past value of another signal, like $y = z$ (same as $y_{new} = z_{old}$).

The two fundamental decisions that must be made before starting a design are then the following:

1) The state machine category (regular, timed, or recursive).
2) The state machine type (Moore or Mealy).

It is important to recall, however, that regardless of the machine category and type, the state transition diagram must fulfill three fundamental requisites (seen in section 1.3):

1) It must include all possible system states.
2) All state transition conditions must be specified (unless a transition is unconditional) and must be truly complementary.
3) The list of outputs must be exactly the same in all states (standard architecture).

11.2 Recursive (Category 3) State Machines

Figure 11.2 shows two examples of very special circuits. In figure 11.2a a simplified flowchart for a memory-write procedure is shown in which an address is set, the data to be stored at that address is presented, then a write-enable pulse is applied to store the data. Note the presence of an incrementer (gray block), responsible for setting the

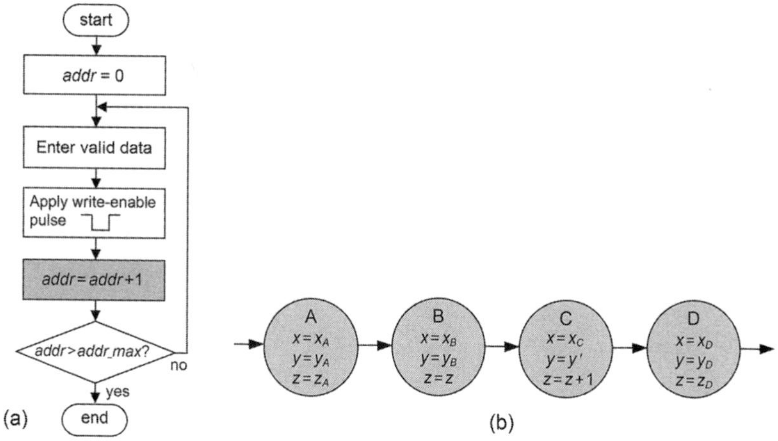
start
addr = 0
Enter valid data
Apply write-enable pulse
addr = addr + 1
addr > addr_max?
no
yes
end
(a)
A
x = x_A
y = y_A
z = z_A
B
x = x_B
y = y_B
z = z
C
x = x_C
y = y'
z = z + 1
D
x = x_D
y = y_D
z = z_D
(b)

Figure 11.2
Examples of category 3 state machines.

next memory address. Because the expression *addr* = *addr* + 1 is not a constant but, rather, depends on the previous value of *addr*, this flowchart cannot be implemented in hardware without some sort of auxiliary memory (to hold the value of *addr*), which must be provided along with the corresponding FSM (note that this is different—and more complex—than a "similar" implementation in software).

The second example (figure 11.2b) consists of a state machine with three outputs. Note that the list of outputs is exactly the same in all states (as required for hardware implementations using the standard architecture; otherwise latches would be inferred), but again not all output values are deterministic: in state B, z must keep the same value that it had when the machine left state A; in state C, y must exhibit the complement of the value that it had in the previous state, while z must be incremented. Recall that we cannot simply write $z = z_A$ in state B because z_A might have changed; for the same reason, we cannot write $y = y_B'$ and $z = z_A + 1$ in state C. Consequently, an extra memory (to hold the values of y and z) is again needed.

11.3 Architectures for Recursive (Category 3) Machines

The general architecture for category 3 machines is summarized in figure 11.3a. This representation follows the style of figures 3.1b and 3.1d, but the style of figures 3.1a and 3.1c could be used equivalently. Note that the timer is optional, but at least one auxiliary register is necessary.

In this illustration, only for the signal that produces *output2* an auxiliary register is needed, so for that output the optional output register (figure 11.3b) is never required (the dashed lines indicate that *output2* can be either the unregistered or the registered

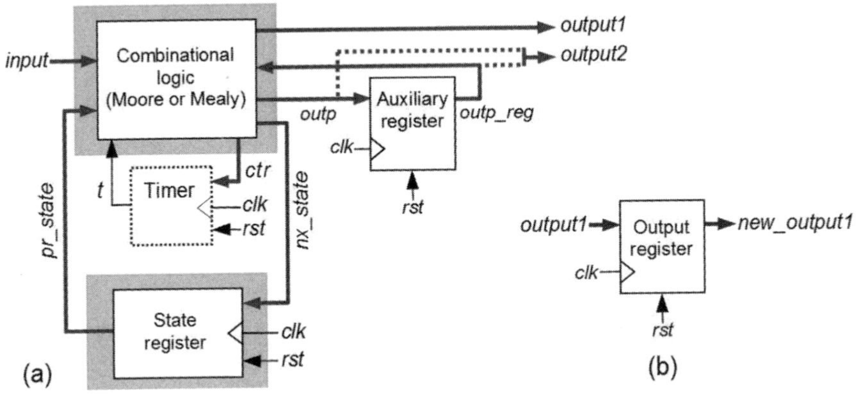
output1
output2
input
Combinational
logic
(Moore or Mealy)
outp
Auxiliary
register
outp_reg
clk
rst
ctr
Timer
t
clk
rst
pr_state
nx_state
State
register
clk
rst
(a)
output1
Output
register
new_output1
clk
rst
(b)

Figure 11.3
(a) General architecture for category 3 machines (timer is optional, but auxiliary register is compulsory). (b) Optional output register (only for outputs not processed by an auxiliary register).

version of *outp*). On the other hand, *output1* is not registered, so depending on the application, for it the optional output register might be needed. The resulting implementations are described below.

Recursive Moore machine: The circuit of figure 11.3a is used, with the input (if it exists) connected only to the logic block for the next state, as in figure 5.2a, and with unregistered output. Regarding the options for the output, see the comments above.
Recursive Mealy machine: Again, the circuit of figure 11.3a is used, but this time with the input connected to both logic blocks (for output and for next state), as in figure 5.2b. Regarding the options for the output, see the comments above.

11.4 Category 3 to Category 1 Conversion

We said in section 1.3 that for an FSM to be implemented in hardware it must obey three fundamental principles, the last one being that the list of outputs be exactly the same in all states. This is indispensable because the outputs are generated by the combinational logic section, which, being combinational, has no memory, so if an output is not specified in a certain state, the compiler usually infers a latch (to hold the output's last value), which is undesirable.

There is, however, an (apparent) exception, which occurs when the outputs are *registered* (that is, when the optional output register seen in all templates is used), because then the outputs are stored anyway (so latches are not needed). In such cases one might not list all outputs in all states, but that simply means that unlisted outputs will exhibit the value previously stored in the corresponding flip-flops. Consequently, for any physical purpose the list of outputs is in fact the same in all states.

The reasoning above allows us to conclude that if a circuit was modeled as a category 3 machine (because it has recursive outputs), with all outputs requiring an auxiliary register, then it can be implemented as if it were a category 1 circuit, with the optional output register included. In practical terms, in such cases the "dangerous" VHDL template of section 6.4.4 can be used (although not recommended).

11.5 Repetitively Looped Category 3 Machines

This section highlights the particular case in which multiple pointers (counters) are needed to implement an FSM. As is shown later in the examples, this can occur particularly when one is dealing with serial data communications (e.g., serial data receiver/transmitter, I^2C interface, SPI interface). Note that this section is the counterpart of section 8.8, in which similar machines were implemented using the category 2 model.

The general problem is stated in figure 11.4a. The machine must stay only one clock period in each state, but the loop must be repeated N_{AB} times, where N_{AB} is the number of times that the AB transition occurs (N_{BA} and $N_{AB} + N_{BA}$ would be fine too, but an extra DFF would be required in the counter for the latter). The solution proper is in figure 11.4b. Note that the counter (k) is incremented only in state B, holding its value while in state A.

A more general case is stated in figure 11.4c. Here, not only must the loop be repeated N_{AB} times, but also the machine must stay N_A clock periods in A and N_B clock periods in B (note N_A and N_B over the state circles). The solution proper is in figure 11.4d. Three counters (i, j, k) are needed. Counter i, which controls the stay in state A, is incremented in A and zeroed in B. Counter j, which controls the stay in state B,

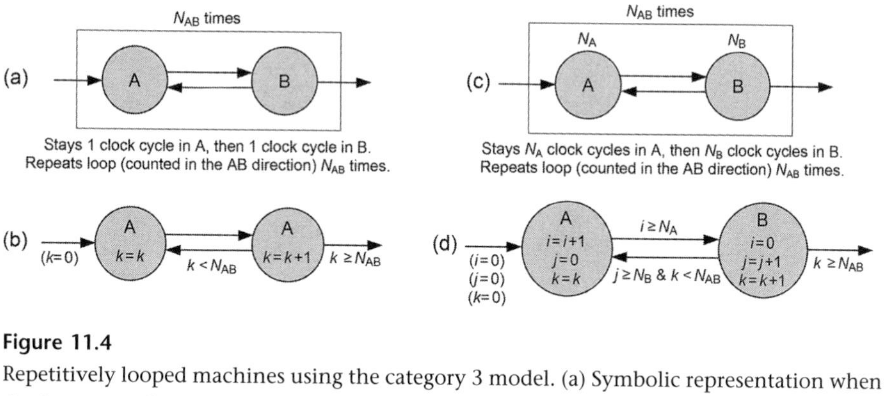

(a)
NAB times
A
B
Stays 1 clock cycle in A, then 1 clock cycle in B.
Repeats loop (counted in the AB direction) NAB times.
(c)
NAB times
NA
NB
A
B
Stays NA clock cycles in A, then NB clock cycles in B.
Repeats loop (counted in the AB direction) NAB times.
(b)
(k=0)
A
k=k
k < NAB
A
k=k+1
k ≥ NAB
(d)
(i=0)
(j=0)
(k=0)
A
i=i+1
j=0
k=k
i ≥ NA
j ≥ NB & k < NAB
B
i=0
j=j+1
k=k+1
k ≥ NAB

Figure 11.4

Repetitively looped machines using the category 3 model. (a) Symbolic representation when only the loop must be repeated and (b) corresponding details (k is incremented only in state B). (c) Symbolic representation with the loop and the individual states repeated and (d) corresponding details (three counters are needed; again, k is incremented only in state B).

is incremented in B and zeroed in A. Finally, counter k, which controls the number of loops, is incremented in B but is not zeroed in A.

11.6 Number of Flip-Flops

As mentioned earlier, it is difficult to estimate the number of logic gates that will be needed in a large design, but it is always possible to determine, and exactly, the number of flip-flops.

In the particular case of sequential circuits implemented as category 3 state machines, there are four demands for DFFs, as follows:

1) For the state register (below, M_{FSM} is the number of states):
   For sequential or Gray encoding: $N_{FSM} = \lceil \log_2 M_{FSM} \rceil$. Example: $M_{FSM} = 25 \rightarrow N_{FSM} = 5$.
   For Johnson encoding: $N_{FSM} = \lceil M_{FSM}/2 \rceil$. Example: $M_{FSM} = 25 \rightarrow N_{FSM} = 13$.
   For One-hot encoding: $N_{FSM} = M_{FSM}$. Example: $M_{FSM} = 25 \rightarrow N_{FSM} = 25$.
2) For the auxiliary register (compulsory, for at least one output, total b_{aux} bits):
   $N_{aux} = b_{aux}$. Example: $b_{aux} = 8 \rightarrow N_{aux} = 8$.
3) For the output register (optional, never needed for outputs processed by auxiliary registers, total b_{output} bits):
   $N_{output} = b_{output}$. Example: $b_{output} = 16 \rightarrow N_{output} = 16$.
4) For the timer (optional; category 3 can have all four types of transitions):
   $N_{timer} = \lceil \log_2 T_{max} \rceil$, where T_{max} is the largest transition time, expressed in "number of clock cycles"; that is, $T_{max} = t_{state\_max} \times f_{clk}$, where t_{state_max} is the largest transition time, in seconds, and f_{clk} is the clock frequency, in hertz.

Therefore, the total number of DFFs is $N_{total} = N_{FSM} + N_{aux} + N_{output} + N_{timer}$. In the examples that follow, as well as in the actual designs with VHDL and SystemVerilog, the number of flip-flops will be examined often.

11.7 Examples of Recursive (Category 3) State Machines

A series of recursive FSMs are presented next. Several of these examples will be designed later using VHDL (chapter 12) and SystemVerilog (chapter 13).

11.7.1 Generic Counters

As mentioned in section 5.4.1, counters are well-known circuits, easily designed without the FSM approach. Nevertheless, because they illustrate the state machine technique well, an example was included in that section using a regular FSM. A limitation seen there is that only small counters can be represented as regular state machines. In this section we are interested in examining how the FSM model can be extended to represent counters of any size. Even though one does not need the FSM approach to implement a counter when using an EDA tool (such as VHDL or SystemVerilog), the formal extension presented here will help in understanding the examples that follow, which often contain an embedded counter.

Two examples of counters modeled as category 3 FSMs are examined in this section (where N is the number of bits): (a) free-running (meaning that once the last value is reached it returns and restarts automatically from the initial value) with modulo 2^N; (b) free-running with modulo $<2^N$.

A modulo 2^N counter is one that has 2^N states, thus spanning all possible N-bit values. A regular modulo 2^N sequential counter will count from 0 to $2^N - 1$, restarting then automatically from zero. This type of counter is depicted in figure 11.5a. As usual, *ena* = '1' allows the counter to run, whereas *ena* = '0' causes it to stop. Note the presence of reset, which acts directly on the *hold* ($x = x$) state, thereby being able to set $x = 0$ (or any other value) as the starting value.

A modulo $<2^N$ counter is one that has fewer than 2^N states, thus not spanning all possible N-bit values. Therefore, a mechanism for starting/stopping the counter at the desired values is needed. A category 3 solution for this kind of counter is presented in figure 11.5b, where x_{min} and x_{max} represent the counter's initial and final values, respectively.

The examples above show that there is a big difference between category 1 and category 3 representations for counters. In the former all states are required to appear in the state transition diagram (section 5.4.1), whereas in the latter only very few states are needed (figure 11.5), regardless of the counter's number of states (thus, only the latter allows large counters to be conveniently represented as state machines). There is a price to pay, however: even though the resulting circuits in category 1 and category 3 are quite similar, only the former can lead to optimal implementations (similar to

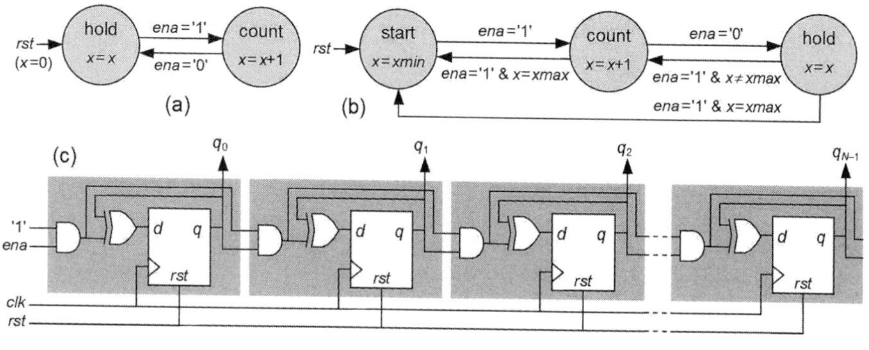

hold
x=x
count
x=x+1
rst
(x=0)
ena='1'
ena='0'
(a)
start
x=xmin
ena='1'
ena='1' & x=xmax
count
x=x+1
ena='0'
ena='1' & x≠xmax
hold
x=x
ena='1' & x=xmax
(b)
(c)
'1'
ena
clk
rst
d
q
rst
q0
q1
q2
qN-1

Figure 11.5

(a, b) Generic counters modeled as category 3 FSMs, free running in the range 0 to $2^N - 1$ or x_{min} to x_{max}, respectively. (c) Usual (optimal) construction for large synchronous counters.

figure 11.5c; this is called a synchronous counter with serial enable; for small counters—typically up to four or five bits—parallel enable can be employed, but then no longer with a standard logic cell).

The nonoptimality mentioned above can be verified, for example, by counting the number of DFFs needed to build the category 3 circuit. Based on section 11.6, and assuming that x is an eight-bit value and that regular sequential encoding (section 3.7) is used for the FSMs, the number of DFFs is as follows: in figure 11.5a: 1 for the two states + 8 for x = 9 DFFs; in figure 11.5b: 2 for the three states + 8 for x = 10 DFFs; with category 1: 8 DFFs in either case.

11.7.2 Long-String Comparator

This section deals with an FSM capable of sequentially comparing two arbitrarily long serial bit streams. The machine must determine whether the last N bits are pairwise equal (this means that the effect of the oldest pair of bits must be discarded when a new pair is received). Note that this is very different from determining whether two sequential blocks of N bits each are equal (in the latter, N bits are compared, then the next N bits are compared, and so on, without overlapping). The former is described in this section, and the latter is treated in exercise 11.5.

The circuit ports are depicted in figure 11.6a. The inputs (serial bit streams) are a and b, while the output is y (= '1' if all last N pairs of bits are equal). The comparator in this case is just a two-input XNOR gate, also depicted in the figure, which produces x = '1' when the inputs are equal. This signal (x) will be the actual input to the FSM.

A corresponding Moore-type solution is presented in figure 11.6b. Note that besides the actual output (y), it also produces an auxiliary output (i) that is a counter

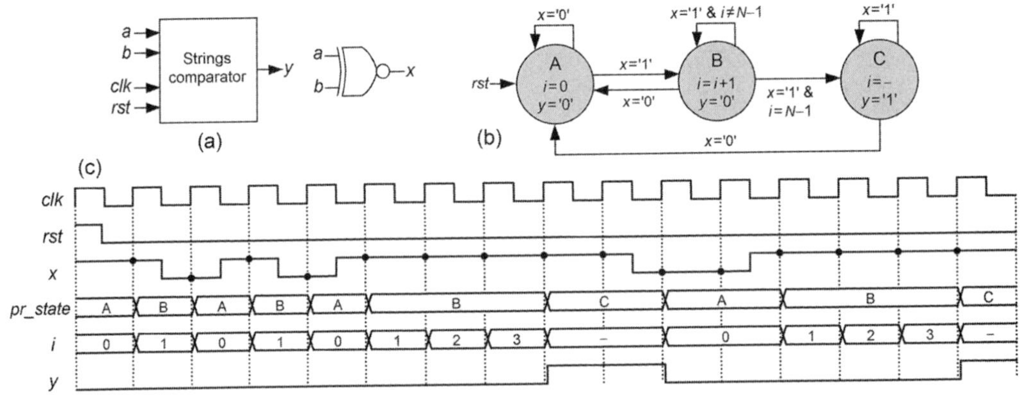
a
b
clk
rst
Strings comparator
y
x
(a)
(b)
A
i=0
y='0'
B
i=i+1
y='0'
C
i=-
y='1'
x='0'
x='1'
x='1' & i≠N–1
x='1' & i=N–1
(c)
pr_state

Figure 11.6
Two-string comparator that produces y = '1' if the last N bits are pairwise equal. (a) Circuit ports and bit comparator. (b) State transition diagram. (c) Illustrative timing diagram, for N = 4.

needed to control some of the machine transitions. The machine only reaches state C, which is the only state with y = '1', if the last N values of x are '1' (no mismatches). Observe the recursive expression $i = i + 1$ in state B, which characterizes a category 3 FSM. Again, to better illustrate the solution, a detailed state transition diagram is presented, but simpler representations can obviously be used as well (as in figure 1.4).

An illustrative timing diagram for this circuit is included in figure 11.6c for $N = 4$. The inputs were considered to be updated at the negative clock edge, whereas the FSM operates at the positive clock transition (note the dots marked on the x waveform, highlighting the values of x as perceived by the state machine). The reader is invited to apply the values of x given in figure 11.6c to the state machine in figure 11.6b to check the correctness of the plots for *pr_state*, i, and y.

Based on section 11.6, the number of flip-flops needed to build the FSM of figure 11.6b is as follows. For the state register: M_{FSM} = 3 states, so N_{FSM} = 2 (assuming sequential encoding). For the auxiliary register: needed for signal i, which ranges from 0 to $N - 1$; assuming N = 64 bits, N_{aux} = 6 DFFs. For the optional output register: not needed, so N_{output} = 0. For the timer: not needed, so N_{timer} = 0. Therefore, N_{total} = 8 DFFs.

11.7.3 Reference-Value Definer

In section 8.11.4 we started a discussion on a very important class of circuits, found particularly in control applications, capable of setting reference values. An example mentioned there was a temperature controller for an air conditioning system, which must have a way of letting the user choose the desired (*reference*) room temperature. As seen in that section, such circuits can be easily implemented without the FSM approach when the increments are constant, or with category 1 FSMs otherwise, but in the latter only if the number of reference values is small. When additional features are required, category 3 can be an interesting alternative because it poses no restrictions.

Let us start by examining two basic building blocks, shown in figures 11.7a,b and 11.7c,d. The circuit of figure 11.7a has only one control input (*up*), which must cause the output (*ref*, the reference value) to be incremented by one unit every time *up* is asserted (by means of a pushbutton, for example). The output must range from ref_{min} to ref_{max}, restarting automatically from ref_{min} after ref_{max} has been reached (or a reset pulse has been applied to the circuit).

A possible solution for this problem is depicted in figure 11.7b, requiring only four states regardless of the number of reference values. The machine must stay in state C during only one clock cycle (otherwise the incrementer would keep incrementing), so CD is an unconditional transition. Note the presence of recursive equations in almost all states, typical of category 3 FSMs.

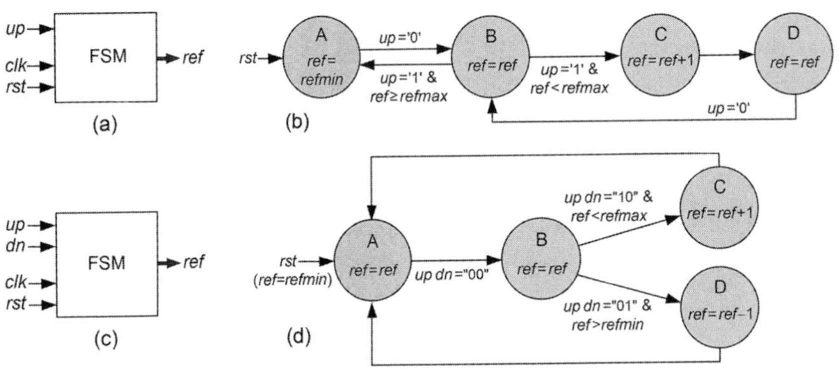
up
clk
rst
FSM
ref
(a)
rst
A
ref=
refmin
up='0'
up='1' &
ref≥refmax
B
ref=ref
up='1' &
ref<refmax
C
ref=ref+1
D
ref=ref
up='0'
(b)
up
dn
clk
rst
FSM
ref
(c)
rst
(ref=refmin)
A
ref=ref
up dn="00"
B
ref=ref
up dn="10" &
ref<refmax
C
ref=ref+1
up dn="01" &
ref>refmin
D
ref=ref-1
(d)

Figure 11.7
Setting a reference value (for any set size). (a, b) Up only. (c, d) Up and down.

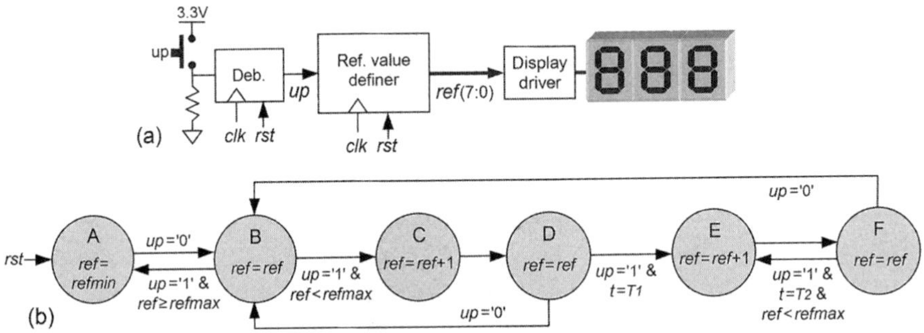
3.3V
up
Deb.
up
clk rst
Ref. value
definer
clk rst
ref(7:0)
Display
driver
(a)
rst
A
ref=
refmin
up='0'
up='1' &
ref≥refmax
B
ref=ref
up='1' &
ref<refmax
C
ref=ref+1
D
ref=ref
up='0'
up='1' &
t=T1
E
ref=ref+1
up='1' &
t=T2 &
ref<refmax
F
ref=ref
up='0'
(b)

Figure 11.8
Practical application for a reference-value definer with a large number of states and timed transitions.

The second case, presented in figures 11.7c,d, has up and down controls. Again, the output must range from ref_{min} to ref_{max}, with *up* and *dn* causing *ref* to be incremented or decremented, respectively. When one of these limits is reached, the machine must remain there until a movement (with *up* or *dn*) in the opposite direction is provoked. A possible solution is depicted in figure 11.7d. Again, the number of states is just four, regardless of the number of reference values. Similarly to the previous case, here too there are states (C and D) that must last only one clock period.

A practical application is presented in figure 11.8a, where *up* is produced by a pushbutton (after a debouncing circuit—see sections 8.11.3 and 11.7.4) and *ref* (reference value) is an eight-bit value, thus capable of operating anywhere in the 0-to-255 range. Every time the pushbutton is pressed (and released), *ref* must be incremented by one unit; however, if the pushbutton is kept pressed for t_1 = 2 s (T_1 clock periods) or longer, the increment must occur automatically and at every t_2 = 0.5 s (T_2 clock periods). If the maximum value is reached, the machine must stop and hold that value, only returning to the initial state if the pushbutton is released and pressed again.

A solution (without the debouncer) is depicted in figure 11.8b, which is simply the basic building block of figure 11.7b plus two extra states (E, F), added to take care of the time-related specifications. Note that the initial and final values can be chosen freely by the designer and that the CD and EF transitions are unconditional. Again, the machine size is independent of the number of reference values and of the time values used in the timed transitions (the time values only affect the size of the counter that implements the timer).

Even though *up* is an asynchronous input in figure 11.8, a synchronizer (section 2.3) is not needed because a debouncer was included in the circuit (and the application might not be critical anyway).

Based on section 11.6, the number of flip-flops needed to build the FSM of figure 11.8b is as follows. For the state register: M_{FSM} = 6 states, so N_{FSM} = 3 if sequential, Gray, or Johnson encoding is used, or 6 for one-hot. For the auxiliary register: needed for *ref*; because it is an eight-bit value, N_{aux} = 8. For the optional output register: not needed, so N_{output} = 0. (If needed, the auxiliary register could be used for that because it contains *ref* anyway.) For the timer: because t_{state_max} = 2 s, and assuming f_{clk} = 50 MHz, $T_{max} = 10^7$ clock cycles results, so N_{timer} = 27. Therefore, N_{total} = 38 or 41 DFFs.

11.7.4 Reference-Value Definer with Embedded Debouncer

This section is an extension to the section above. Because in many control applications reference values are set by means of mechanical switches, which might require some sort of debouncer (section 8.11.3), we want to examine the possibility of embedding the debouncer directly into the reference-value definer circuit.

Three possible situations are depicted in figure 11.9: (a) with debouncers implemented as two separate circuits; (b) with the debouncers combined into a single circuit; (c) with the debouncers embedded in the FSM that implements the reference-value definer. The case in a was seen in section 8.11.3; that in b was treated in exercises 8.11 and 8.12; and that in c is discussed in this section.

The general debouncing principle seen in section 8.11.3 is summarized in figure 11.10a (with a simplified representation—see figure 1.4), which says that for the output to change from '0' to '1' the input must remain high during T consecutive clock cycles (recall that the timer is zeroed every time the machine changes its state, so if a '0' occurs before the time has been completed, the machine returns to the initial state, restarting the timer).

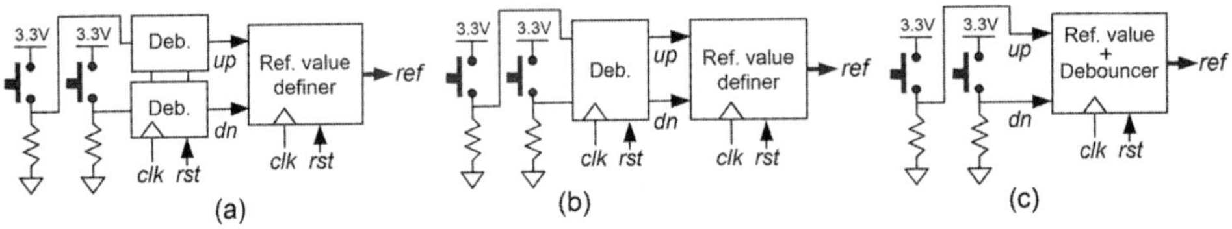
3.3V
3.3V
Deb.
Deb.
up
dn
Ref. value definer
ref
clk rst
clk rst
(a)
Deb.
up
dn
Ref. value definer
ref
clk rst
clk rst
(b)
Ref. value + Debouncer
up
dn
ref
clk rst
(c)

Figure 11.9
Reference-value definer with up and down controls set by two pushbuttons having the debouncers (a) implemented as two separate circuits, (b) implemente as a combined circuit, and (c) embedded into the main FSM.

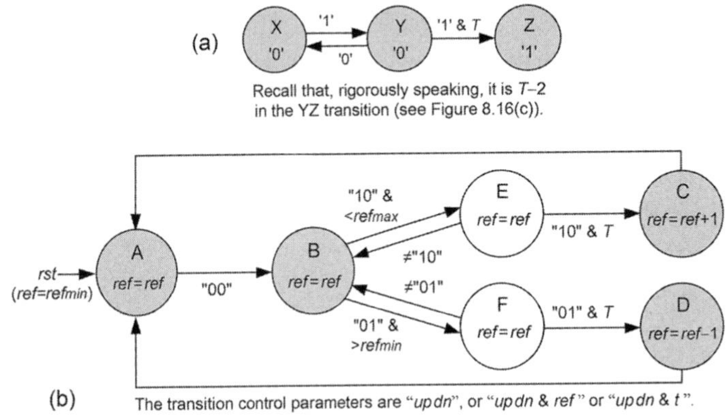
(a)
X '0'
Y '0'
Z '1'
'1'
'0'
'1' & T
Recall that, rigorously speaking, it is T–2 in the YZ transition (see Figure 8.16(c)).
rst
(ref=refmin)
A ref=ref
"00"
B ref=ref
"10" & <refmax
≠"10"
≠"01"
"01" & >refmin
E ref=ref
F ref=ref
"10" & T
"01" & T
C ref=ref+1
D ref=ref–1
(b)
The transition control parameters are "updn", or "updn & ref" or "updn & t".

Figure 11.10
(a) Review of the general debouncing principle. (b) Machine of figure 11.7d with embedded debouncer (for the '0'-to-'1' transition only).

This principle was applied to the '0'-to-'1' ('1'-to-'0' not included) transitions of figure 11.7d, resulting in the state diagram of figure 11.10b. Note the white circles between states BC and BD, related to the debouncing procedure.

For an analysis of the number of flip-flops, see exercise 11.6. For another implementation, concerning the case of figure 11.7b, see exercise 11.7.

11.7.5 Datapath Control for a Sequential Multiplier

Before we examine this example, a review of Section 3.13 is useful. Particular attention should be paid to comment 4 at the end of that section, which is helpful here.

Figure 11.11a presents an algorithm for unsigned sequential multiplication using only add and shift operations. It computes the product in N iterations (after a data-load operation), where N is the number of bits in the multiplier and multiplicand, and $2N$ is the number of bits in the product. Note that the product is divided into

Iteration	Procedure	Multiplicand *mult*	Product *carry prodL prodR*
–	Load data	1100	0 0000 101 1
1	*prod*(0) = 1 → *prodL* = *prodL* + *mult* Shift right *prod* with carry		0 1100 101 1 0 0110 010 1
2	*prod*(0) = 1 → *prodL* = *prodL* + *mult* Shift right *prod* with carry		1 0010 010 1 0 1001 001 0
3	*prod*(0) = 0 → no operation Shift right *prod* (with carry)		 0 0100 100 1
4	*prod*(0) = 1 → *prodL* = *prodL* + *mult* Shift right *prod* (with carry)		1 0000 100 1 0 1000 010 0

(a)

(b)

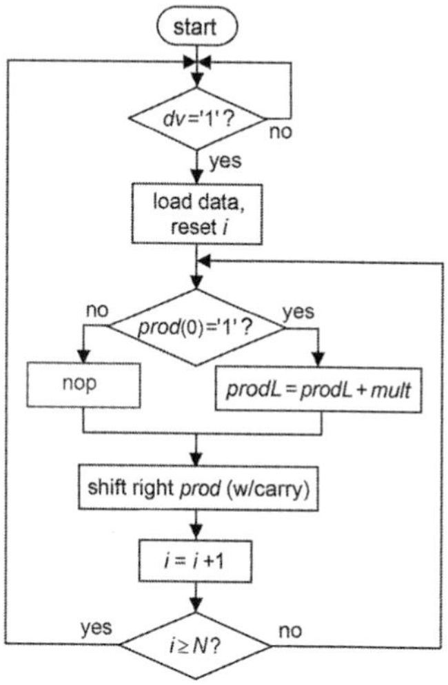
start
dv='1'?
no
yes
load data, reset i
no
prod(0)='1'?
yes
nop
prodL = prodL + mult
shift right prod (w/carry)
i = i +1
yes
i ≥ N?
no

Figure 11.11
Sequential add-and-shift multiplier. (a) Algorithm. (b) Flowchart.

two halves, called *prodL* (product left) and *prodR* (product right). In this example the inputs are "1100" (multiplicand = 12) and "1011" (multiplier = 11), so the expected result is "10000100" (product = 132).

Initially, the multiplicand is stored in a (fixed) register, and the multiplier is loaded into *prodR*, with *prodL* loaded with zeros. The algorithm checks the LSB (least significant bit) of the product; if it is '0', the product register is simply shifted to the right one position (empty position filled with the carry bit); if, however, it is '1', *mult* is added (with carry) to *prodL* before the shift operation is executed. After N iterations the product will be available in the product register.

The algorithm is described in ASM form in figure 11.11b. A data-valid bit (dv = '1' during one clock period) is used to tell the circuit when the computation should start. The algorithm runs N times (for $i = 0$ to $N - 1$), so when $i = N$ occurs the algorithm returns to the beginning, ready to start a new computation when dv is asserted again. Note that a nop (no operation) stage was included in the left branch to consume one clock cycle, so the computations will always take a fixed amount of time (depending on the application, the nop stage can be suppressed). Observe in the flowchart the recursive equation $i = i + 1$, which characterizes a category 3 FSM.

Figure 11.12a shows the parts of a datapath used to implement this multiplier, consisting of an ALU, two registers (REG1, REG2), and a multiplexer (MUX). It is assumed that it is a 16-bit system. The control unit (FSM) must generate the signals *wrR1* and *wrR2* (to enable writing into REG1 and REG2, respectively), *sel* (for mux input selection), *ALUop* (to control the ALU operation), and *shift* (to shift REG2 to the

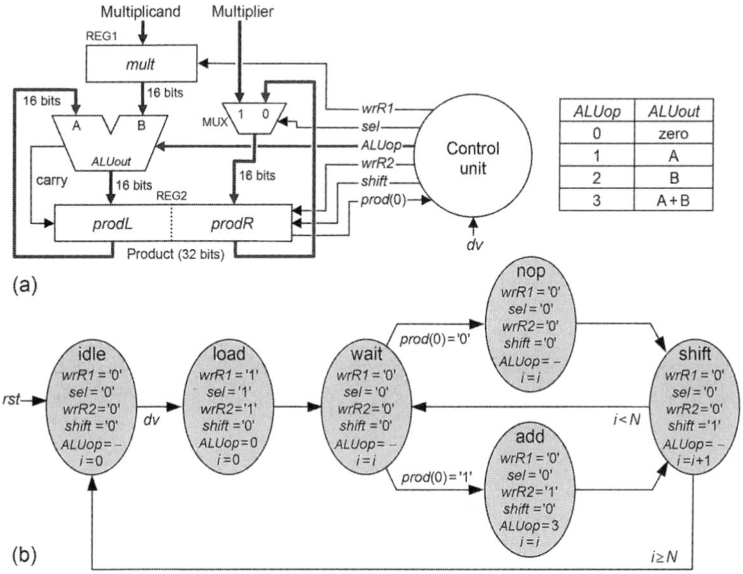
Multiplicand
Multiplier
REG1
mult
16 bits
16 bits
A
B
MUX
1 0
ALUout
carry
16 bits
REG2
16 bits
prodL
prodR
Product (32 bits)
wrR1
sel
ALUop
wrR2
shift
prod(0)
Control unit
dv
ALUop ALUout
0 zero
1 A
2 B
3 A+B
(a)
idle
wrR1 = '0'
sel = '0'
wrR2='0'
shift ='0'
ALUop= –
i=0
rst
dv
load
wrR1 = '1'
sel = '1'
wrR2='1'
shift ='0'
ALUop=0
i=0
wait
wrR1 = '0'
sel = '0'
wrR2='0'
shift ='0'
ALUop= –
i=i
prod(0)='0'
nop
wrR1 = '0'
sel = '0'
wrR2='0'
shift ='0'
ALUop= –
i=i
prod(0)='1'
add
wrR1 = '0'
sel = '0'
wrR2='1'
shift ='0'
ALUop=3
i=i
shift
wrR1 = '0'
sel = '0'
wrR2='0'
shift ='1'
ALUop= –
i=i+1
i<N
i≥N
(b)

Figure 11.12
Sequential add-and-shift multiplier. (a) Datapath. (b) Control unit implementation.

right one position, with carry). The ALU opcode table is assumed to be that included in the figure.

The multiplication starts when the control unit receives *dv* = '1' (during one clock period), at which time it enables REG1 (by means of *wrR1* = '1') to store (at the next positive clock edge) the multiplicand, and REG2 (by means of *ALUop* = 0, *sel* = '1', and *wrR2* = '1') to store zero in *prodL* and the multiplier in *prodR*. After this, *wrR1* stays low until the end of the computations, while *wrR2* is asserted at the end of each iteration to enable the storage of *ALUout* into *prodL*, after which *shift* = '1' is produced to shift REG2 one position to the right. After *N* of such iterations, the product will be available in REG2.

A Moore machine that implements the control unit of figure 11.12(a) is presented in figure 11.12(b), which is a direct translation of the algorithm described above. Observe the inclusion of a wait state, needed for the reason explained in comment 4 at the end of section 3.13.

VHDL and SystemVerilog implementations for this multiplier are presented in sections 12.4 and 13.4, respectively.

11.7.6 Sequential Divider

This section describes a state machine capable of sequentially computing the division *num*/*den* (numerator/denominator), producing the corresponding quotient (*quot*) and

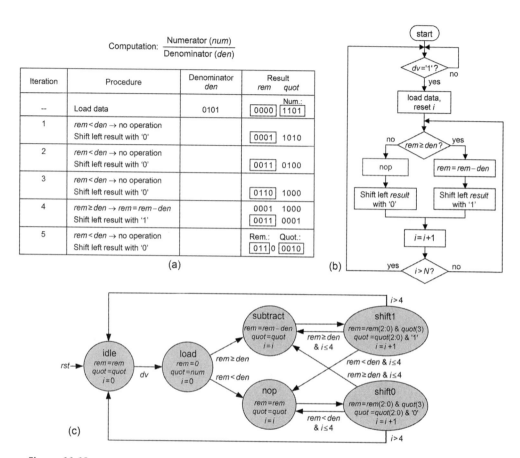

Computation: $\frac{\text{Numerator } (num)}{\text{Denominator } (den)}$

Iteration	Procedure	Denominator *den*	Result *rem* *quot*
--	Load data	0101	Num.: 0000 1101
1	*rem*<*den* → no operation Shift left result with '0'		0001 1010
2	*rem*<*den* → no operation Shift left result with '0'		0011 0100
3	*rem*<*den* → no operation Shift left result with '0'		0110 1000
4	*rem*≥*den* → *rem*=*rem*−*den* Shift left result with '1'		0001 1000 0011 0001
5	*rem*<*den* → no operation Shift left result with '0'		Rem.: Quot.: 0110 0010

(a)

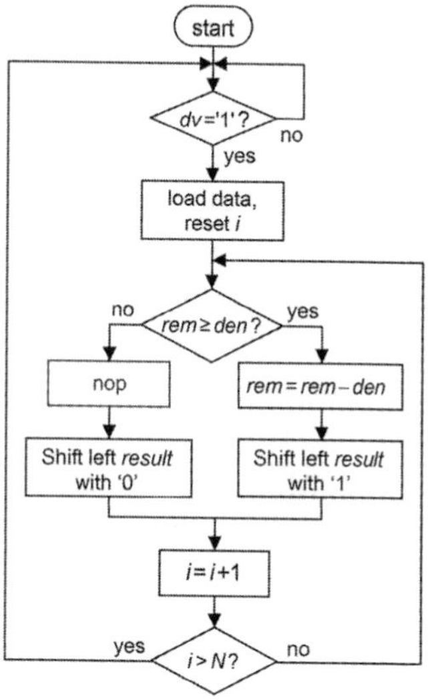
start
dv='1'?
no
yes
load data, reset i
rem≥den?
no
yes
nop
rem=rem−den
Shift left result with '0'
Shift left result with '1'
i=i+1
yes
i>N?
no

(b)

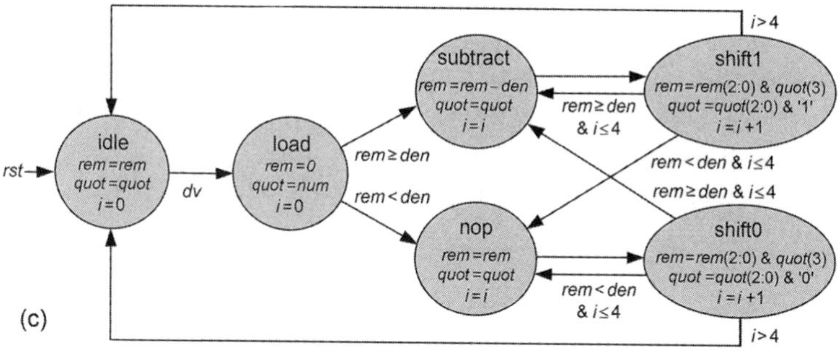
i>4
subtract
rem=rem−den
quot=quot
i=i
shift1
rem=rem(2:0) & quot(3)
quot=quot(2:0) & '1'
i=i+1
rem≥den
& i≤4
idle
rem=rem
quot=quot
i=0
rst
dv
load
rem=0
quot=num
i=0
rem≥den
rem<den
rem<den & i≤4
rem≥den & i≤4
nop
rem=rem
quot=quot
i=i
shift0
rem=rem(2:0) & quot(3)
quot=quot(2:0) & '0'
i=i+1
rem<den
& i≤4
i>4

(c)

Figure 11.13
Complete sequential divider. (a) Algorithm. (b) Flowchart. (c) A possible implementation (for $N = 4$).

remainder (*rem*) values. Contrary to the previous section, here a datapath is not employed, so the machine is a complete divider, not a control unit.

The division algorithm, for unsigned inputs and employing only subtract and shift operations, is illustrated in figure 11.13(a). The computations take $N+1$ iterations (after a data-load operation), where N is the number of bits in all four signals (*num, den, quot, rem*). In this example the inputs are *num* = "1101" (= 13) and *den* = "0101" (= 5), so the expected results are *quot* = "0010" (= 2) and *rem* = "0011" (= 3).

Initially, the denominator is stored in a (fixed) register, while the numerator is loaded into the quotient register, and the remainder is loaded with zeros. The algorithm checks whether *rem*≥*den*; if yes, *den* is subtracted from *rem*, and the entire result (*rem* and *quot*) is shifted to the left one position with the empty position filled with a '1'; otherwise, no subtraction occurs, and the result is shifted to the left with a '0' in the empty position. After $N + 1$ iterations, the final result will be available. Note that the actual value of *rem* does not include its LSB.

The algorithm is described in ASM form in figure 11.13b. A data-valid bit (*dv* = '1' during one clock period) is used to tell the circuit when the computation should start. The algorithm runs $N + 1$ times (for $i = 0$ to N), so when $i = N + 1$ occurs, the algorithm returns to the beginning, ready to start a new computation when *dv* is asserted again. As in the previous section, an optional nop (no operation) stage was included in the left branch to consume one clock cycle, so the computations will always take a fixed amount of time.

A Moore machine that implements the complete divider is presented in figure 11.13c (note that $N = 4$ in this example). In the *load* state, *rem* is zeroed and *quot* is loaded with *num*. If *rem*≥*den*, the machine moves to state *subtract*, in which *rem–den* occurs, followed by state *shift1*, responsible for shifting the data one position to the left with a '1' included in the rightmost position (following VHDL notation, "&" means concatenation in the expression *rem* = *rem*(2:0) & *quot*(3), meaning that *rem*(3:1) = *rem*(2:0) and *rem*(0) = *quot*(3); the expression *quot* = *quot*(2:0) & '1' has a similar meaning). On the other hand, if *rem*<*den* when the machine is in *load*, it goes through the *nop* state, followed by state *shift0*, responsible for shifting the result one position to the left with a '0' included in the rightmost position. Observe the presence of recursive equations (*quot* = *quot*, $i = i + 1$, etc.) in several states, which characterize a category 3 FSM.

11.7.7 Serial Data Receiver

This section shows another application that can be solved using a category 3 machine. It consists of a serial data receiver, which must store the received (one bit at a time) data in a multibit register. Even though this kind of circuit is simple, so it can be implemented without the FSM approach, we want to see how it can be modeled as a state machine (recall that we should be able to model any sequential circuit as an FSM).

The circuit ports are depicted in figure 11.14a. The inputs are *x* (serial bit stream), *dv* (data-valid bit, high during only one clock cycle, informing that data storage should start), plus the conventional clock and reset. The received data must be stored in *y*, which is an *N*-flip-flop register. A signal called *done* is also shown, which informs when the machine is free to receive/store another serial vector.

It will be assumed that the first bit of *x* is made available at the same time that *dv* is asserted, which is more difficult to implement. Because this kind of problem was already treated in section 3.10, a review of that section is recommended before proceeding. Indeed, two solutions for this problem were already presented in figures 3.16c,e, using a timed machine.

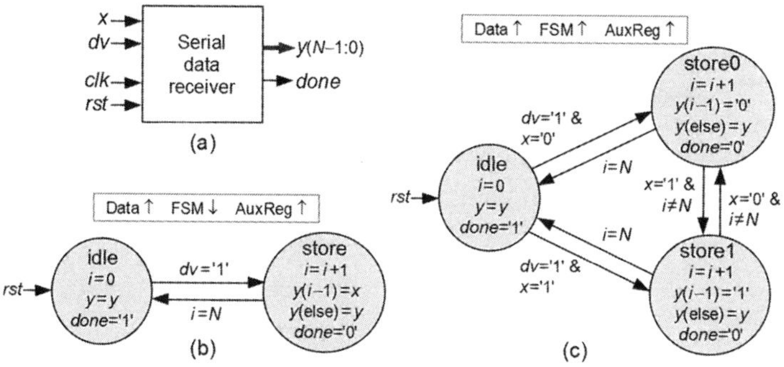
x
dv
clk
rst
Serial data receiver
y(N–1:0)
done
(a)
Data ↑ FSM ↓ AuxReg ↑
rst
idle
i=0
y=y
done='1'
dv='1'
i=N
store
i=i+1
y(i–1)=x
y(else)=y
done='0'
(b)
Data ↑ FSM ↑ AuxReg ↑
rst
idle
i=0
y=y
done='1'
dv='1' &
x='0'
i=N
store0
i=i+1
y(i–1)='0'
y(else)=y
done='0'
x='1' &
i≠N
x='0' &
i≠N
i=N
dv='1' &
x='1'
store1
i=i+1
y(i–1)='1'
y(else)=y
done='0'
(c)

Figure 11.14
Serial data receiver. (a) Circuit ports. (b) Solution based on figure 3.16c (FSM operating at the negative clock edge). (c) Solution with all units operating at the same clock edge.

Two additional solutions are presented in figure 11.14, this time using category 3 machines, both of Moore type and without timed transitions. The FSM of figure 11.14b was based directly on that of figure 3.16c, with the timer (t) replaced with a pointer (i). As indicated in the rectangle above the state machine, the data is updated at the positive clock edge, which is also the edge that causes the storage of i and y in auxiliary registers, whereas the FSM operates at the negative clock transition.

A final solution is presented in figure 11.14c, operating with the default clocking scheme (everybody operating at the same clock edge). In this case the first bit of x is not lost because it is part of the transition conditions (observe the *idle-store0* and *idle-store1* transitions, the first for x = '0', the second for x = '1').

11.7.8 Memory Interface

We want to develop a circuit for the memory interface of figure 11.15a, which must write data to an asynchronous SRAM chip. The only nonoperational input is *dv* (data valid), and the outputs are *A* (address at which the data must be stored), *OEn* (output enable, active low), *CEn* (chip enable, active low), and *WEn* (write enable, also active low). The actual memory-write command, internal to the SRAM, normally corresponds to the overlap between *CEn* and *WEn*. As in the previous example, this too is a simple circuit, but it is important to understand how it can be modeled as a finite state machine.

Figure 11.15b shows a possible (conservative) memory-write sequence. All signals are updated/produced at positive clock edges. As usual, a small propagation delay is included between clock transitions and corresponding responses in order to portray a more realistic situation. In this example it is assumed that writing occurs only while *dv* is high. When *dv* is raised, the circuit lowers *CEn* and *WEn*, causing *D0* to be

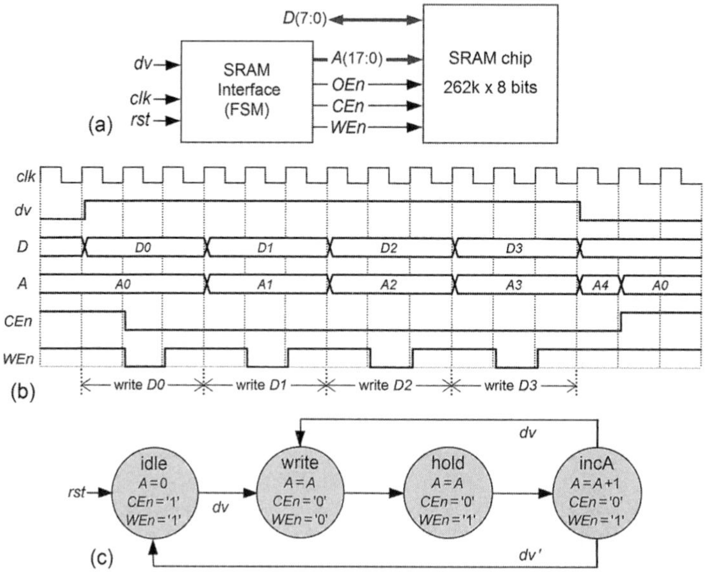
D(7:0)
dv
clk
rst
(a)
SRAM
Interface
(FSM)
A(17:0)
OEn
CEn
WEn
SRAM chip
262k x 8 bits
clk
dv
D
A
CEn
WEn
D0
D1
D2
D3
A0
A1
A2
A3
A4
A0
(b)
write D0
write D1
write D2
write D3
rst
idle
A=0
CEn='1'
WEn='1'
dv
write
A=A
CEn='0'
WEn='0'
hold
A=A
CEn='0'
WEn='1'
incA
A=A+1
CEn='0'
WEn='1'
dv
dv'
(c)

Figure 11.15
FSM implementing a memory-write procedure for an asynchronous SRAM. (a) Circuit ports. (b) Illustrative timing diagram (note that here writing occurs while *dv* is high). (c) Corresponding state machine.

stored at the initial memory address, after which *WEn* is raised, disabling further writing. Each subsequent iteration consists of three clock cycles, during which the memory address is updated and then another write-enable pulse (*WEn* = '0') is applied to the circuit. When *dv* returns to '0', the address is reset to zero (or to any other initial value). These operations can be done with *OEn* permanently high (thus not shown).

A Moore-type state machine capable of implementing this sequence of events is presented in figure 11.15c, which is a direct translation of the timing diagram of figure 11.15b. The address is updated in state *incA*, which increments the value of *A*. Note the recursive expressions $A = A$ and $A = A + 1$, which characterize a category 3 FSM.

An example involving an actual SRAM chip is depicted in figure 11.16. The SRAM (IS61LV25616 device, from ISSI) is shown in figure 11.16a. It can store 262 kwords of 16 bits each, hence requiring an 18-bit address bus, *A*(17:0), and a 16-bit data bus, *D*(15:0). It also contains five control signals, all active low, called *CEn*, *WEn*, *OEn*, *UBn* (upper byte enable), and *LBn* (lower byte enable).

A memory-write procedure based on this device's truth table and time parameters is presented in the left half of figure 11.16b. Note that it is less conservative than that in figure 11.15b (the end of the *WEn* pulse coincides with the beginning of a new

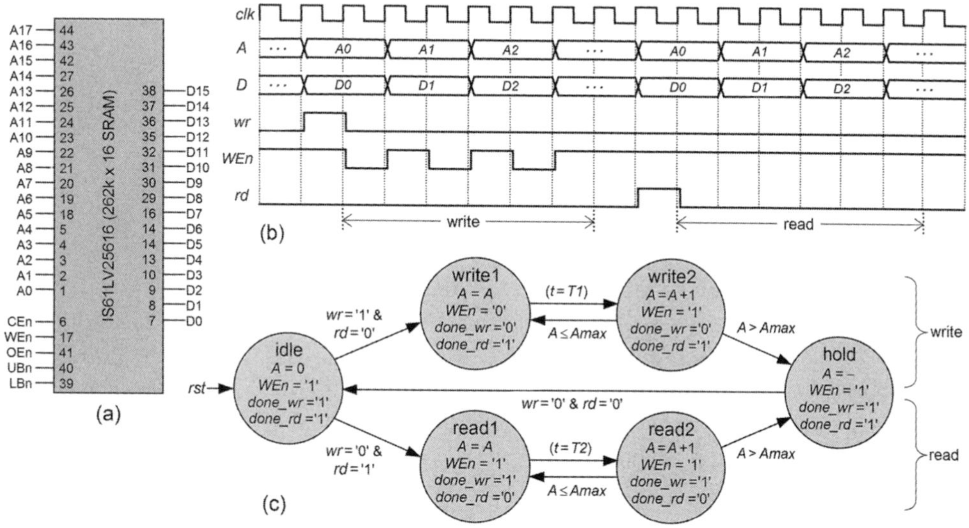
IS61LV25616 (262k x 16 SRAM)
A17 44
A16 43
A15 42
A14 27
A13 26
A12 25
A11 24
A10 23
A9 22
A8 21
A7 20
A6 19
A5 18
A4 5
A3 4
A2 3
A1 2
A0 1
CEn 6
WEn 17
OEn 41
UBn 40
LBn 39
38 D15
37 D14
36 D13
35 D12
32 D11
31 D10
30 D9
29 D8
16 D7
14 D6
14 D5
13 D4
10 D3
9 D2
8 D1
7 D0
(a)
clk
A
D
wr
WEn
rd
A0
A1
A2
D0
D1
D2
write
read
(b)
idle
A = 0
WEn = '1'
done_wr ='1'
done_rd ='1'
rst
wr='1' &
rd='0'
write1
A = A
WEn = '0'
done_wr ='0'
done_rd ='1'
(t=T1)
A≤Amax
write2
A=A+1
WEn = '1'
done_wr ='0'
done_rd ='1'
A > Amax
hold
A = –
WEn = '1'
done_wr ='1'
done_rd ='1'
wr='0' & rd='0'
wr='0' &
rd='1'
read1
A = A
WEn = '1'
done_wr ='1'
done_rd ='0'
(t=T2)
A≤Amax
read2
A=A+1
WEn = '1'
done_wr ='1'
done_rd ='0'
A > Amax
write
read
(c)

Figure 11.16

FSM implementing memory-write and memory-read procedures for an actual 262k × 16 SRAM. (a) Chip pinout. (b) Illustrative timing diagram (here, *wr* and *rd* are short pulses). (c) Corresponding state machine.

memory-write cycle). The largest read/write time parameter is 10 ns, so a clock of up to 100 MHz can be used in the procedure shown in the figure. Finally, note that the *wr* command lasts just one clock cycle, so the end of writing is determined by a predefined maximum address value.

A memory-read procedure is presented in the right half of figure 11.16b. When the device is not in write mode (write is done with *WEn* low), it is automatically in read mode, so when the *rd* command (which also lasts only one clock cycle) occurs, all that is needed is to do the address sweep.

A complete FSM for writing to (upper branch) and reading from (lower branch) this device is presented in figure 11.16c. If a *wr* = '1' pulse occurs, data is written to the SRAM from address $A = 0$ (or any other initial value) up to $A = A_{max}$. A similar situation occurs for reading when an *rd* = '1' pulse is received. Note that state *hold* is important to prevent overwriting or overreading in case *wr* or *rd* is too long. The signals *done_wr* and *done_rd* were included to inform the user when writing or reading has been completed. Note also the inclusion of $t = T_1$ and $t = T_2$ in two of the transitions, which indicate a way of reducing the write/read speed if that is desired.

A complete design for this memory interface, using VHDL and SystemVerilog, is presented in sections 12.6 and 13.6, respectively. The number of flip-flops is treated in exercise 11.15.

11.8 Exercises

Exercise 11.1: Machines Category

a) Why are the state machines in figures 11.5, 11.6, and 11.7 (among others) said to be of category 3?
b) What types of transitions (section 1.6) can category 3 machines have?
c) What differentiates category 3 from categories 1 and 2?

Exercise 11.2: Generic Counter with a Stop Value

Say that we must design a counter that starts at x_{min} and stops (and remains there) when x_{max} is reached, only returning to the initial value and running again after a reset pulse is applied to the circuit. As in section 11.7.1, the counter must have an enable input (*ena*) that allows the counter to run when asserted or holds it otherwise.

a) Draw a Moore-type state transition diagram for this counter modeled as a category 3 machine.
b) Does the number of states depend on the counting range?
c) Does the number of flip-flops depend on the counting range? How many are needed to build your machine with $x_{min} = 1$ and $x_{max} = 200$?
d) Is it advantageous or necessary to use the FSM approach to design counters in general?

Exercise 11.3: Hamming-Weight Calculator

The circuit of figure 11.17 must determine the Hamming weight (number of '1's) of a *serial* bit vector *x*. The vector is delimited by a data-valid bit (the counting must occur during all the time while dv = '1'). Study the illustrative timing diagram included in the figure. Observe that *dv* and *x* (= "100110101," so $N = 9$) are updated at positive clock edges and that the FSM too operates at positive clock edges (see the plot for *y*). As usual, small propagation delays were included to portray a more realistic situation.

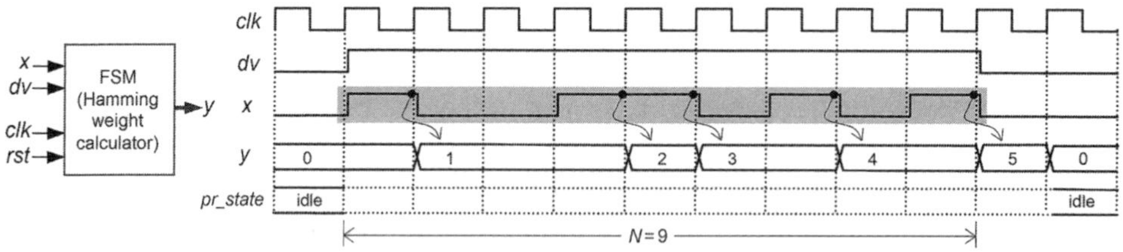
x
dv
clk
rst
FSM
(Hamming
weight
calculator)
y
clk
dv
x
y
0
1
2
3
4
5
0
pr_state
idle
idle
N=9

Figure 11.17

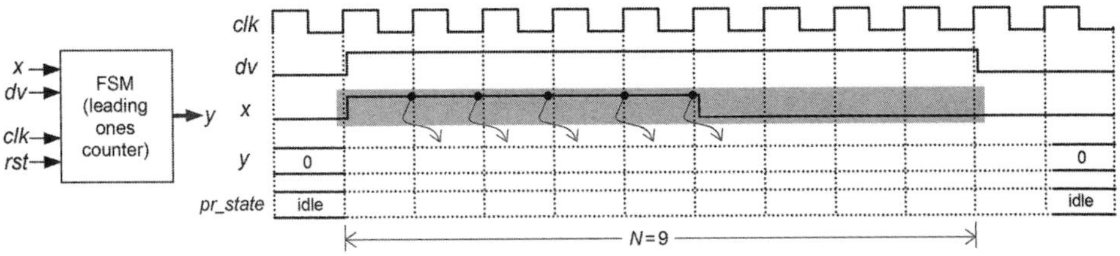
x
dv
clk
rst
FSM
(leading
ones
counter)
y
clk
dv
x
y
pr_state
0
idle
0
idle
N=9

Figure 11.18

a) Based on the given data, draw a Moore-type state transition diagram for this problem. Include a reset signal but assume that it can be asserted only at power-up.
b) Based on your state diagram, fill in the waveform for *pr_state* in the figure.
c) Redo part a assuming now that a reset pulse is applied to the FSM before each new computation starts. Can you find a solution with fewer states than in a?
d) Draw an illustrative timing diagram, similar to that in figure 11.17, for the FSM developed in part c.
e) How many DFFs are needed to build each machine developed above, assuming that sequential encoding is used and that *x* is a 32-bit vector (so *y* can go from 0 to 32)?

Exercise 11.4: Leading-Ones Counter

The circuit of figure 11.18 must count the number of '1's before a '0' is found in a *serial* bit vector *x*. The vector is delimited by a data-valid bit (the counting must occur during all the time while *dv* = '1'). Study the illustrative timing diagram included in the figure. Observe that *dv* and *x* (= "111110000," so *N* = 9) are updated at positive clock edges, which are the same edges at which the FSM must operate.

a) Draw a state transition diagram for this machine.
b) Based on your machine, complete the plots for *y* and *pr_state* in the figure.
c) Say that we want the output value to remain stable (constant) during the computations, with the current value replaced only when a new value is ready. How can that be done? (Suggestion: see section 3.11.)

Exercise 11.5: Long-String Comparator

Develop an FSM that detects if two *serial* bit streams *a* and *b* of length *N* are pair-wise equal. This is an extension to the example of section 11.7.2 in which the FSM had to detect if the *last* N bits were equal. The circuit ports are depicted in the upper part of figure 11.19, which also shows an XNOR gate (*x* = '1' when *a* = *b*). The desired behavior is also illustrated in the figure for *N* = 4. Note in the *y* and *done* waveforms that after every four bits, starting right after the reset pulse, *done* must be asserted, informing that a complete block has been inspected, with *y* high during that pulse if the four pairs of bits were equal (*x* = '1' in all four time slots) or low otherwise.

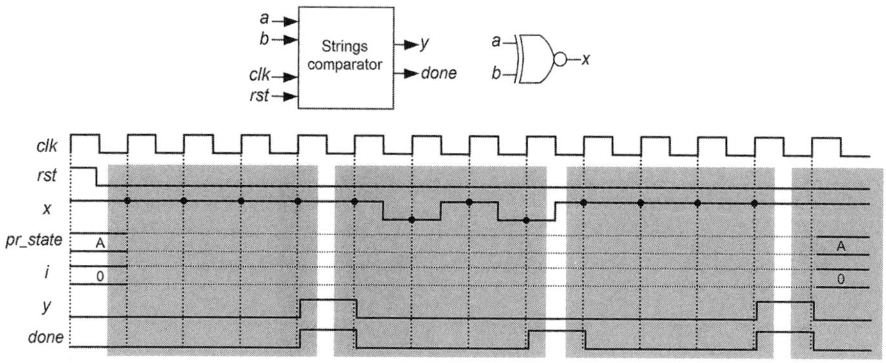
a
b
clk
rst
Strings comparator
y
done
a
b
x
clk
rst
x
pr_state
i
y
done
A
0
A
0

Figure 11.19

a) Draw a state transition diagram for a machine that solves this problem for any (arbitrarily long) value of N.

b) Based on your solution, fill in the missing plots in figure 11.19.

c) How many DFFs are needed to build your machine, assuming that sequential encoding is used and that $N = 256$ bits?

Exercise 11.6: Reference-Value Definer with Embedded Debouncer #1

This exercise concerns the reference-value definer with embedded debouncer seen in figure 11.10b.

a) Assuming that *ref* is an eight-bit signal, regular sequential encoding is used for the FSM, the debouncing time interval is 1 ms, and f_{clk} = 50 MHz, calculate the number of flip-flops needed to build that circuit.

b) The inputs *up* and *dn* are asynchronous. Is a synchronizer (section 2.3) needed in this application?

Exercise 11.7: Reference-Value Definer with Embedded Debouncer #2

We saw in figure 11.10b an FSM that embeds, in the reference-value definer of figure 11.7d, a pair of debouncers for the *up* and *dn* pushbuttons.

a) Using the same principle, modify the state transition diagram of figure 11.7b in order to include in it a debouncer for the '0'-to-'1' transitions of *up*.

b) Determine the number of DFFs needed to implement your FSM, assuming that sequential encoding is used, *ref* is an eight-bit signal, the debouncing interval is 1 ms, and f_{clk} = 50 MHz.

Exercise 11.8: Greatest Common Divisor

The algorithm and a corresponding flowchart for calculating the greatest common divisor (GCD) between two integers *a* and *b* are presented in figure 5.12. A data-valid

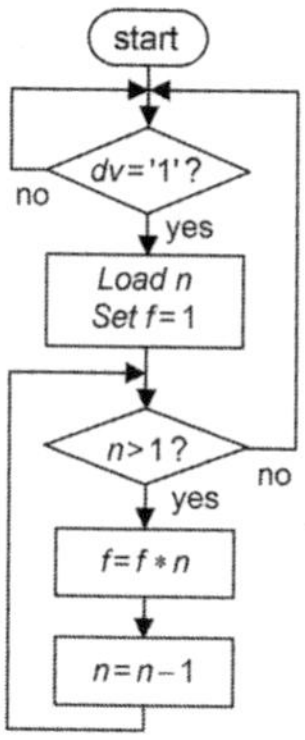
start
dv='1'?
no
yes
Load n
Set f=1
n>1?
no
yes
f=f*n
n=n-1

Figure 11.20

(*dv*) pulse, lasting only one clock period, informs when the computations must start. We want to redesign that machine, now without the datapath (so this is a *complete* GCD calculator). Note that the "load data" block of figure 5.12 is not indispensable here, but then the inputs must remain stable during the whole computations. Because the circuit will take a variable amount of time to compute the GCD (it depends on the input values), an output called *done* must be provided, which should remain high while the machine is idle. Draw a state transition diagram for an FSM capable of solving this problem.

Exercise 11.9: Factorial Calculator

An algorithm for calculating $f = n!$ ($n \geq 0$, integer) is described in the flowchart of figure 11.20. Assume that *dv* (data valid) is asserted during one clock cycle, indicating when the data (n) is ready, so the computation should commence. Because the circuit will take a variable amount of time to compute f (it depends on the value of n), an output called *done* must be provided, which should remain high while the machine is idle. Draw a state transition diagram for a Moore-type machine that solves this problem.

Exercise 11.10: Datapath Control for a Sequential Multiplier

This exercise concerns the datapath and corresponding control unit for multiplication using add-and-shift operations seen in figure 11.12.

a) How many flip-flops are needed to build the machine of figure 11.12b for $N = 4$ and for $N = 32$ bits?

b) Draw a timing diagram that illustrates its operation (as done in figure 5.13d, for example), for $N = 4$. Consider that the first four values of *prod*(0), after *dv* = '1' occurs, are *prod*(0) = {'1', '0', '1', '0'}.

Exercise 11.11: Datapath Control for a Sequential Divider

The algorithm and a corresponding flowchart for calculating the division *num*/*den* between two integers were presented in figure 11.13. In that case a complete divider was developed, whereas in this case we are interested in the same division but using a datapath (in other words, a *control unit* is needed here).

a) Based on the algorithm of figure 11.13 (and on the several examples using a datapath shown in chapters 3, 5, and 11), sketch a datapath that seems adequate for this problem.
b) Draw a state transition diagram for a (Moore) control unit such that the desired division is produced by your datapath.

Exercise 11.12: Serial Data Receiver

Two category 3 solutions for a serial data receiver were presented in figure 11.14.

a) Present an illustrative timing diagram for the solution in figure 11.14b, considering for *clk*, *dv*, and *x* the same waveforms of figure 3.16.
b) Do the same for the solution in figure 11.14c.

Exercise 11.13: Serial Data Transmitter

Two category 3 solutions for a serial data receiver (a deserializer) were presented in section 11.7.7.

a) Develop a category 3 solution for a serial data transmitter (a serializer).
b) Present an illustrative timing diagram for your FSM, for $N = 4$ and $x(3:0)$ = "1101."

Exercise 11.14: Memory Interface

Calculate the number of flip-flops needed to build the memory interface of figure 11.16c. Assume that sequential encoding is used for the FSM and that no timer is employed.

12 VHDL Design of Recursive (Category 3) State Machines

12.1 Introduction

This chapter presents several VHDL designs of category 3 state machines. It starts by presenting two VHDL templates, for Moore- and Mealy-based implementations, which are used subsequently to develop a series of designs related to the examples introduced in chapter 11.

The codes are always complete (not only partial sketches) and are accompanied by comments and often also simulation results illustrating the design's main features. All circuits were synthesized using Quartus II (from Altera) or ISE (from Xilinx). The simulations were performed with Quartus II or ModelSim (from Mentor Graphics). The default encoding scheme for the states of the FSMs was regular sequential encoding (see encoding options in section 3.7; see ways of selecting the encoding scheme at the end of section 6.3).

The same designs are presented in chapter 13 using SystemVerilog, so the reader can make a direct comparison between the codes.

Note: See suggestions of VHDL books in the bibliography.

12.2 VHDL Template for Recursive (Category 3) Moore Machines

The template is presented below. Because it is an extension to the Moore templates for categories 1 and 2, described in sections 6.3 and 9.2, respectively, a review of those templates is suggested before this one is examined because only the differences are described. Review also some possible code variations in section 6.4.

The only differences are those needed for the inclusion of an auxiliary register, compulsory in category 3 machines. As seen in section 6.2, the architecture is composed of two parts, the declarative part (before **begin**) and the statements part (from **begin** on); both have new elements in order to accommodate the auxiliary register.

In the architecture's declarative part (lines 14–21), the difference is in line 21, in which two signals are created to deal with the auxiliary register. It is assumed that there is only one output and that it must be stored, but recall that the circuit might have several outputs, not all registered. The actual number of auxiliary registers is determined by the number of outputs that depend on past values.

In the architecture's statements part (lines 23–75), two differences are seen: the inclusion of a process to infer the auxiliary register and the replacement of *outp* with *outp_reg* on the right-hand side of the recursive equations. The latter removes the recursiveness, thus allowing the output to be computed by a combinational circuit.

Lines 29–36 show the process that implements the auxiliary register. If one prefers, this process can be combined with that for the FSM's state register (a shorter code results, but less didactic, with no effect on the result).

Lines 42–68 show the process that implements the machine's combinational logic section. The only difference here is that *outp_reg*, instead of *outp* itself, appears on the right-hand side of the (originally) recursive equations (lines 46 and 56).

As explained in section 11.3, an interesting aspect of category 3 FSMs is that the auxiliary register can also play the role of output register (for glitch-free and/or pipelined construction). To do so, we simply send *outp_reg* out instead of *outp* in line 73.

The code is concluded in line 73, in which the value of *outp* is passed to the actual output. In fact, the actual output could be used in lines 34, 46, and 56, in which case the mode of *output* (in the entity) should be changed from *out* to *buffer* (see example in section 12.5).

```
1    ------------------------------------------------------------
2    library ieee;
3    use ieee.std_logic_1164.all;
4    ------------------------------------------------------------
5    entity circuit is
6       generic (
7          (same as for category 2 Moore, section 9.2)
8       port (
9          (same as for category 1 Moore, section 6.3)
10   end entity;
11   ------------------------------------------------------------
12   architecture moore_fsm of circuit is
13
14      --FSM-related declarations:
15      (same as for category 1 Moore, section 6.3)
16
17      --Timer-related declarations:
18      (same as for category 2 Moore, section 9.2)
19
20      --Auxiliary-register-related declarations:
```

```
   signal outp, outp_reg: std_logic_vector(...);

begin

   --Timer:
   (same as for category 2 Moore, section 9.2)

   --Auxiliary register:
   process (clk, rst)
   begin
      if (rst='1') then
         outp_reg <= <initial value>;
      elsif rising_edge(clk) then
         outp_reg <= outp;
      end if;
   end process;

   --FSM state register:
   (same as for category 2 Moore, section 9.2)

   --FSM combinational logic:
   process (all) --list proc. inputs if "all" not supported
   begin
      case pr_state is
         when A =>
            outp <= outp_reg;
            tmax <= T1-1;
            if <condition> then
               nx_state <= B;
            elsif <condition> then
               nx_state <= ...;
            else
               nx_state <= A;
            end if;
         when B =>
            outp <= outp_reg + 1;
            tmax <= T2-1;
            if <condition> then
               nx_state <= C;
            elsif <condition> then
               nx_state <= ...;
            else
               nx_state <= B;
            end if;
         when C =>
            ...
      end case;
   end process;

   --Optional output register:
   (same as for category 1 Moore, section 6.3)

   output <= outp;

end architecture;
------------------------------------------------------------
```

12.3 VHDL Template for Recursive (Category 3) Mealy Machines

The template is presented below. The only difference with respect to the Moore template just described is in the process for the combinational logic (lines 23–57) because the output is specified differently here. Recall that in a Mealy machine the output depends not only on the FSM's state but also on the input, so **if** statements are expected for the output in one or more states because the output value might not be unique.

Please review the following comments, which can be easily adapted from the Moore case to the Mealy case:

—On the Moore template for category 1, in section 6.3, especially comment 10.
—On the *enum_encoding* and *fsm_encoding* attributes, also in section 6.3.
—On possible code variations, in section 6.4.
—On the Mealy template for category 1, in section 6.5.
—On the Moore template for category 2, in section 9.2.
—On the Mealy template for category 2, in section 9.3.
—Finally, on the Moore template for category 3, in section 12.2.

```
1    -----------------------------------------------------------
2    library ieee;
3    use ieee.std_logic_1164.all;
4    -----------------------------------------------------------
5    entity circuit is
6       (same as for Moore, section 12.2)
7    end entity;
8    -----------------------------------------------------------
9    architecture mealy_fsm of circuit is
10      (same as for Moore, section 12.2)
11   Begin
12
13      --Timer:
14      (same as for Moore, section 9.2)
15
16      --Auxiliary register:
17      (same as for Moore, section 12.2)
18
19      --FSM state register:
20      (same as for Moore, section 9.2)
21
22      --FSM combinational logic:
23      process (all) --list proc. inputs if "all" not supported
24      begin
25         case pr_state is
26            when A =>
27               if <condition> then
28                  outp <= outp_reg;
29                  tmax <= <value>;
30                  nx_state <= B;
31               elsif <condition> then
```

```
               outp <= outp_reg + 1;
               tmax <= <value>;
               nx_state <= ...;
            else
               outp <= outp_reg;
               tmax <= <value>;
               nx_state <= A;
            end if;
         when B =>
            if <condition> then
               outp <= outp_reg;
               tmax <= <value>;
               nx_state <= C;
            elsif <condition> then
               outp <= outp_reg - 1;
               tmax <= <value>;
               nx_state <= ...;
            else
               outp <= outp_reg;
               tmax <= <value>;
               nx_state <= B;
            end if;
         when C =>
            ...
      end case;
   end process;

   --Optional output register:
   (same as for Moore, Section 12.2)

   output <= outp;

end architecture;
--------------------------------------------------------------
```

12.4 Design of a Datapath Controller for a Multiplier

This section presents a VHDL-based design for the control unit introduced in section 11.7.5, which controls a datapath to produce a sequential add-and-shift multiplier. The Moore template for category 3 machines seen in section 12.2 is used to implement the FSM of figure 11.12b.

The entity, called *control_unit_for_multiplier*, is in lines 5–11. The number of bits (N) in the multiplier and multiplicand was entered as a generic parameter (line 6); a small value ($N = 4$) was used to ease the inspection of the simulation results. Note that all ports (lines 8–10) are of type *std_logic* or *std_logic_vector* (industry standard).

The architecture, called *moore_fsm*, is in lines 13–93. As usual, it contains a declarative part and a statements part, with three processes in the latter.

The declarative part of the architecture (lines 15–20) contains FSM- and auxiliary-register-related declarations. In the former the enumerated type *state* is created to represent the machine's present and next states. In the latter the signals *i* and *i_reg* are created to deal with the auxiliary register (note that in this case none of the actual outputs is stored in an auxiliary register).

The first process (lines 25–32) implements the auxiliary register, exactly as in the template.

The second process (lines 35–42) implements the FSM's state register, again exactly as in the template.

The third and final process (lines 45–91) implements the entire combinational logic section. It is just a list of all states, each containing the output values and the next state. Note that because some of the output values get repeated several times, default values were entered in lines 48–53, so they only need to be included in the **case** statement when different values are required (see section 6.4.3). Observe that in the (originally) recursive equations (lines 68, 75, 80, and 84), *i_reg* appears on the right-hand side instead of *i* itself (as seen in the template). As usual, in each state the output value is unique because in a Moore machine the output depends only on the state in which the machine is.

In datapath-related designs, possible glitches at the output of the control unit during clock transitions are normally not a problem, so the optional output register is not employed.

Observe the correct use of registers and the completeness of the code, as described in comment 10 of section 6.3. Note in particular the following:

1) Regarding the use of registers: The circuit is not overregistered. This can be observed in the **elsif rising_edge(clk)** statement of line 39 (responsible for the inference of flip-flops), which is closed in line 41, guaranteeing that only the machine state (line 40) gets stored (the auxiliary register is a separate circuit, built in the preceding process). The outputs are in the next process, which is purely combinational (thus not registered).
2) Regarding the outputs: The list of outputs (*wrR1, sel, wrR2, shft, ALUop, i*) is exactly the same in all states, and the corresponding values/expressions are always properly declared (note that some values are declared in the default list of lines 48–53).
3) Regarding the next state: Again, the coverage is complete because all states are included (see lines 56, 62, 67, . . .), and in each state the conditional declarations for the next state are always finalized with an **else** statement (lines 59, 71, 87), guaranteeing that no condition is left unchecked.

```
1    -------------------------------------------------------------
2    library ieee;
3    use ieee.std_logic_1164.all;
4    -------------------------------------------------------------
5    entity control_unit_for_multiplier is
6       generic (N: natural := 4);    --number of bits (any >0)
7       port (
8          dv, prod, clk, rst: in std_logic;
```

```
      wrR1, sel, wrR2, shft: out std_logic;
      ALUop: out std_logic_vector(1 downto 0));
end entity;
-------------------------------------------------------------
architecture moore_fsm of control_unit_for_multiplier is

   --FSM-related declarations:
   type state is (idle, load, waitt, nop, add, shift);
   signal pr_state, nx_state: state;

   --Auxiliary-register-related declarations:
   signal i, i_reg: natural range 0 to N;

begin

   --Auxiliary register:
   process (clk, rst)
   begin
      if rst='1' then
         i_reg <= 0;
      elsif rising_edge(clk) then
         i_reg <= i;
      end if;
   end process;

   --FSM state register:
   process (clk, rst)
   begin
      if rst='1' then
         pr_state <= idle;
      elsif rising_edge(clk) then
         pr_state <= nx_state;
      end if;
   end process;

   --FSM combinational logic:
   process (all)
   begin
      --Default values:
      wrR1 <= '0';
      sel <= '0';
      wrR2 <= '0';
      shft <= '0';
      ALUop <= "00";
      i <= 0;
      --Case statement:
      case pr_state is
         when idle =>
            if dv='1' then
               nx_state <= load;
            else
               nx_state <= idle;
            end if;
         when load =>
            wrR1 <= '1';
            sel <= '1';
            wrR2 <= '1';
            nx_state <= waitt;
```

```
67           when waitt =>
68              i <= i_reg;
69              if prod='0' then
70                 nx_state <= nop;
71              else
72                 nx_state <= add;
73              end if;
74           when nop =>
75              i <= i_reg;
76              nx_state <= shift;
77           when add =>
78              wrR2 <= '1';
79              ALUop <= "11";
80              i <= i_reg;
81              nx_state <= shift;
82           when shift =>
83              shft <= '1';
84              i <= i_reg + 1;
85              if i<N then
86                 nx_state <= waitt;
87              else
88                 nx_state <= idle;
89              end if;
90        end case;
91     end process;
92
93  end architecture;
94  ------------------------------------------------------------
```

The number of flip-flops inferred by the compiler on synthesizing the code above, with regular sequential encoding (section 3.7), was six for $N = 4$ and nine for $N = 32$ bits. Compare these results against your predictions made in exercise 11.10.

Simulation results are shown in figure 12.1. Observe in the plot for *prod* that the circuit was tested for the sequence *prod* = {'1', '0', '1', '0'}, so the expected sequence of states is *pr_state* = {0, 1, 2, 4, 5, 2, 3, 5, 2, 4, 5, 2, 3, 5, 0}, which indeed occurs (recall that the states are enumerated in the order that they appear in line 16; however, some compilers reserve the value zero for the reset state, but that is not a concern here because that is the first state in our list anyway). Note that the values produced at the output in each state are exactly as in figure 11.12b. Finally, compare these simulation results against your sketch in exercise 11.10 to see whether they match.

12.5 Design of a Serial Data Receiver

This section presents a VHDL-based design for the serial data receiver introduced in section 11.7.7. The Moore template for category 3 machines seen in section 12.2 is used to implement the solution of figure 11.14c.

The entity, called *serial_data_receiver*, is in lines 5–11. The number of bits (N) is entered as a generic parameter (line 6). All ports (lines 8–10) are of type *std_logic* or

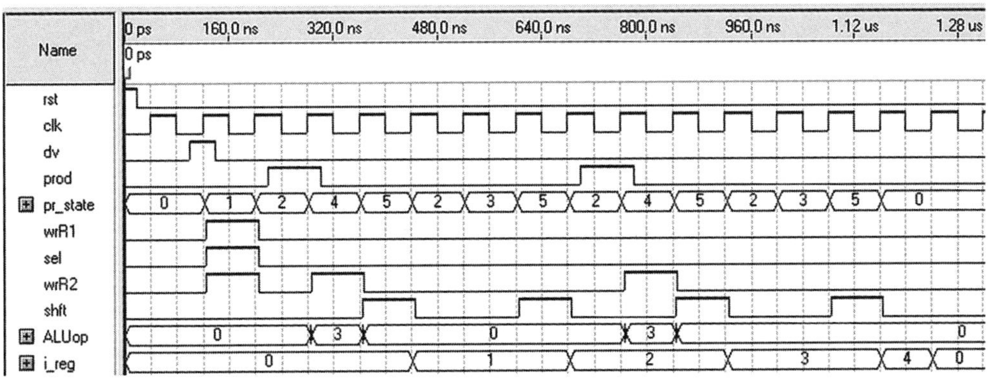
Name
0 ps
160,0 ns
320,0 ns
480,0 ns
640,0 ns
800,0 ns
960,0 ns
1.12 us
1.28 us
rst
clk
dv
prod
pr_state
wrR1
sel
wrR2
shft
ALUop
i_reg

Figure 12.1
Simulation results from the VHDL code for the control unit of figure 11.12b, for $N = 4$, which controls a multiplying datapath.

std_logic_vector (industry standard). Note that mode *buffer* is used this time for *y*, so *y* can be associated directly with *y_reg*.

The architecture, called *moore_fsm*, is in lines 13–89. As usual, it contains a declarative part and a statements part, with three processes in the latter.

The declarative part of the architecture (lines 15–21) contains FSM- and auxiliary-register-related declarations. In the former the enumerated type *state* is created to represent the machine's present and next states. In the latter the signals *y_reg*, *i*, and *i_reg* are created to deal with the auxiliary registers. Note that two auxiliary registers are needed in this example: for the main (actual) output (*y*) and for the output that operates as an auxiliary pointer (*i*) to the FSM.

The first process (lines 26–35) implements the auxiliary register, similarly to the template, except for the fact that there are now two auxiliary registers.

The second process (lines 38–45) implements the FSM's state register, exactly as in the template.

The third and final process (lines 48–87) implements the entire combinational logic section. It is just a list of all states, each containing the output values and the next state. Observe that in the (originally) recursive equations (lines 53, 63–64, and 75–76), *i_reg* and and *y_reg* appear on the right-hand side instead of *i* and *y* themselves (as proposed in the template). As usual, note that in each state the output values are unique because in a Moore machine the outputs depend only on the state in which the machine is. Another important aspect can be observed in lines 64–65 and 76–77; note that first a value is assigned to the entire vector *y* (lines 64 and 76), then one of its bits, *y*(*i*–1), is overwritten (lines 65 and 77).

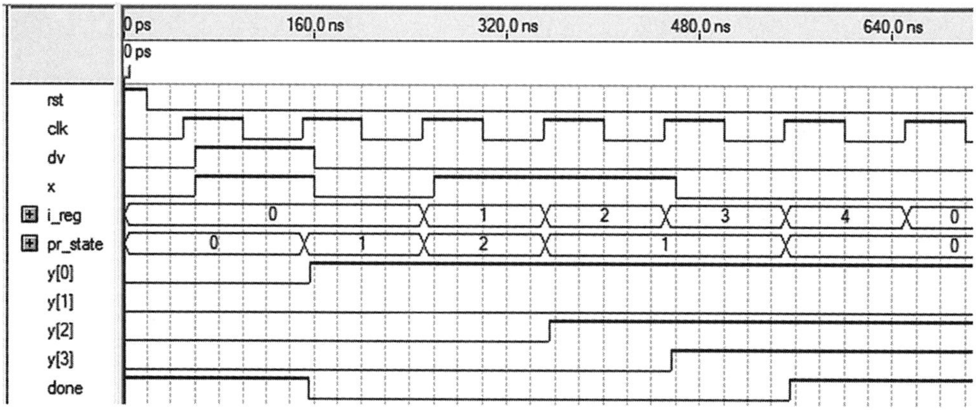
0 ps
160.0 ns
320.0 ns
480.0 ns
640.0 ns
0 ps
rst
clk
dv
x
i_reg
pr_state
y[0]
y[1]
y[2]
y[3]
done
0
1
2
3
4
0
0
1
2
1
0

Figure 12.2
Simulation results from the VHDL code for the serial data receiver of figure 11.14d, with $N = 8$.

In this kind of application glitches during clock transitions are generally not a problem. In any case, because *y* is one of the signals that go through an auxiliary register, if a glitch-free/pipelined output is required we can simply send out *y_reg* instead of *y*.

Observe the correct use of registers and the completeness of the code, as described in comment number 10 of section 6.3.

The number of flip-flops inferred by the compiler on synthesizing the code below, with regular sequential encoding (section 3.7), was 14 for $N = 8$ and 40 for $N = 32$.

Simulation results are shown in figure 12.2, for *x*="1011".

```
---------------------------------------------------------
library ieee;
use ieee.std_logic_1164.all;
---------------------------------------------------------
entity serial_data_receiver is
   generic (N: natural := 4); --number of bits (any >0)
   port (
      x, dv, clk, rst: in std_logic;
      done: out std_logic;
      y: buffer std_logic_vector(N-1 downto 0));
end entity;
---------------------------------------------------------
architecture moore_fsm of serial_data_receiver is

   --FSM-related declarations:
   type state is (idle, store0, store1);
   signal pr_state, nx_state: state;

   --Auxiliary-register-related declarations:
   signal y_reg: std_logic_vector(N-1 downto 0);
```

```
   signal i, i_reg: natural range 0 to N;

begin

   --Auxiliary register:
   process (clk, rst)
   begin
      if rst='1' then
         i_reg <= 0;
         y_reg <= (others => '0');
      elsif rising_edge(clk) then
         i_reg <= i;
         y_reg <= y;
      end if;
   end process;

   --FSM state register:
   process (clk, rst)
   begin
      if rst='1' then
         pr_state <= idle;
      elsif rising_edge(clk) then
         pr_state <= nx_state;
      end if;
   end process;

   --FSM combinational logic:
   process (all)
   begin
      case pr_state is
         when idle =>
            i <= 0;
            y <= y_reg;
            done <= '1';
            if dv='1' and x='0' then
               nx_state <= store0;
            elsif dv='1' and x='1' then
               nx_state <= store1;
            else
               nx_state <= idle;
            end if;
         when store0 =>
            i <= i_reg + 1;
            y <= y_reg;
            y(i-1) <= '0';
            done <= '0';
            if i=N then
               nx_state <= idle;
            elsif x='1' then
               nx_state <= store1;
            else
               nx_state <= store0;
            end if;
         when store1 =>
            i <= i_reg + 1;
            y <= y_reg;
```

```
            y(i-1) <= '1';
            done <= '0';
            if i=N then
               nx_state <= idle;
            elsif x='0' then
               nx_state <= store0;
            else
               nx_state <= store1;
            end if;
      end case;
   end process;

end architecture;
------------------------------------------------------
```

12.6 Design of a Memory Interface

This section presents a VHDL-based design for the memory interface introduced in section 11.7.8 (figure 11.16). The SRAM used in the experiments is the IS61LV25616 device, from ISSI, which is capable of storing 262k 16-bit words. The corresponding FSM was presented in figure 11.16c, and the circuit ports are depicted in figure 12.3 (note that a test circuit has been included).

The entity, called *sram_interface*, is in lines 7–25. Note that several parameters were declared as generic (lines 8–14), so they can be easily modified and overridden. Note also that the port names are from figure 12.3 and that all ports (lines 15–24) are of type *std_logic* or *std_logic_vector* (industry standard).

The signals in lines 21–24 are for the test circuit (see figure 12.3). The signal *seed* (line 22), set by two switches, is added to the actual address (*A*, line 19) to produce the test data (*D*, line 21), which is displayed on a seven-segment display by means of the signal *ssd* (line 23). Two LEDs are lit by the signals *done_wr* and *done_rd* (line 24)

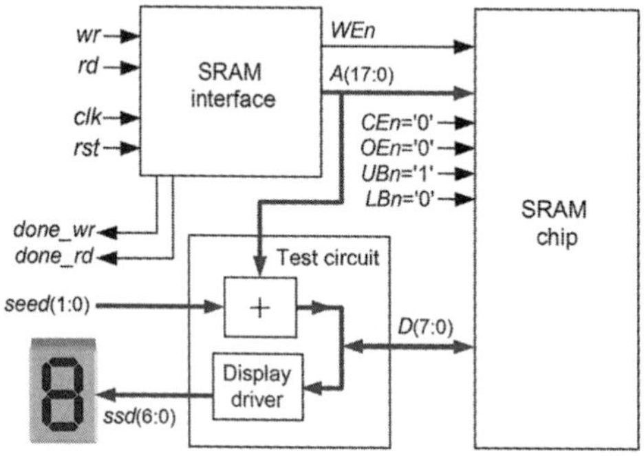
wr
rd
clk
rst
SRAM
interface
WEn
A(17:0)
CEn='0'
OEn='0'
UBn='1'
LBn='0'
SRAM
chip
done_wr
done_rd
Test circuit
seed(1:0)
+
D(7:0)
Display
driver
ssd(6:0)

Mode	$\overline{WE}$	$\overline{CE}$	$\overline{OE}$	$\overline{LB}$	$\overline{UB}$	I/O0-I/O7	I/O8-I/O15
Not Selected	X	H	X	X	X	High-Z	High-Z
Output	H	L	H	X	X	High-Z	High-Z
Disabled	X	L	X	H	H	High-Z	High-Z
Read	H	L	L	L	H	DOUT	High-Z
	H	L	L	H	L	High-Z	DOUT
	H	L	L	L	L	DOUT	DOUT
Write	L	L	X	L	H	DIN	High-Z
	L	L	X	H	L	High-Z	DIN
	L	L	X	L	L	DIN	DIN

Figure 12.3
Setup for the experiments with the SRAM memory interface introduced in figure 11.16, including a test circuit as well. The device's truth table is also shown.

to indicate when the test circuit has finished writing to or reading from the memory, respectively.

The architecture, called *moore_fsm*, is in lines 27–181. As usual, it contains a declarative part and a statements part, with six code sections in the latter.

The architecture's declarative part is in lines 29–62. In the FSM-related declarations (lines 30–31), the enumerated type *state* is created to represent the machine's present and next states. In the auxiliary-register-related declarations (line 34), the signals *addr* and *addr_reg* are created to deal with the auxiliary register (observe that the address is the signal that appears in the recursive expressions, so that is the signal to be stored in that register). In the timer-related declarations (line 37), the signals needed to build a 0.5-s timer are created to be used in the *read1-read2* transition (see $t = T_2$) of figure 11.16c, so the user will have enough time to observe the value presented on the display during the tests. Finally, a function is created in lines 39–62 to later implement the SSD driver (integer-to-SSD conversion).

The first code section (line 67) in the architecture's statements part is a list of static signals to be connected to the SRAM chip during the tests. Note that they are all enabled (because they are active low) except for the upper byte of the data word, which is not used here.

The second code section (lines 70–81) contains a process that implements the timer (needed in the *read1-read2* transition; the *write1-write2* transition is made at full clock speed). This code is similar to the template of section 9.2. Both timer control strategies (section 8.5) are allowed for this FSM.

The third code section (lines 84–91) in the architecture's statements part is a process that implements the auxiliary register, exactly as in the template.

The fourth code section (lines 94–101) is another process, which implements the FSM's state register, again exactly as in the template.

The fifth code section (lines 104–172) contains a process that implements the entire combinational logic section. It is just a list of all states, each containing the output and time parameter values plus the next state. Observe that in the (originally) recursive equations (lines 121, 128, 139, and 150), *addr_reg* appears on the right-hand side instead of *addr* itself (as proposed in the template). As usual, in each state the output values are unique because in a Moore machine the outputs depend only on the state in which the machine is.

The sixth and final code section (lines 174–179) in the architecture's statements part passes the value of *addr* to *A* (in *std_logic_vector* form) and also builds the test circuit. The test circuit is important because it illustrates how we can deal with a bidirectional bus. Note that during the writing procedure the FPGA sends data to the SRAM, but when data is being read from the SRAM the FPGA's output must go into high-impedance (floating) mode because they (FPGA and SRAM) are physically connected to the same wires (data bus *D*). Observe also that the generated data consist simply of *seed* + *A* (line 177—see also figure 12.3), which is written to the SRAM when a *wr* = '1' pulse occurs and is read from the SRAM and displayed on the SSD when a *rd* = '1' pulse occurs.

In this kind of application, glitches during clock transitions are generally not a problem, so the optional output register is not needed.

Observe the correct use of registers and the completeness of the code, as described in comment 10 of section 6.3.

```
1     ------------------------------------------------------------------
2     library ieee;
3     use ieee.std_logic_1164.all;
4     use ieee.std_logic_unsigned.all;
5     use ieee.std_logic_arith.all;
6     ------------------------------------------------------------------
7     entity sram_interface is
8        generic (
9           --Main-circuit parameters:
10          Abus: natural := 18; --Address bus width
11          Dbus: natural := 16; --Data bus width
12          --Test-circuit parameters:
13          Tread: natural := 25_000_000; --Time=0.5s @fclk=50MHz
14          Amax: natural := 12); --Max address in test circuit
15       port (
16          --Main-circuit ports:
17          rd, wr, clk, rst: in std_logic;
18          CEn, WEn, OEn, UBn, LBn: out std_logic;
19          A: out std_logic_vector(Abus-1 downto 0);
20          --Test-circuit ports:
21          D: inout std_logic_vector(7 downto 0); --Lower-byte only
22          seed: in std_logic_vector(1 downto 0);
23          ssd: out std_logic_vector(6 downto 0);
24          done_wr, done_rd: buffer std_logic);
25    end entity;
26    ------------------------------------------------------------------
27    architecture moore_fsm of sram_interface is
28
29       --FSM-related declarations:
30       type state is (idle, write1, write2, read1, read2, hold);
31       signal pr_state, nx_state: state;
32
33       --Auxiliary-register-related declarations:
34       signal addr, addr_reg: natural range 0 to 2**Abus-1;
35
36       --Timer-related declarations:
37       signal t, tmax: natural range 0 to Tread-1; --range≥Tread
38
39       function int_to_ssd(signal input: natural) return std_logic_vector
40          is variable output: std_logic_vector(6 downto 0);
41       begin
42          case input is
43             when 0 => output:="0000001";
44             when 1 => output:="1001111";
45             when 2 => output:="0010010";
```

```
         when 3 => output:="0000110";
         when 4 => output:="1001100";
         when 5 => output:="0100100";
         when 6 => output:="0100000";
         when 7 => output:="0001111";
         when 8 => output:="0000000";
         when 9 => output:="0000100";
         when 10 => output:="0001000";
         when 11 => output:="1100000";
         when 12 => output:="0110001";
         when 13 => output:="1000010";
         when 14 => output:="0110000";
         when 15 => output:="0111000";
         when others => output:="1111110"; --"-"
      end case;
      return output;
   end integer_to_ssd;

begin

   --Static SRAM signals:
   CEn<='0'; OEn<='0'; UBn<='1'; LBn<='0';

   --Timer (using strategy #2, section 8.5.3):
   process (clk, rst)
   begin
      if (rst='1') then
         t <= 0;
      elsif rising_edge(clk) then
            if t < tmax then
            t <= t + 1;
         else
            t <= 0;
      end if;
   end if;
end process;

   --Auxiliary register:
   process (clk, rst)
   begin
      if rst='1' then
         addr_reg <= 0;
      elsif rising_edge(clk) then
         addr_reg <= addr;
      end if;
   end process;

   --FSM state register:
   process (clk, rst)
   begin
      if rst='1' then
         pr_state <= idle;
      elsif rising_edge(clk) then
         pr_state <= nx_state;
      end if;
   end process;

```

```
   --FSM combinational logic:
   process (all)
   begin
      case pr_state is
         when idle =>
            addr <= 0;
            WEn <= '1';
            done_wr <= '1';
            done_rd <= '1';
            tmax <= 0;
            if wr='1' and rd='0' then
               nx_state <= write1;
            elsif wr='0' and rd='1' then
               nx_state <= read1;
            else
               nx_state <= idle;
            end if;
         when write1 =>
            addr <= addr_reg;
            WEn <= '0';
            done_wr <= '0';
            done_rd <= '1';
            tmax <= 0;
            nx_state <= write2;
         when write2 =>
            addr <= addr_reg + 1;
            WEn <= '1';
            done_wr <= '0';
            done_rd <= '1';
            tmax <= 0;
            if addr <= Amax then
               nx_state <= write1;
            else
               nx_state <= hold;
            end if;
         when read1 =>
            addr <= addr_reg;
            WEn <= '1';
            done_wr <= '1';
            done_rd <= '0';
            tmax <= Tread;
            if t>=tmax then
               nx_state <= read2;
            else
               nx_state <= read1;
            end if;
         when read2 =>
            addr <= addr_reg + 1;
            WEn <= '1';
            done_wr <= '1';
            done_rd <= '0';
            tmax <= 0;
            if addr <= Amax then
               nx_state <= read1;
            else
               nx_state <= hold;
            end if;
```

```
          when hold =>
             addr <= 0;
             WEn <= '1';
             done_wr <= '1';
             done_rd <= '1';
             tmax <= 0;
             if wr='0' and rd='0' then
                nx_state <= idle;
             else
                nx_state <= hold;
             end if;
       end case;
    end process;

    A <= conv_std_logic_vector(addr, Abus);

    --Test circuit:
    D <= seed + conv_std_logic_vector(addr, 8) when done_wr='0' else
         (others => 'Z');
    ssd <= int_to_ssd(conv_integer(D));

 end architecture;
 -----------------------------------------------------------------
```

12.7 Exercises

Exercise 12.1: Long-String Comparator #1

This exercise concerns the long-string comparator of figure 11.6, which must detect whether the last N bits in two serial bit streams are equal.

a) Implement it using VHDL. Compile it for $N = 64$ bits and sequential encoding and check if the number of DFFs inferred by the compiler matches the estimate made in section 11.7.2.
b) Recompile it for $N = 4$; then simulate it using the same stimuli of figure 11.6c and check if the same waveforms result.

Exercise 12.2: Long-String Comparator #2

This exercise concerns the long-string comparator of exercise 11.5.

a) Solve exercise 11.5 if not done yet.
b) Implement the resulting FSM using VHDL. Check if the number of DFFs inferred by the compiler matches your estimate.
c) Simulate it using the same stimuli of figure 11.19, checking if the same waveforms result.

Exercise 12.3: Hamming-Weight Calculator

This exercise concerns the Hamming-weight calculator of exercise 11.3.

a) Solve parts a and b of exercise 11.3 if not done yet.
b) How many DFFs are needed to build the resulting FSM, with sequential encoding and *dv* lasting 64 clock periods (so *y* can go from 0 to 64)?
c) Implement your machine using VHDL. Check if the number of DFFs inferred by the compiler matches your estimate.
d) Recompile the code for $N = 9$ (hence with four bits for *y*) and simulate it using the same stimuli of figure 11.17, checking if the same waveforms result.
e) Even though exercise 11.3 is important to understand how that kind of circuit can be modeled as an FSM, it was said in sections 5.4.1 and 11.7.1 that counters are well-known circuits, easily designed without the FSM approach. Therefore, because a Hamming calculator is a kind of counter, it can be designed directly in VHDL. Do it. Check the number of DFFs and combinational elements needed to implement it for *dv* lasting 64 clock periods and compare the results against those obtained in part c above.

Exercise 12.4: Leading-Ones Counter

This exercise concerns the leading-ones counter of exercise 11.4.

a) Solve parts a and b of exercise 11.4 if not done yet.
b) How many DFFs are needed to build the resulting FSM, with sequential encoding and *dv* lasting 64 clock periods (so *y* can go from 0 to 64)?
c) Implement your machine using VHDL. Check if the number of DFFs inferred by the compiler matches your estimate.
d) Recompile the code for $N = 9$ (hence with four bits for *y*) and simulate it using the same stimuli of figure 11.18, checking if the same waveforms result.
e) Even though exercise 11.4 is important to understand how that kind of circuit can be modeled as an FSM, it is said in sections 5.4.1 and 11.7.1 that counters are well-known circuits, easily designed without the FSM approach. Therefore, because a leading-ones counter is a kind of counter, it can be designed directly in VHDL. Do it. Check the number of DFFs and combinational elements needed to implement it for *dv* lasting 64 clock periods and compare the results against those obtained in part c above.

Exercise 12.5: Complete Reference-Value Definer

Figure 12.4 shows an initial block diagram for the experiment to be developed in this exercise. It consists of a reference-value definer with up-down controls, which must also include some type of debouncer for the pushbuttons. The output (reference value) must range from 00 to 60 and must be displayed on two SSDs or an LCD. Note that *ref* is a six-bit signal, while each display digit (*dig0* for units, *dig1* for tens of units) is a seven-bit value if SSDs are employed. A special feature desired for this circuit is the

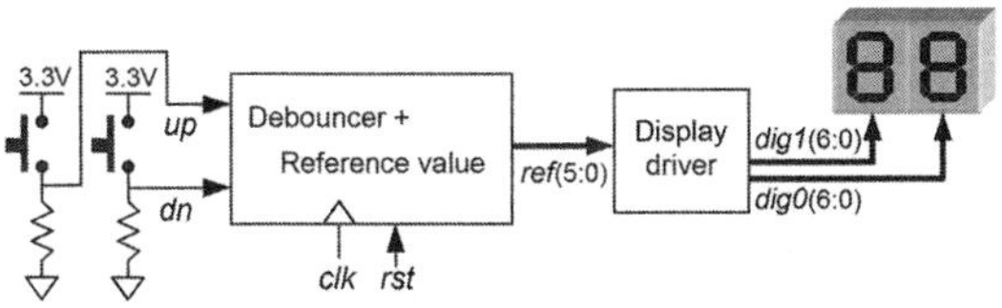
3.3V
3.3V
up
dn
Debouncer +
Reference value
clk
rst
ref(5:0)
Display
driver
dig1(6:0)
dig0(6:0)

Figure 12.4

following: if either pushbutton is kept pressed for $t_1 \geq 2$ s (T_1 clock periods), the increment must occur automatically and at every $t_2 = 0.5$ s (T_2 clock periods).

a) Carefully review sections 8.11.4, 11.7.3, and 11.7.4 and decide what should go inside the main block of figure 12.4. Then draw an expanded block diagram with proper details.
b) Draw a state transition diagram for each FSM to be used in this problem.
c) Estimate the number of DFFs that will be needed to build the complete circuit. Assume sequential encoding for the FSM(s) and a 1-ms debouncing interval (check the clock frequency in your development board).
d) Implement the circuit using VHDL. Check whether the number of DFFs inferred by the compiler matches your prediction.
f) Physically demonstrate your design in the FPGA development board.

Exercise 12.6: Factorial Calculator
This exercise concerns the factorial calculator of exercise 11.9.

a) Solve exercise 11.9 if not done yet.
b) Implement the resulting FSM using VHDL. Show meaningful simulation results.

Exercise 12.7: Divider
This exercise concerns the sequential divider of figure 11.13.

a) How many DFFs are needed to build it for $N = 32$ bits and sequential encoding?
b) Implement it using VHDL. Check whether the number of DFFs inferred by the compiler matches your estimate.
c) Recompile the code for $N = 4$ and simulate it using the same stimuli of figure 11.13a, checking whether the same results are obtained here.

13 SystemVerilog Design of Recursive (Category 3) State Machines

13.1 Introduction

This chapter presents several SystemVerilog designs of category 3 state machines. It starts by presenting two SystemVerilog templates, for Moore- and Mealy-based implementations, which are used subsequently to develop a series of designs related to the examples introduced in chapter 11.

The codes are always complete (not only partial sketches) and are accompanied by comments and often also simulation results, illustrating the design's main features. All circuits were synthesized using Quartus II (from Altera) or ISE (from Xilinx). The simulations were performed with Quartus II or ModelSim (from Mentor Graphics). The default encoding scheme for the states of the FSMs was regular sequential encoding (see encoding options in section 3.7).

The same designs were developed in chapter 12 using VHDL, so the reader can make a direct comparison between the codes.

Note: See suggestions of SystemVerilog books in the bibliography.

13.2 SystemVerilog Template for Recursive (Category 3) Moore Machines

The template is presented below. Because it is an extension to the Moore templates for categories 1 and 2, described in sections 7.3 and 10.2, respectively, a review of those templates is suggested before this one is examined because only the differences are described.

The only differences are those needed for the inclusion of an auxiliary register, compulsory in category 3 machines. In summary, the following must be added/done to the previous template: declarations concerning the auxiliary register; an **always_ff** block to infer the auxiliary register; and proper adjustments in the recursive equations to invoke the auxiliary register. These modifications are described next.

The auxiliary-register-related declarations are in line 13. It is assumed that there is only one output and that it must be stored, but recall that the circuit might have several outputs, not all registered. The actual number of auxiliary registers is determined by the number of outputs that depend on past output values.

To implement the auxiliary register, an **always_ff** block is employed in lines 21–23.

Finally, note in the **always_comb** block of lines 29–49 that *outp* (lines 32 and 39) is no longer a function of itself but rather a function of *outp_reg*. This removes the recursiveness, allowing the output to be computed by a combinational circuit.

As explained in section 11.3, an interesting aspect of category 3 FSMs is that the auxiliary register can also play the role of output register (for glitch-free and/or pipelined construction) when the output is one of the signals stored in an auxiliary register. To do so, simply send *outp_reg* out instead of *outp*.

```
1    //Part 1: Module header:----------------------------
2       (same as for categ. 1 and 2, sections 7.3 and 10.2)
3
4    //Part 2: Declarations:-----------------------------
5
6       //FSM-related declarations:
7       (same as for category 1 Moore, Section 7.3)
8
9       //Timer-related declarations:
10      (same as for category 2 Moore, section 10.2)
11
12      //Auxiliary-register-related declarations:
13      logic [N-1:0] outp, outp_reg;
14
15   //Part 3: Statements:-------------------------------
16
17      //Timer:
18      (same as for category 2 Moore, section 10.2)
19
20      // Auxiliary register:
21      always_ff @(posedge clk, posedge rst)
22         if (rst) outp_reg <= <initial_value>;
23         else outp_reg <= outp;
24
25      //FSM state register:
26      (same as for category 2 Moore, section 10.2)
27
28      //FSM combinational logic:
29      always_comb
30         case (pr_state)
31            A: begin
32               outp <= outp_reg;
33               tmax <= T1-1; //if using strategy #2
34               if (condition) nx_state <= B;
35               else if (condition) nx_state <= ...;
36               else nx_state <= A;
```

```
37          end
38          B: begin
39             outp <= outp_reg + 1;
40             tmax <= T2-1; //if using strategy #2
41             if (condition) nx_state <= C;
42             else if (condition) nx_state <= ...;
43             else nx_state <= B;
44          end
45          C: begin
46             ...
47          end
48          ...
49       endcase
50
51    //Optional output register:
52    (same as for category 1 Moore, section 7.3)
53
54 endmodule
55 //------------------------------------------------------
```

13.3 SystemVerilog Template for Recursive (Category 3) Mealy Machines

The template is presented below. The only difference with respect to the Moore template just described is in the **always_comb** block for the combinational logic (lines 27–63) because the output is specified differently now. Recall that in a Mealy machine the output depends not only on the FSM's state but also on its input, so **if** statements are expected for the output in one or more states because the output (and t_{max}) values might not be unique. This is achieved by including such values *within* the conditional statements for *nx_state*. For example, observe in lines 29–44, relative to state A, that the output (and t_{max}) values are now conditional. Compare these lines against lines 31–37 in the previous template.

```
1  //Part 1: Module header:----------------------------
2     (same as for category 3 Moore, section 13.2)
3
4  //Part 2: Declarations:-----------------------------
5
6     //FSM-related declarations:
7     (same as for category 3 Moore, section 13.2)
8
9     //Timer-related declarations:
10    (same as for category 3 Moore, section 13.2)
11
12    //Auxiliary-register-related declarations:
13    (same as for category 3 Moore, section 13.2)
14
15 //Part 3: Statements:-------------------------------
16
17    //Timer:
18    (same as for category 3 Moore, section 13.2)
19
```

```
    // Auxiliary register:
    (same as for category 3 Moore, section 13.2)

    //FSM state register:
    (same as for category 3 Moore, section 13.2)

    //FSM combinational logic:
    always_comb
       case (pr_state)
          A:
             if (condition) begin
                outp <= outp_reg;
                tmax <= <value>; //if using strategy #2
                nx_state <= B;
             end
             else if (condition) begin
                outp <= outp_reg + 1;
                tmax <= <value>; //if using strategy #2
                nx_state <= ...;
             end
             else begin
                outp <= <value>;
                tmax <= <value>;
                nx_state <= A;
             end
          B:
             if (condition) begin
                outp <= outp_reg + 1;
                tmax <= <value>;
                nx_state <= C;
             end
             else if (condition) begin
                outp <= outp_reg;
                tmax <= <value>;
                nx_state <= ...;
             end
             else begin
                outp <= <value>;
                tmax <= <value>;
                nx_state <= B;
             end
          C: ...
          ...
    endcase

    //Optional output register:
    (same as for category 3 Moore, section 13.2)

endmodule
//----------------------------------------------------
```

13.4 Design of a Datapath Controller for a Multiplier

This section presents a SystemVerilog-based design for the control unit introduced in section 11.7.5, which controls a datapath to produce a sequential add-and-shift multiplier. The Moore template for category 3 machines seen in section 13.2 is used to implement the FSM of figure 11.12b.

The first part of the code (*module header*) is in lines 1–7. The module's name is *contol_unit_for_multiplier*. Note that all ports are of type **logic**.

The second part of the code (*declarations*) is in lines 9–17. In the FSM-related declarations (lines 12–14), the enumerated type *state* is created to represent the machine's present and next states. In the auxiliary-register-related declarations (line 17), *i* and *i_reg* are created to deal with the auxiliary register. Note that in this example none of the actual outputs is stored in an auxiliary register (the auxiliary registers are always for the variables that appear in the recursive equations).

The third and final part of the code (*statements*) is in lines 19–75. It contains three **always** blocks, described next.

The first **always** block (lines 22–24) is an **always_ff** that implements the auxiliary register, exactly as in the template.

The second **always** block (lines 27–29) is another **always_ff**, which implements the machine's state register, again exactly as in the template.

The third and final **always** block (lines 32–73) is an **always_comb**, which implements the entire combinational logic section. It is just a list of all states, each containing the output values and the next state. Note that because some of the output values get repeated several times, default values were entered in lines 35–40, so they only need to be included in the **case** statement when different values are required. Observe that in the (originally) recursive equations (lines 53, 58, 64, and 69), *i_reg* appears on the right-hand side instead of *i* itself (as proposed in the template). As usual, in each state the output values are unique because in a Moore machine the outputs depend only on the state in which the machine is.

In datapath-related designs, possible glitches at the output of the control unit following clock transitions are normally not a problem, so the optional output register was not employed.

Finally, and very importantly, observe the correct use of registers and the completeness of the code, as described in comment 8 of section 7.3. Observe in particular the following: 1) all states are included; 2) the list of outputs is exactly the same in all states, and the corresponding values/expressions are always properly declared; 3) any conditional specification for *nx_state* is finalized with an **else** statement, so no condition is left unchecked.

The number of flip-flops inferred by the compiler on synthesizing the code below, with regular sequential encoding (section 3.7), was six for $N = 4$ and nine for $N = 32$ bits. Compare these results against your predictions made in exercise 11.10.

Simulation results from this code are exactly the same as those obtained using VHDL, shown in figure 12.1.

```
//Module header:-----------------------------------------
module control_unit_for_multiplier
   #(parameter N=4) //number of bits
   (
   input logic dv, prod, clk, rst,
   output logic wrR1, sel, wrR2, shft,
   output logic [1:0] ALUop);

//Declarations:------------------------------------------

   //FSM-related declarations:
   typedef enum logic [2:0]
      {idle, load, waitt, nop, add, shift} state;
   state pr_state, nx_state;

   //Auxiliary-register-related declarations:
   logic [$clog2(N):0] i, i_reg; //function ceiling(log2(N))

//Statements:--------------------------------------------

   // Auxiliary register:
   always_ff @(posedge clk, posedge rst)
      if (rst) i_reg <= 0;
      else i_reg <= i;

   //FSM state register:
   always_ff @(posedge clk, posedge rst)
      if (rst) pr_state <= idle;
      else pr_state <= nx_state;

   //FSM combinational logic:
   always_comb begin

      //Default values:
      wrR1 <= 'b0;
      sel <= 'b0;
      wrR2 <= 'b0;
      shft <= 'b0;
      ALUop <= 2'b00;
      i <= 0;

      case (pr_state)
         idle:
            if (dv) nx_state <= load;
            else nx_state <= idle;
         load: begin
            wrR1 <= 'b1;
            sel <= 'b1;
            wrR2 <= 'b1;
            nx_state <= waitt;
            end
         waitt: begin
            i <= i_reg;
            if (~prod) nx_state <= nop;
            else nx_state <= add;
```

```
               end
            nop: begin
               i <= i_reg;
               nx_state <= shift;
               end
            add: begin
               wrR2 <= 'b1;
               ALUop <= 2'b11;
               i <= i_reg;
               nx_state <= shift;
               end
            shift: begin
               shft <= 'b1;
               i <= i_reg + 1;
               if (i<N) nx_state <= waitt;
               else nx_state <= idle;
               end
         endcase

endmodule
//-----------------------------------------------------------
```

13.5 Design of a Serial Data Receiver

This section presents a SystemVerilog-based design for the serial data receiver introduced in section 11.7.7. The Moore template for category 3 machines seen in section 13.2 is used to implement the solution of figure 11.14c.

The first part of the code (*module header*) is in lines 1–7. The module's name is *serial_data_receiver*. Note that all ports are of type **logic**.

The second part of the code (*declarations*) is in lines 9–17. In the FSM-related declarations (lines 12–13), the enumerated type *state* is created to represent the machine's present and next states. In the auxiliary-register-related declarations (lines 16–17), *y_reg*, *i*, and *i_reg* are created to deal with the auxiliary registers. Note that two auxiliary registers are needed in this example: for the main (actual) output (*y*) and for the output that operates as an auxiliary pointer (*i*) to the FSM.

The third and final part of the code (*statements*) is in lines 19–68. It contains three **always** blocks, described next.

The first **always** block (lines 22–30) is an **always_ff** that implements the auxiliary register, similarly to the template.

The second **always** block (lines 33–35) is another **always_ff**, which implements the machine's state register, exactly as in the template.

The third and final **always** block (lines 38–66) is an **always_comb**, which implements the entire combinational logic section. It is just a list of all states, each containing the output values and the next state. Observe that in the (originally) recursive equations (lines 42, 49–50, and 58–59), *i_reg* and *y_reg* appear on the right-hand side instead of *i* and *y* themselves (as proposed in the template). As usual, in each state the output values are unique because in a Moore machine the outputs depend only on the state in which the machine is. Another important aspect can be observed in lines 50–51 and 59–60; note that first a value is assigned to the entire vector *y* (lines 50 and 59); then one of its bits, *y*(*i* – 1), is overwritten (lines 51 and 60).

In this kind of application, glitches during clock transitions are generally not a problem. Anyway, because *y* is one of the signals that go through an auxiliary register, if a glitch-free (pipelined) output is required, we can simply send out *y_reg* instead of *y*.

Finally, and very importantly, observe the correct use of registers and the completeness of the code, as described in comment 8 of section 7.3. Observe in particular the following: 1) all states are included; 2) the list of outputs is exactly the same in all states, and the corresponding values/expressions are always properly declared; 3) the specifications for *nx_state* are always finalized with an **else** statement, so no condition is left unchecked.

The number of flip-flops inferred by the compiler on synthesizing the code below, with regular sequential encoding (section 3.7) and $N = 8$, was 14.

Simulation results from this code are exactly the same as those obtained using VHDL, shown in figure 12.2.

```
1     //Module header:------------------------------------------
2     module serial_data_receiver
3        #(parameter N=8) //number of bits (any >0)
4        (
5        input logic x, dv, clk, rst,
6        output logic done,
7        output logic [N-1:0] y);
8
9     //Declarations:------------------------------------------
10
11       //FSM-related declarations:
12       typedef enum logic [1:0] {idle, load0, load1} state;
13       state pr_state, nx_state;
14
15       //Auxiliary-register-related declarations:
16       logic [N-1:0] y_reg;
17       logic [$clog2(N):0] i, i_reg; //function ceiling(log2(N))
18
19    //Statements:--------------------------------------------
20
21       //Auxiliary register:
22       always_ff @(posedge clk, posedge rst)
23          if (rst) begin
24             i_reg <= '0;
25             y_reg <= '0;
26             end
27          else begin
28             i_reg <= i;
29             y_reg <= y;
30             end
```

```

   //FSM state register:
   always_ff @(posedge clk, posedge rst)
      if (rst) pr_state <= idle;
      else pr_state <= nx_state;

   //FSM combinational logic:
   always_comb
      case (pr_state)
         idle: begin
            i <= 1'b0;
            y <= y_reg;
            done <= 1'b1;
            if (dv & ~x) nx_state <= load0;
            else if (dv & x) nx_state <= load1;
            else nx_state <= idle;
            end
         load0: begin
            i <= i_reg + 1;
            y <= y_reg;
            y[i-1] <= 1'b0;
            done <= 1'b0;
            if (i=N) nx_state <= idle;
            else if (x) nx_state <= load1;
            else nx_state <= load0;
            end
         load1: begin
            i <= i_reg + 1;
            y <= y_reg;
            y[i-1] <= 1'b1;
            done <= 1'b0;
            if (i=N) nx_state <= idle;
            else if (~x) nx_state <= load0;
            else nx_state <= load1;
            end
      endcase

endmodule
//----------------------------------------------------------
```

13.6 Design of a Memory Interface

This section presents a SystemVerilog-based design for the memory interface introduced in section 11.7.8 (figure 11.16). The SRAM used in the experiments is the IS61LV25616 device, from ISSI, which is capable of storing 262k 16-bit words. The corresponding FSM was presented in figure 11.16c, and the circuit ports are depicted in figure 13.1 (note that a test circuit has been included).

The first part of the code (*module header*) is in lines 1–19. The module's name is *sram_interface*. Several global parameters were included for both the main circuit and a test circuit. The port names are from figure 13.1. All ports are of type **logic**.

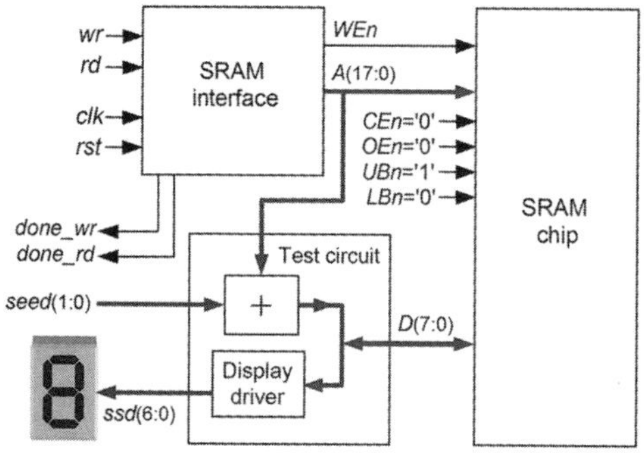
wr
rd
clk
rst
SRAM interface
WEn
A(17:0)
CEn='0'
OEn='0'
UBn='1'
LBn='0'
SRAM chip
done_wr
done_rd
Test circuit
seed(1:0)
+
D(7:0)
Display driver
ssd(6:0)

Mode	$\overline{WE}$	$\overline{CE}$	$\overline{OE}$	$\overline{LB}$	$\overline{UB}$	I/O0-I/O7	I/O8-I/O15
Not Selected	X	H	X	X	X	High-Z	High-Z
Output	H	L	H	X	X	High-Z	High-Z
Disabled	X	L	X	H	H	High-Z	High-Z
Read	H	L	L	L	H	DOUT	High-Z
	H	L	L	H	L	High-Z	DOUT
	H	L	L	L	L	DOUT	DOUT
Write	L	L	X	L	H	DIN	High-Z
	L	L	X	H	L	High-Z	DIN
	L	L	X	L	L	DIN	DIN

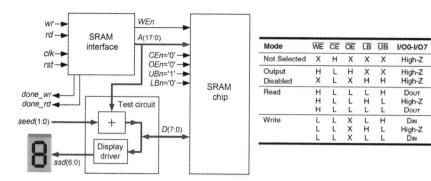

Figure 13.1
Setup for the experiments with the SRAM memory interface introduced in figure 11.16, also including a test circuit. The device's truth table is also shown.

The signals in lines 16–19 are for the test circuit. *seed* (line 17), set by two switches, is added to the actual address (*A*, line 14) to produce the test data (*D*, line 16), which is displayed on a seven-segment display by means of the signal *ssd* (line 18). Two LEDs are lit by the signals *done_wr* and *done_rd* (line 19) to indicate when the test circuit has finished writing to or reading from the memory, respectively.

The second part of the code (*declarations*) is in lines 21–56. In the FSM-related declarations (lines 24–26), the enumerated type *state* is created to represent the machine's present and next states. In the auxiliary-register-related declarations (line 29), *A_reg* is created to deal with the auxiliary register (observe that the address is the signal that appears in the recursive expressions, so that is the signal to be stored in that register). In the timer-related declarations (line 32), the signals needed to build a 0.5-s timer are created, to be used in the *read1-read2* transition (see $t = T_2$) of figure 11.16c, so the user has enough time to observe the value presented on the display during the tests. Finally, a function is created in lines 35–56 to implement later the SSD driver (integer-to-SSD conversion).

The third and final part of the code (*statements*) is in lines 58–149. It contains six sections, described next.

The first section (lines 61–64) of the statements produces the static signals to be connected to the SRAM chip during the tests. Note that they are all enabled (because they are active low), except for the upper byte of the data word, which is not used here.

The second section (lines 67–69) of the statements contains an **always_ff** block, which implements the timer (needed in the *read1-read2* transition; the *write1-write2* transition is made at full clock speed). This code is similar to the template of section 10.2. Both timer control strategies (section 8.5) are allowed for this FSM.

The third code section (lines 73–75) is another **always_ff** block, which implements the auxiliary register, exactly as in the template.

The fourth code section (lines 78–80) is an **always_ff** block that implements the FSM's state register, again exactly as in the template.

The fifth portion (lines 83–139) of the statements is part of an **always_comb** block that implements the entire FSM's combinational logic section. It is just a list with all states, each containing the output and time parameter values, plus the next state. Observe that in the (originally) recursive equations (lines 96, 104, 113, and 122), *A_reg* appears on the right-hand side instead of *A* itself (as proposed in the template). As usual, in each state the output values are unique because in a Moore machine the outputs depend only on the state in which the machine is.

Finally, the code in lines 142–147 implements the test circuit. This code is important because it illustrates one way (similar to VHDL—see section 12.6) of dealing with a bidirectional bus. Note that during the writing procedure the FPGA sends data to the SRAM, but when data is being read from the SRAM, the FPGA's output must go into high-impedance (floating) mode because they (FPGA and SRAM) are physically connected to the same wires (data bus *D*). Observe also that the generated data consist simply of *A* + *seed* (line 142; see also figure 13.1), which is written to the SRAM when a *wr* = '1' pulse occurs and is read from the SRAM and displayed on the SSD when a *rd* = '1' pulse occurs.

In this kind of application, glitches during clock transitions are generally not a problem, so the optional output register is not needed.

Finally, and very importantly, observe the correct use of registers and the completeness of the code, as described in comment 8 of section 7.3. Observe in particular the following: 1) all states are included; 2) the list of outputs is exactly the same in all states, and the corresponding values/expressions are always properly declared; 3) the specifications for *nx_state*, when conditional, are always finalized with an **else** statement, so no condition is left unchecked.

```
1     //Module header:------------------------------------
2     module sram_interface
3        #(parameter
4        //Main-circuit parameters:
5        Abus = 18, //Address bus width
6        Dbus = 16, //Data bus width
7        //Test-circuit parameters:
8        Tread = 25_000_000, //Time=0.5s @fclk=50MHz
9        Amax = 12) //Max address in test circuit
10       (
11       //Main-circuit ports:
12       input logic rd, wr, clk, rst,
13       output logic CEn, WEn, OEn, UBn, LBn,
14       output logic [17:0] A,
15       //Test-circuit ports:
16       inout logic [7:0] D, //Only lower-byte used
17       input logic [1:0] seed,
18       output logic [6:0] ssd,
19       output logic  done_wr, done_rd);
20
```

```
//Declarations:----------------------------------

   //FSM-related declarations:
   typedef enum logic [2:0]
      {idle, write1, write2, read1, read2, hold} state;
   state pr_state, nx_state;

   //Auxiliary-register-related declarations:
   logic [Abus-1:0] A_reg;

   //Timer-related declarations:
   logic [$clog2(Tread)-1:0] t, tmax; //range≥Tread

   //Function construction:
   function [6:0] integer_to_ssd;
      input [3:0] inp;
      case (inp)
         0: integer_to_ssd = 7'b0000001;
         1: integer_to_ssd = 7'b1001111;
         2: integer_to_ssd = 7'b0010010;
         3: integer_to_ssd = 7'b0000110;
         4: integer_to_ssd = 7'b1001100;
         5: integer_to_ssd = 7'b0100100;
         6: integer_to_ssd = 7'b0100000;
         7: integer_to_ssd = 7'b0001111;
         8: integer_to_ssd = 7'b0000000;
         9: integer_to_ssd = 7'b0000100;
         10: integer_to_ssd = 7'b0001000;
         11: integer_to_ssd = 7'b1100000;
         12: integer_to_ssd = 7'b0110001;
         13: integer_to_ssd = 7'b1000010;
         14: integer_to_ssd = 7'b0110000;
         15: integer_to_ssd = 7'b0111000;
         default: integer_to_ssd = 7'b1111110;
      endcase
   endfunction

//Statements:------------------------------------

   //Static SRAM signals:
   assign CEn = 1'b0;
   assign OEn = 1'b0;
   assign UBn = 1'b0;
   assign LBn = 1'b0;

   //Timer (using strategy #2):
   always_ff @(posedge clk, posedge rst)
      if (rst) t <= 0;
      else if (t < tmax) t <= t + 1;
      else t <= 0;

   // Auxiliary register:
   always_ff @(posedge clk, posedge rst)
      if (rst) A_reg <= 0;
      else A_reg <= A;
```

```
    //FSM state register:
    always_ff @(posedge clk, posedge rst)
       if (rst) pr_state <= idle;
       else pr_state <= nx_state;

    //FSM combinational logic:
    always_comb begin
       case (pr_state)
          idle: begin
             A <= 0;
             WEn <= 1'b1;
             done_wr <= 1'b1;
             done_rd <= 1'b1;
             tmax <= 0;
             if (wr & ~rd) nx_state <= write1;
             else if (~wr & rd) nx_state <= read1;
             else nx_state <= idle;
             end
          write1: begin
             A <= A_reg;
             WEn <= 1'b0;
             done_wr <= 1'b0;
             done_rd <= 1'b1;
             tmax <= 0;
             nx_state <= write2;
             end
          write2: begin
             A <= A_reg + 1;
             WEn <= 1'b1;
             done_wr <= 1'b0;
             done_rd <= 1'b1;
             tmax <= 0;
             if (A <= Amax) nx_state <= write1;
             else nx_state <= hold;
             end
          read1: begin
             A <= A_reg;
             WEn <= 1'b1;
             done_wr <= 1'b1;
             done_rd <= 1'b0;
             tmax <= Tread;
             if (t>=tmax) nx_state <= read2;
             else nx_state <= read1;
             end
          read2: begin
             A <= A_reg + 1;
             WEn <= 1'b1;
             done_wr <= 1'b1;
             done_rd <= 1'b0;
             tmax <= 0;
             if (A <= Amax) nx_state <= read1;
             else nx_state <= hold;
             end
          hold: begin
             A <= 0;
             WEn <= 1'b1;
             done_wr <= 1'b1;
```

```
                done_rd <= 1'b1;
                tmax <= 0;
                if (~wr & ~rd) nx_state <= idle;
                else nx_state <= hold;
                end
        endcase

        //In-out port with tri-state:
        if (~done_wr) D <= A[7:0] + seed;
        else D <= 'z;

        //SSD signal produced by function integer_to_ssd:
        ssd <= integer_to_ssd(D[3:0]);
    end

endmodule
//---------------------------------------------
```

13.7 Exercises

Exercise 13.1: Long-String Comparator #1

Solve exercise 12.1 using SystemVerilog instead of VHDL.

Exercise 13.2: Long-String Comparator #2

Solve exercise 12.2 using SystemVerilog instead of VHDL.

Exercise 13.3: Hamming-Weight Calculator

Solve exercise 12.3 using SystemVerilog instead of VHDL.

Exercise 13.4: Leading-Ones Counter

Solve exercise 12.4 using SystemVerilog instead of VHDL.

Exercise 13.5: Complete Reference-Value Definer

Solve exercise 12.5 using SystemVerilog instead of VHDL.

Exercise 13.6: Factorial Calculator

Solve exercise 12.6 using SystemVerilog instead of VHDL.

Exercise 13.7: Divider

Solve exercise 12.7 using SystemVerilog instead of VHDL.

14 Additional Design Examples

This chapter presents three additional FSM-based designs. They are included in a separate chapter because theoretical details and background material are also provided, leading to much longer design examples. Moreover, FSMs from all three categories are involved, depending on the application. The chapter starts with a simple LCD driver, followed by the I^2C and SPI interfaces, which are currently the most popular circuits for serial communication between integrated circuits.

14.1 LCD Driver

Like SSDs (seven segment displays), alphanumeric LCDs (liquid crystal displays) are popular options for displaying readings in all sorts of equipment, from watches to car speedometers, from microwave ovens to medical instruments. Their main advantages over SSDs are a much lower power consumption and the possibility of displaying basically any character and also simple figures, but at a higher price and a more complex driver.

14.1.1 Alphanumeric LCD

A popular alphanumeric LCD is shown in figure 14.1, which contains two lines of 16 characters each. A picture of the display is shown in figure 14.1a. The corresponding pinout is exhibited in figure 14.1b. The internal display layout is illustrated in figure 14.1c, showing 16×2 dot arrays of size 8×5 each. In figure 14.1d, its most frequent exhibition mode is depicted, consisting of 8×5-dot arrays for 7×5 characters. Finally, in figure 14.1e, its other predefined exhibition mode is depicted, consisting of 11×5-dot arrays, for 10×5 characters.

Even though this kind of display can also be found with I^2C and other serial interfaces, for low-cost applications the use of parallel access through an HD44780U microcontroller constitutes the industry standard. Such a controller (from Hitachi, or an equivalent one such as KS0066U from Samsung) is installed on the back of the device, acting as the interface between the LCD and the external world. The device is then

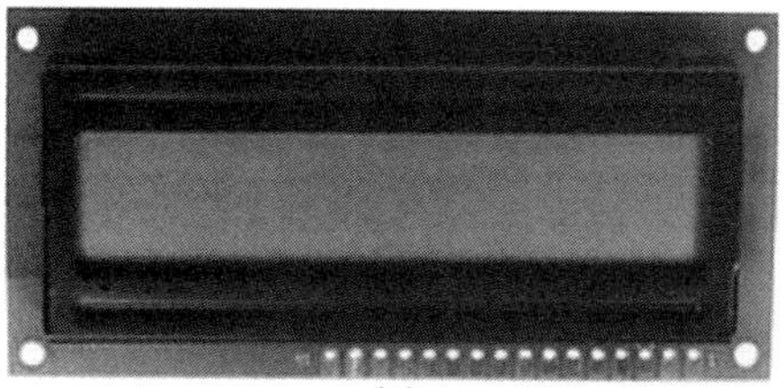

(a)

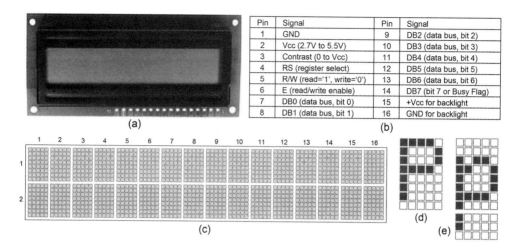

Pin	Signal	Pin	Signal
1	GND	9	DB2 (data bus, bit 2)
2	Vcc (2.7V to 5.5V)	10	DB3 (data bus, bit 3)
3	Contrast (0 to Vcc)	11	DB4 (data bus, bit 4)
4	RS (register select)	12	DB5 (data bus, bit 5)
5	R/W (read='1', write='0')	13	DB6 (data bus, bit 6)
6	E (read/write enable)	14	DB7 (bit 7 or Busy Flag)
7	DB0 (data bus, bit 0)	15	+Vcc for backlight
8	DB1 (data bus, bit 1)	16	GND for backlight

(b)

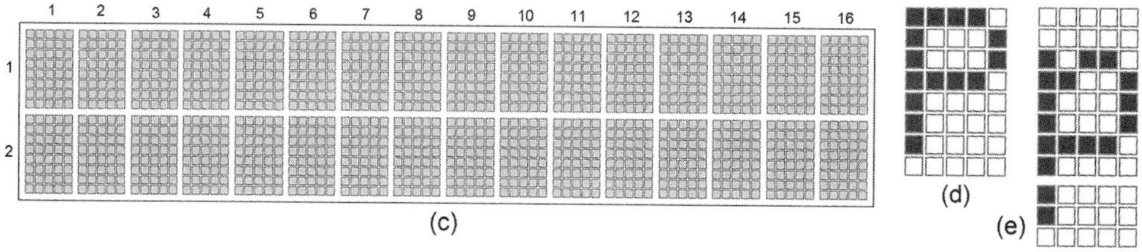
1 2 3 4 5 6 7 8 9 10 11 12 13 14 15 16
1
2
(c)
(d)
(e)

Figure 14.1

Alphanumeric LCD. (a) Popular 16×2 device. (b) Pinout (with HD44780U microcontroller). (c) Internal layout (16×2 8×5-dot blocks). (d) Standard 8×5-dot exhibition mode (two lines of 7×5 characters). (e) Standard 11×5-dot configuration (single line of 10×5 characters).

accessed through the 16 pins listed in figure 14.1b, which include power, contrast, control, and data.

Circuit Ports

To use this kind of display, the first step is to understand its microcontroller. The purposes of the signals listed in figure 14.1b are described below.

—*E* (enable, pin 6): Writing into the LCD controller occurs at the negative edge of *E*, whereas reading occurs at the positive edge.

—*RS* (register select, pin 4): '0' selects the controller's instruction register (for initialization, for example), whereas '1' selects its data register (for the characters to be displayed by the LCD).

—*R/W* (read/write, pin 5): '1' for reading, '0' for writing. If *R/W* = '0', the next negative edge of *E* causes the present instruction or data to be written into the controller's register selected by *RS*. If *R/W* = '1', data is read from the controller's register at the next positive edge of *E*.

—*DB* (data bus, pins 7–14): Bidirectional eight-bit bus for sending/receiving data or instructions to/from the LCD controller.

—*BF* (busy flag, pin 14): The microcontroller sets bit 7 of *DB* to '1' when it is busy, informing that writing is not allowed. In practice, the use of this signal can be avoided by adopting for each instruction a time long enough to guarantee completion.

Instruction	RS	RW	DB7 ... DB0	Description	Max. exec. time (*)
1) Clear Display	0	0	0 0 0 0 0 0 0 1	Clears display and resets DD RAM address to zero	1.52ms
2) Return Home	0	0	0 0 0 0 0 0 1 –	Returns display to origin and resets DD RAM address to zero	1.52ms
3) Entry Mode Set	0	0	0 0 0 0 0 1 I/D S	Sets cursor direction and display shift during read and write I/D=1/0 incr./decrement DD RAM address S=1/0 shift/do not shift display	37μs
4) Display on/off Control	0	0	0 0 0 0 1 D C B	D=1/0 display on/display off C=1/0 cursor on/cursor off B=1/0 blink/do not blink character	37μs
5) Cursor or Display Shift	0	0	0 0 0 1 S/C R/L – –	Moves cursor or display without changing DD RAM contents S/C=1 shift display, =0 shift cursor R/L=1 shift to right, =0 shift to left	37μs
6) Function Set	0	0	0 0 1 DL N F – –	Sets bus size, number of lines, and digit size (font) DL=1/0 8-bit bus/4-bit bus N=1/0 two-line/one-line operation F=1/0 5x10 dots/5x8 dots	37μs
7) Set CG RAM Address	0	0	0 1 A A A A A A	Sets CG RAM address to AAAAAA 1st position in 1st row: address=0 1st position in 2nd row: address=64	37μs
8) Set DD RAM Address	0	0	1 A A A A A A A	Sets DD RAM address to AAAAAAA	37μs
9) Read Busy Flag and Address	0	1	BF A A A A A A A	Reads busy flag and address counter	0μs
10) Write Data to CG or DD RAM	1	0	D D D D D D D D	Writes data into DD RAM or CG RAM (defined by last DD or CG RAM addr set)	41μs
11) Read Data from CG or DD RAM	1	1	D D D D D D D D	Reads data from DD RAM or CG RAM (defined by last DD or CG RAM addr set)	41μs
(*) For 270kHz internal oscillator; for other frequencies (100 to 500 kHz), multiply time given by 270kHz/oscillator freq.					

Figure 14.2
LCD controller (HD44780U or equivalent) instruction set.

Controller Instructions

The controller's instruction set is shown in figure 14.2, along with explanatory comments and worst-case execution times. As mentioned above, a common design practice is to adopt a slow clock to operate the LCD, such that any of the instructions has enough time to be completed (so reading *BF* is not necessary); for example, 500 Hz, hence allowing 2 ms for execution.

The maximum execution times in figure 14.2 are for the controller's internal oscillator operating at 270 kHz. This frequency is set by an external resistor between 75 kΩ (for $V_{DD} = 3$ V) and 91 kΩ (for $V_{DD} = 5$ V). If a different frequency is employed (with different resistor values, the range that can be covered is roughly 100–500 kHz), then the execution times must be multiplied by 270 kHz/$f_{oscillator}$.

Figure 14.3
Predefined characters of size 7×5 available in the LCD controller's ROM.

Character ROM
A total of 192 predefined characters of size 7×5 (see figure 14.3) are available in the LCD controller's character generator ROM (CGROM). The CGROM also contains 32 characters of size 10×5. When the former are used (hence, with 8×5-dot blocks), the LCD can operate with two lines; when the latter (11×5-dot blocks) are used, only single-line operation is possible. User-defined characters are also allowed, so other exhibition modes are possible, such as full-height (16×5-dot) characters.

Initialization by Instructions
An important design consideration is the controller's initialization procedure, which can be done in two ways: automatically, at power up, or by instructions. The latter can be used when the power supply conditions for automatic initialization are not met, or for safety. It consists of the following (adjusted) sequence (always with *R/W* = '0' and *RS* = '0'; as usual, "–" means "don't care"):

1) Turn the power ON.
2) Wait >15 ms after V_{DD} rises to 4.5 V.
3) Execute instruction "Function set," with *DB* = "0011– – – –".
4) Wait >4.1 ms.
5) Execute instruction "Function set," with *DB* = "0011– – – –".
6) Wait >100 μs.
7) Execute instruction "Function set," with *DB* = "0011– – – –".
8) Execute instruction "Function set," with *DB* = "0011 *N F* – –" (choose *N* and *F*).
9) Execute instruction "Clear display," with *DB* = "00000001".
10) Execute instruction "Display on/off control," with *DB* = "00001100".
11) Execute instruction "Entry mode set," with *DB* = "000001 *I/D S*" (choose *I/D* and *S*).

(Some equivalent microcontrollers have a slightly simpler initialization.)

14.1.2 Typical FSM Structure for Alphanumeric LCD Drivers

A typical FSM structure for writing to the LCD of figure 14.1 is shown in figure 14.4a, clocked by *E* at around 500 or 600 Hz. The seven states in the upper row constitute the initialization sequence (compare those states to the sequence described above),

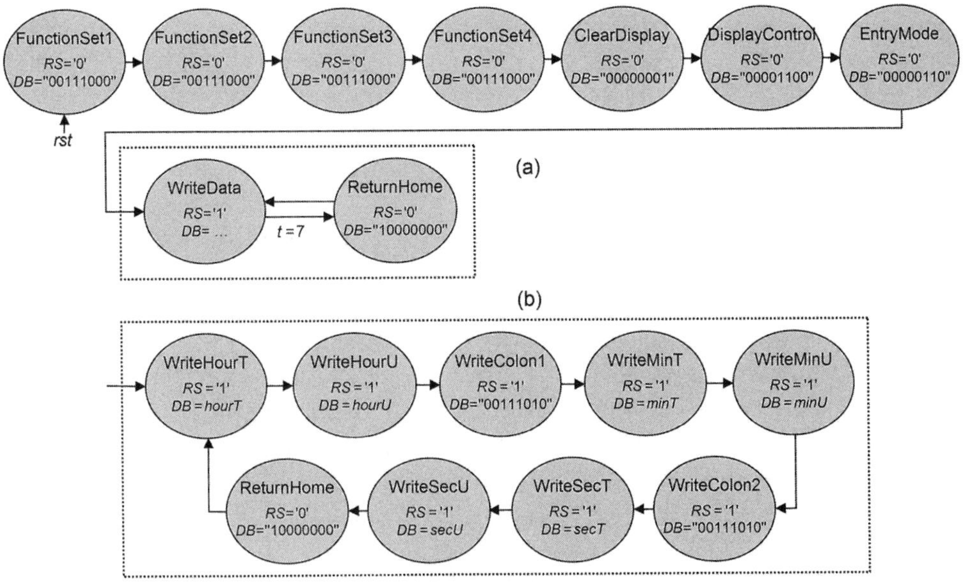
FunctionSet1
RS='0'
DB="00111000"
FunctionSet2
RS='0'
DB="00111000"
FunctionSet3
RS='0'
DB="00111000"
FunctionSet4
RS='0'
DB="00111000"
ClearDisplay
RS='0'
DB="00000001"
DisplayControl
RS='0'
DB="00001100"
EntryMode
RS='0'
DB="00000110"
rst
WriteData
RS='1'
DB= ...
t=7
ReturnHome
RS='0'
DB="10000000"
(a)
(b)
WriteHourT
RS='1'
DB=hourT
WriteHourU
RS='1'
DB=hourU
WriteColon1
RS='1'
DB="00111010"
WriteMinT
RS='1'
DB=minT
WriteMinU
RS='1'
DB=minU
ReturnHome
RS='0'
DB="10000000"
WriteSecU
RS='1'
DB=secU
WriteSecT
RS='1'
DB=secT
WriteColon2
RS='1'
DB="00111010"

Figure 14.4
(a) Typical structure for an FSM that writes characters to the LCD of figure 14.1. (b) Another example for the actual data-writing states, displaying the digits of a clock.

and the states in the lower row (within a dashed rectangle) are responsible for the actual data-writing procedure. Note that *RS* = '0' in the upper row, needed to write to the instruction register, whereas *RS* = '1' in the lower row (except for state *ReturnHome*), so the writing occurs in the data register. *R/W* (not shown) is kept permanently low (writing only).

In the initialization sequence of figure 14.4a the same value was adopted for *DB* in all four repetitions of the "Function set" instruction, with *N* = '1' (two-line operation) and *F* = '0' (5×8-dot characters). In the "Entry mode" instruction the selected values were *I/D* = '1' (DD RAM address incremented automatically) and *S* = '0' (display not shifted).

The actual data-writing sequence (lower row, inside the dashed rectangle) depends on the application. In figure 14.4a a timed (category 2) machine is employed, which writes a total of eight characters (*t* running from 0 to 7) to the LCD, then returns to the initial display address and overwrites those eight characters, repeating this loop indefinitely.

Another data-writing example is presented in figure 14.4b, this time with a regular (category 1) machine (this is the FSM that will be implemented with VHDL in the next section). It displays the digits of a digital clock, with tens of hours and units of hours in the first two positions, then a colon, followed by tens and units of minutes in the next two characters, then another colon, and finally tens and units of seconds in the last two positions, after which the machine repeats the loop, overwriting the characters with the new readings. This FSM will implement the LCD driver, while the values of *DB* (*hourT, hourU, minT, minU, secT, secU*) shown in figure 14.4b are produced by another circuit, responsible for implementing the timer proper. Note that *DB* = "00111010" was used in states *WriteColon1* and *WriteColon2*, which corresponds to the ":" character (check this in figure 14.3).

The FSM of figure 14.4 is simple enough to also be implemented using the pointer-based technique described in chapter 15.

14.1.3 Complete Design Example: Clock with LCD Display

Figure 14.5 shows a digital clock that displays hours, minutes, and seconds on an alphanumeric LCD. The circuit was divided into two blocks, with the first block implementing the clock proper and the second block implementing the LCD driver.

The Clock block is controlled by five pushbuttons, as follows. Powers-of-two (simple shifts) were chosen as speed-up factors to reduce the amount of hardware.

—*sec* (adjustment of seconds): Increases the clock speed by a factor of 8.
—*min* (adjustment of minutes): Increases the clock speed by a factor of 256.
—*hour* (adjustment of hours): Increases the clock speed by a factor of 8192.
—*rst_clock* (clock reset): Resets the clock (and so the display) to zero.
—*rst_lcd* (LCD reset): Resets the FSM that implements the LCD driver.

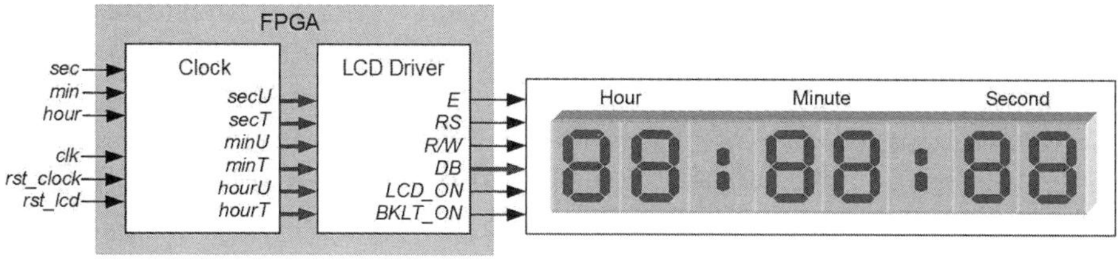
FPGA
Clock
LCD Driver
sec
min
hour
clk
rst_clock
rst_lcd
secU
secT
minU
minT
hourU
hourT
E
RS
R/W
DB
LCD_ON
BKLT_ON
Hour
Minute
Second
88:88:88

Figure 14.5
Digital clock with LCD display.

The outputs of the Clock block are the following:

—*secU*: Units of seconds.
—*secT*: Tens of seconds.
—*minU*: Units of minutes.
—*minT*: Tens of minutes.
—*hourU*: Units of hours.
—*hourT*: Tens of hours.

Finally, the outputs of the LCD Driver block are the signals already described (listed in figure 14.1b):

—*E* (enable): Actual LCD clock.
—*RS* (register select): Selects LCD instruction ('1') or data ('0') register.
—*R/W* (read/write): Selects LCD read ('1') or write ('0') operation.
—*DB* (data bus): Bidirectional eight-bit bus.
—*LCD_ON*: Turns display on ('1') or off ('0').
—*BKLT_ON*: Turns backlight on ('1') or off ('0').

VHDL Code

Because of space limitations, only the VHDL code is presented. However, with that code and the SystemVerilog codes seen in chapters 7, 10, and 13, writing the corresponding SystemVerilog code is relatively simple.

A complete VHDL code for the FSM of figure 14.4b, with the initialization sequence of figure 14.4a, is presented below. Because it is a category 1 machine, it was based on the template of section 6.3.

The entity, called *clock_with_LCD_display*, is in lines 5–12. The clock frequency was entered as a generic constant (line 6), so the speed-up factors (lines 50–53), the 1-s time base for the clock (line 57), and the frequency of the LCD clock (500 Hz, lines 120 and 124) will adjust automatically when this parameter changes. The circuit ports (lines 8–11) are exactly as in figure 14.5 and are all of type *std_logic* or *std_logic_vector* (industry standard).

The architecture, called *moore_fsm*, is in lines 14–214. As usual, it contains a declarative part and a statements part, both commented below.

The declarative part of the architecture (lines 16–43) starts with a function that converts binary-coded decimal (BCD) values into LCD characters (according to figure 14.3). It also contains FSM-related and other signal declarations. In the FSM declarations (lines 35–39), the enumerated type *state* is created to represent the machine's present and next states. The other declarations contain in line 42 the signals needed to interface the Clock block with the LCD block (see figure 14.5) and in line 43 a signal that will allow the adoption of different limits in the first clock counter so the clock can be sped up during seconds, minutes, or hours adjustments.

The code proper (statements part, lines 45–214) is divided into two parts. Part I (lines 47–110) implements the Clock block of figure 14.5, while part II (lines 112–212) implements the LCD Driver block of figure 14.5.

Part I (Clock block) contains just definitions for the speed-up factors (lines 50–53) and a basic process (lines 56–110) that implements the clock proper.

Part II (LCD Driver block) starts with definitions for the LCD static signals (lines 114–116), followed by a process (lines 119–129) that creates the 500 Hz clock for the LCD controller. The last two processes implement the FSM that runs the LCD, with the sequential section (FSM state register) in lines 132–139 and the combinational logic section in lines 142–212, based directly on the template of section 6.3. Note that this last process follows figure 14.4 exactly.

Finally, observe the correct use of registers and the completeness of the code, as described in comment 10 of section 6.3.

The reader is invited to set up this (or an equivalent) experiment and play with it in the FPGA board.

```
1     ---------------------------------------------------------------
2     library ieee;
3     use ieee.std_logic_1164.all;
4     ---------------------------------------------------------------
5     entity clock_with_LCD_display is
6        generic (fclk: natural := 50_000_000);  --freq in Hz
7        port (
8           clk, rst_clock, rst_lcd, sec, min, hour: in std_logic;
9           RS, RW, LCD_ON, BKLT_ON: out std_logic;
10          E: buffer std_logic;
11          DB: out std_logic_vector(7 downto 0));
12    end entity;
13    ---------------------------------------------------------------
14    architecture moore_fsm of clock_with_LCD_display is
15
16       --BCD-to-LCD character conversion function:
17       function bcd_to_lcd (input: natural) return std_logic_vector is
18       begin
19          case input is
```

```
         when 0 => return "00110000"; --"0" on LCD
         when 1 => return "00110001"; --"1" on LCD
         when 2 => return "00110010"; --"2" on LCD
         when 3 => return "00110011"; --"3" on LCD
         when 4 => return "00110100"; --"4" on LCD
         when 5 => return "00110101"; --"5" on LCD
         when 6 => return "00110110"; --"6" on LCD
         when 7 => return "00110111"; --"7" on LCD
         when 8 => return "00111000"; --"8" on LCD
         when 9 => return "00111001"; --"9" on LCD
         when others => return "00111111"; --"?" on LCD
      end case;
   end bcd_to_lcd;

   --FSM-related declarations:
   type state is (FunctionSet1, FunctionSet2, FunctionSet3,
      FunctionSet4, ClearDisplay, DisplayControl, EntryMode,
      WriteHourT, WriteHourU, WriteColon1, WriteMinT, WriteMinU,
      WriteColon2, WriteSecT, WriteSecU, ReturnHome);
   signal pr_state, nx_state: state;

   --Other signal declarations:
   signal secU, secT, minU, minT, hourU, hourT: natural range 0 to 9;
   signal limit: natural range 0 to fclk;

begin

   --PART I: CLOCK BLOCK-------------------------

   --Speed-up factors:
   limit <= fclk/8192 when hour='1' else
            fclk/256 when min='1' else
            fclk/8 when sec='1' else
            fclk;

   --Clock design:
   process (clk, rst_clock)
      variable counter1: natural range 0 to fclk;
      variable counter2: natural range 0 to 10;
      variable counter3: natural range 0 to 6;
      variable counter4: natural range 0 to 10;
      variable counter5: natural range 0 to 6;
      variable counter6: natural range 0 to 10;
      variable counter7: natural range 0 to 3;
   begin
      if rst_clock='1' then
         counter1 := 0;
         counter2 := 0;
         counter3 := 0;
         counter4 := 0;
         counter5 := 0;
         counter6 := 0;
         counter7 := 0;
      elsif rising_edge(clk) then
         counter1 := counter1 + 1;
         if counter1=limit then
```

```
         counter1 := 0;
         counter2 := counter2 + 1;
         if counter2=10 then
            counter2 := 0;
            counter3 := counter3 + 1;
            if counter3=6 then
               counter3 := 0;
               counter4 := counter4 + 1;
               if counter4=10 then
                  counter4 := 0;
                  counter5 := counter5 + 1;
                  if counter5=6 then
                     counter5 := 0;
                     counter6 := counter6 + 1;
                     if (counter7/=2 and counter6=10) OR
                        (counter7=2 and counter6=4) then
                        counter6 := 0;
                        counter7 := counter7 + 1;
                        if counter7=3 then
                           counter7 := 0;
                        end if;
                     end if;
                  end if;
               end if;
            end if;
         end if;
      end if;
   end if;
   secU <= counter2;
   secT <= counter3;
   minU <= counter4;
   minT <= counter5;
   hourU <= counter6;
   hourT <= counter7;
end process;

--PART II: LCD DRIVER BLOCK--------------------

LCD_ON <= '1';
BKLT_ON <= '1';
RW <= '0';

--Generate 500Hz enable (E):
process (clk)
   variable counter1: natural range 0 to fclk/1000;
begin
   if rising_edge(clk) then
      counter1 := counter1 + 1;
      if counter1=fclk/1000 then
         counter1 := 0;
         E <= not E;
      end if;
   end if;
end process;

--FSM state register:
process (E, rst_lcd)
```

```
   begin
      if rst_lcd='1' then
         pr_state <= FunctionSet1;
      elsif rising_edge(E) then
         pr_state <= nx_state;
      end if;
   end process;

   --FSM combinational logic:
   process (all)
   begin
      case pr_state is
         --Initialization:
         when FunctionSet1 =>
            RS <= '0';
            DB <= "00111000";
            nx_state <= FunctionSet2;
         when FunctionSet2 =>
            RS <= '0';
            DB <= "00111000";
            nx_state <= FunctionSet3;
         when FunctionSet3 =>
            RS <= '0';
            DB <= "00111000";
            nx_state <= FunctionSet4;
         when FunctionSet4 =>
            RS <= '0';
            DB <= "00111000";
            nx_state <= ClearDisplay;
         when ClearDisplay =>
            RS <= '0';
            DB <= "00000001";
            nx_state <= DisplayControl;
         when DisplayControl =>
            RS <= '0';
            DB <= "00001100";
            nx_state <= EntryMode;
         when EntryMode =>
            RS <= '0';
            DB <= "00000110";
            nx_state <= WriteHourT;
         --Write data:
         when WriteHourT =>
            RS <= '1';
            DB <= bcd_to_lcd(hourT);
            nx_state <= WriteHourU;
         when WriteHourU =>
            RS <= '1';
            DB <= bcd_to_lcd(hourU);
            nx_state <= WriteColon1;
         when WriteColon1 =>
            RS <= '1';
            DB <= "00111010";
            nx_state <= WriteMinT;
         when WriteMinT =>
            RS <= '1';
            DB <= bcd_to_lcd(minT);
```

```
            nx_state <= WriteMinU;
         when WriteMinU =>
            RS <= '1';
            DB <= bcd_to_lcd(minU);
            nx_state <= WriteColon2;
         when WriteColon2 =>
            RS <= '1';
            DB <= "00111010";
            nx_state <= WriteSecT;
         when WriteSecT =>
            RS <= '1';
            DB <= bcd_to_lcd(secT);
            nx_state <= WriteSecU;
         when WriteSecU =>
            RS <= '1';
            DB <= bcd_to_lcd(secU);
            nx_state <= ReturnHome;
         when ReturnHome =>
            RS <= '0';
            DB <= "10000000";
            nx_state <= WriteHourT;
      end case;
   end process;

end architecture;
--------------------------------------------------------------
```

14.2 I^2C Interface

I^2C (Inter Integrated Circuit) is a synchronous eight-bit oriented serial bus for communication between integrated circuits installed next to each other (normally on the same board). Created by Philips in the 1980s, it is a two-wire bus with five standardized speed modes, called *standard* (100 kbps), *fast* (400 kbps), *fast-plus* (1 Mbps), *high-speed* (3.4 Mbps), and *ultrafast* (5 Mbps).

14.2.1 I^2C Bus Structure

The I^2C bus general structure is depicted in figure 14.6. Its two wires are called *SCL* (serial clock) and *SDA* (serial data), which interconnect a master unit to a number of

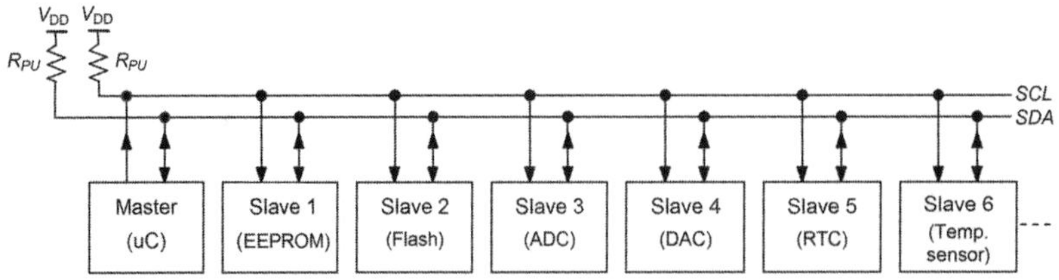
VDD
VDD
RPU
RPU
SCL
SDA
Master
(uC)
Slave 1
(EEPROM)
Slave 2
(Flash)
Slave 3
(ADC)
Slave 4
(DAC)
Slave 5
(RTC)
Slave 6
(Temp. sensor)

Figure 14.6
General I^2C bus structure.

slave units. A common ground wire (not shown) is obviously also needed for the system to function. Examples of IC families currently fabricated with I^2C support are also shown in the figure, including microcontrollers, EEPROM and Flash memories, A/D and D/A converters, RTC (real time clock) circuits, temperature sensors, and accelerometers, among others.

As indicated in figure 14.6, the clock (*SCL*) is unidirectional, always generated by the master (usually a microcontroller), whereas data (*SDA*) transmission is bidirectional. Because *SCL* and *SDA* are open-drain lines (the 5-Mbps version also allows push-pull logic), external pull-up resistors (R_{PU}) must be connected between these wires and the power supply.

The number of devices sharing the same bus can be up to 128 (seven-bit address) or 1024 (10-bit address). More than one master is allowed, in which case the I^2C protocol provides bus arbitration. Other advanced features include clock stretching, general call, reset by software, and others.

14.2.2 Open-Drain Outputs

As mentioned, *SCL* and *SDA* are open-drain pins. Details on open-drain connections for *SDA* (which is bidirectional) are shown in figure 14.7. Note that for an individual stage (the master, for example) to have its output high, the corresponding nMOS transistor must be cut off (so its gate voltage must be low), because then the output voltage will be elevated to V_{DD} by the pull-up resistor. Because all stages are hardwired to the same *SDA* node, the only way to have *SDA* high is to have all nMOS transistors off; in other words, all individual outputs must be high. Consequently, the *SDA* node behaves as an AND gate, so any nonactive unit must keep its output high (i.e., internal nMOS transistor gate voltage low) in order not to interfere with the communication between any other units.

Because of the open-drain arrangement, the high-impedance state ('Z', in VHDL) provided by tristate buffers is actually not needed. However, the design example shown ahead is tested with the master on an FPGA (without open-drain pads), so in that case the 'Z' state is required.

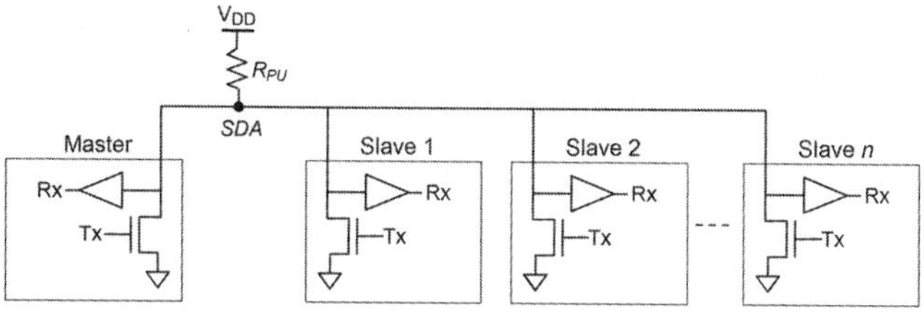
VDD
RPU
SDA
Master
Rx
Tx
Slave 1
Rx
Tx
Slave 2
Rx
Tx
Slave n
Rx
Tx

Figure 14.7
Open-drain connections for the *SDA* wire.

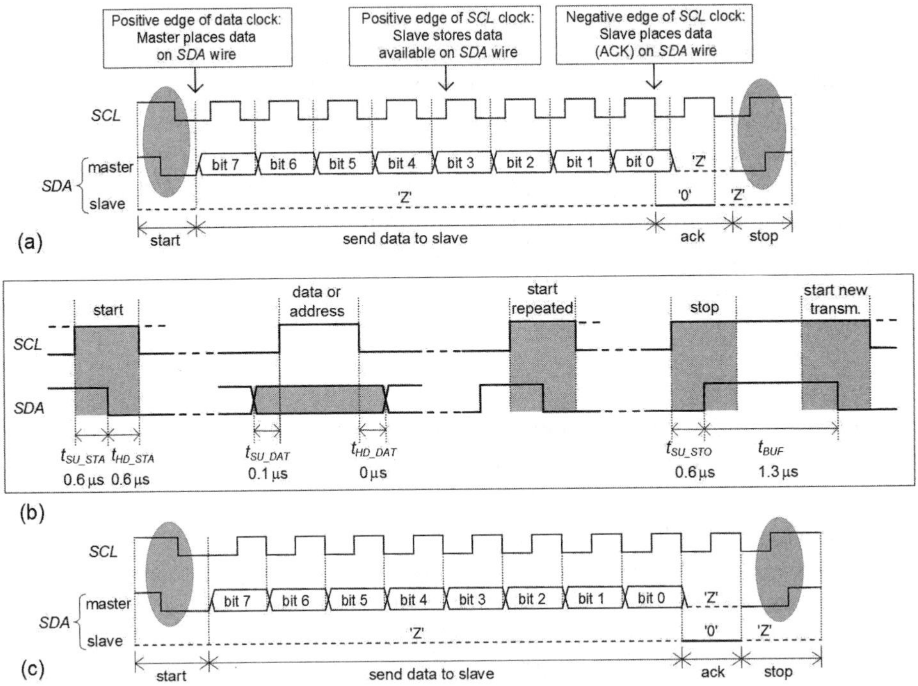

Positive edge of data clock: Master places data on SDA wire
Positive edge of SCL clock: Slave stores data available on SDA wire
Negative edge of SCL clock: Slave places data (ACK) on SDA wire
SCL
SDA
master
slave
bit 7
bit 6
bit 5
bit 4
bit 3
bit 2
bit 1
bit 0
'Z'
'0'
(a)
start
send data to slave
ack
stop
data or address
start repeated
start new transm.
tSU_STA tHD_STA
0.6 µs 0.6 µs
tSU_DAT
0.1 µs
tHD_DAT
0 µs
tSU_STO
0.6 µs
tBUF
1.3 µs
(b)
(c)

Figure 14.8
(a) I²C communication principle. (b) Time parameters. (c) Resulting (allowed) operation.

The value of R_{PU} is typically in the 1-kΩ to 33-kΩ range. It depends on the total *SCL* or *SDA* wire capacitance; if it is large (long bus with many slaves), then the resistor must be small to achieve the rise time defined in the I²C specifications. The value of V_{DD} was 5 V in initial I²C-driven devices, but voltages as low as 1.8 V are now common.

14.2.3 I²C Bus Operation

Data transfers are always done one byte at a time, after which an acknowledgment bit is issued by the receiving end. The general principle is depicted in figure 14.8a, which shows a data transmission from the master to a slave. The start sequence consists of lowering *SDA* with *SCL* high, whereas the stop sequence consists of raising *SDA* with *SCL* high. This means that during data transmission *SDA* must remain stable while *SCL* is high; otherwise start/stop commands will occur (note in the figure that the data is always updated while *SCL* is low). While the master is transmitting (always MSB first), the slave remains with its output high (nMOS transistor cut off—represented by 'Z' in the figure), so the master has control over the *SDA* wire. After the eighth bit is received, the slave issues an ack bit, which (for obvious reasons) can only be a '0'. As also depicted in figure 14.8a, the slave stores the data available on the *SDA* wire at the positive edge of *SCL* and places data on that wire at the negative edge.

The main time parameters are summarized in figure 14.8b, which have the following meaning:

t_{SU_STA} (setup time for start): *SCL* stable high before *SDA* high-to-low transition.
t_{HD_STA} (hold time for start): *SCL* still stable high after *SDA* high-to-low transition.
t_{SU_DAT} (setup time for data): data or address stable before *SCL* pulse.
t_{HD_DAT} (hold time for data): data or address still stable after *SCL* pulse.
t_{SU_STO} (setup time for stop): *SCL* stable high before *SDA* low-to-high transition.
t_{BUF} (bus-free time): bus-free time before another data transmission.

Figure 14.8b also shows examples of numerical values for these time parameters. A very important observation is that $t_{HD\_DAT} = 0$ (this is generally true for I^2C-interfaced devices), which causes the simpler timing diagram of figure 14.8c to be allowed.

The overall I^2C protocol is summarized in figure 14.9. In figure 14.9a the master writes data to a slave, whereas in figure 14.9b it reads data from a slave. White blocks represent actions taken by the master, and gray blocks indicate actions taken by the slave.

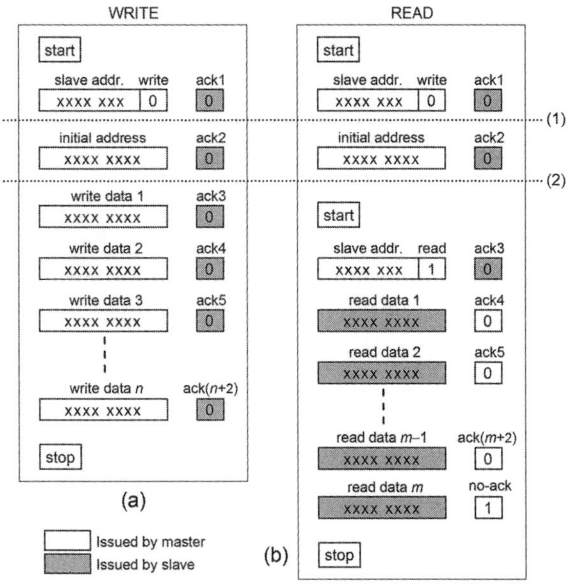
WRITE
start
slave addr.
write
ack1
XXXX XXX
0
0
initial address
ack2
XXXX XXXX
0
write data 1
ack3
write data 2
ack4
write data 3
ack5
write data n
ack(n+2)
stop
(a)
Issued by master
Issued by slave
READ
(1)
(2)
read
ack3
1
read data 1
ack4
read data 2
ack5
read data m–1
ack(m+2)
read data m
no-ack
(b)

Figure 14.9
Summary of I^2C operation for (a) writing and (b) reading.

The write procedure (figure 14.9a) begins with a start pulse, followed by the first byte, which contains the slave's seven-bit address plus a '0' appended to its right-end (this '0' informs the slave that a write procedure will occur). The corresponding slave responds with an acknowledgment bit (= '0'). The second byte is then issued by the master, containing the initial memory address where the writing must occur, to which the slave responds with another ack bit. After this, data writing begins, which can consist of any number of bytes, until the master ends the operation with a stop pulse.

The read procedure (figure 14.9b) is exactly the same as the write procedure up to line 1, or up to line 2 if the same initial address is used for writing and for reading. After line 2 another start pulse is issued by the master, followed by the seven-bit slave's address, this time with a '1' appended to its right-end (informing that a read operation will occur), to which the slave responds with a final ack bit. After this point the slave issues the data and the master issues the ack bit. Again, any number of bytes can be transferred, until a no-ack (= '1') bit is issued by the master, followed by a stop pulse.

The repetition seen in figure 14.9b (before line 2) is sometimes referred to as *dummy write*. It is necessary because I²C also permits reading from wherever the address pointer sits, in which case the entire portion before line 2 is omitted. In other words, the dummy write resets the address pointer to a specific position.

The diagram of figure 14.9 is shown in a temporal fashion in figure 14.10, now with all waveforms to be used in the experiments explicitly shown. This diagram is based directly on figure 14.8c. Because $t_{HD\_DAT} = 0$ (figure 14.8b), only a single clock is actually needed. Observe the safe construction of the restart sequence, needed for reading, which takes two clock cycles (left and right portions). Note also the inclusion of a hold state at the end, which waits for wr = '0' (and also rd = '0' if sequential

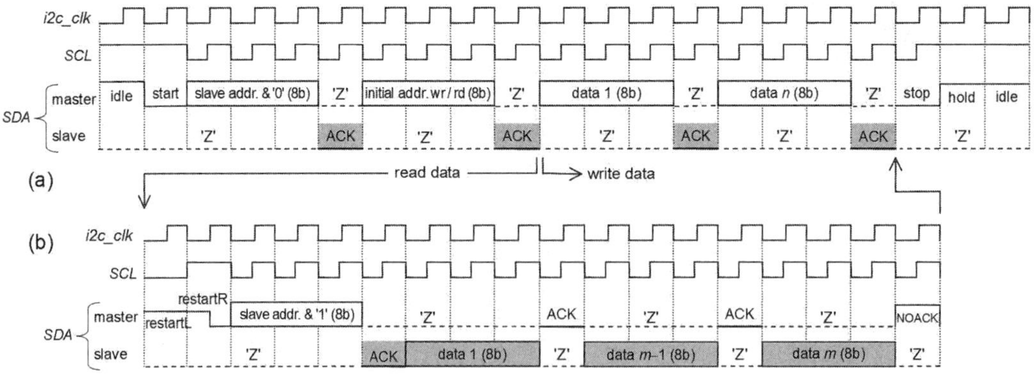
i2c_clk
SCL
master
SDA
slave
idle
start
slave addr. & '0' (8b)
'Z'
initial addr. wr / rd (8b)
data 1 (8b)
data n (8b)
stop
hold
idle
ACK
(a)
read data
write data
(b)
restartR
restartL
slave addr. & '1' (8b)
data 1 (8b)
data m-1 (8b)
data m (8b)
NOACK

Figure 14.10

Detailed I²C signals for (a) writing and (b) reading (compare to the sequences in figure 14.9).

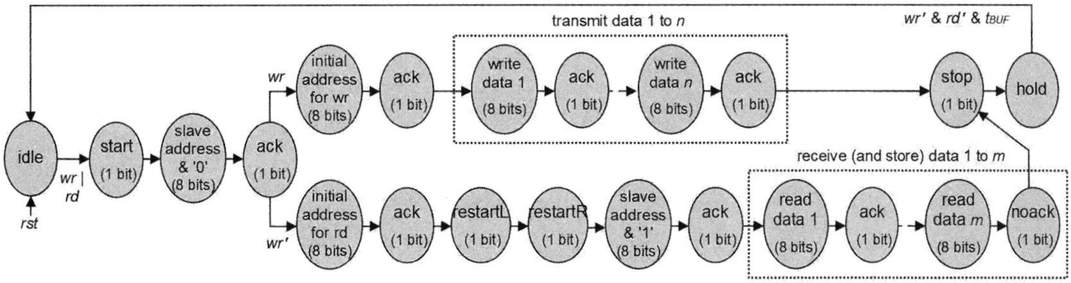
transmit data 1 to n
wr' & rd' & tBUF
idle
wr | rd
rst
start (1 bit)
slave address & '0' (8 bits)
ack (1 bit)
wr
wr'
initial address for wr (8 bits)
ack (1 bit)
write data 1 (8 bits)
ack (1 bit)
write data n (8 bits)
ack (1 bit)
stop (1 bit)
hold
receive (and store) data 1 to m
initial address for rd (8 bits)
ack (1 bit)
restartL (1 bit)
restartR (1 bit)
slave address & '1' (8 bits)
ack (1 bit)
read data 1 (8 bits)
ack (1 bit)
read data m (8 bits)
noack (1 bit)

Figure 14.11
Typical FSM structure for I^{2}C master implementations.

reading is not wanted) and for a time equal to t_{BUF} before a new transmission can take place. Because this diagram is completely generic, it can be used as the basis for any circuit that writes/reads data to/from any I^{2}C-interfaced device.

14.2.4 Typical FSM Structure for I^{2}C Applications

Figure 14.11 shows a typical FSM structure for implementing an I^{2}C master circuit. The sequence of states is based directly on figures 14.9 and 14.10. The process starts when a *wr* = '1' or *rd* = '1' pulse is received, with the first three states after the idle state corresponding to the initialization sequence (down to line 1 of figure 14.9). After this point the upper branch is pursued if writing is wanted, or the lower branch if reading is intended.

Understanding this state transition diagram well is very important because basically the only changes from one I^{2}C application to another are in the data-write and data-read sequences inside the dashed rectangles. Two very important aspects of this machine are commented on below.

The first point regards the duration of the *wr* and *rd* signals. Note that these signals are used to make decisions in two points along the FSM, so at least one of them must last up to the point where the second decision must be made. If *wr* and *rd* are short pulses, then a stretcher (section 8.11.10) can be used; another (simpler) solution is to repeat the three initial states for writing and for reading. If, on the other hand, *wr* and *rd* are long pulses, then the *hold* state (which waits also for t_{BUF}) shown after *stop* can solve the problem.

The second point regards the two blocks within dashed rectangles. If the number of bytes to be transmitted or received is small, then one pair "write-data + ack" or "read-data + ack" can be used for each byte. However, for a large number of bytes, it is more practical to build a loop to have the same pair repeated a number of times (except for the last pair when reading data, because then no-ack must be used in place of ack). Both solutions can be implemented with a category 2 machine (based on the

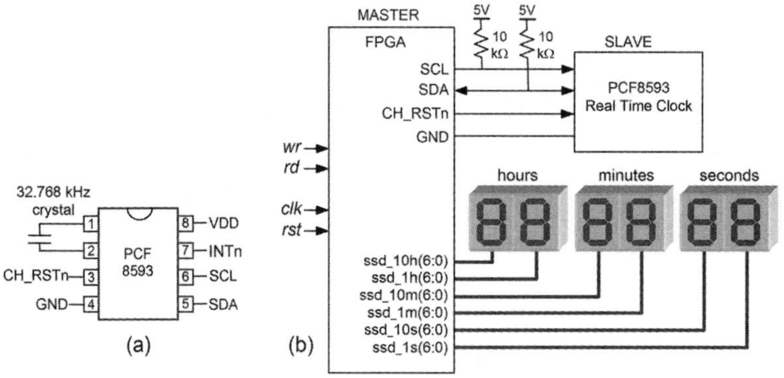
32.768 kHz
crystal
PCF
8593
VDD
INTn
SCL
SDA
CH_RSTn
GND
(a)
(b)
MASTER
FPGA
SCL
SDA
CH_RSTn
GND
wr
rd
clk
rst
5V
5V
10 kΩ
10 kΩ
SLAVE
PCF8593
Real Time Clock
hours
minutes
seconds
ssd_10h(6:0)
ssd_1h(6:0)
ssd_10m(6:0)
ssd_1m(6:0)
ssd_10s(6:0)
ssd_1s(6:0)

Figure 14.12
(a) PCF8593 pinout. (b) Setup for the experiment.

material in section 8.7), or with a category 3 machine (based on the material in section 11.5), or still using the pointer-based technique described in chapter 15. In the design example shown next, a category 3 machine is used, with individual pairs for reading and a looped pair for writing.

14.2.5 Complete Design Example: RTC (Real-Time Clock) Interface

This section shows a complete design example for an I^2C master that interfaces with an RTC (real-time clock) circuit. The master first writes the current time and date to the RTC; then it reads the clock-related data continuously, hence having the RTC operate as a high-precision clock. The circuit is implemented with VHDL and physically tested in the FPGA development board.

The RTC employed in this example is the PFC8593 (see figure 14.12a), from NXP, which contains clock, calendar, timer, and alarm features (see the device's manual).

The setup for the experiment is shown in figure 14.12b. The inputs are the *wr* and *rd* commands plus the traditional clock and reset. The output is divided into two sets; the first set contains *SCL* and *SDA* (plus a chip-reset signal), thus constituting the actual I^2C bus; the second set of outputs is for testing the circuit, displaying on six SSDs (or on an LCD) the data (clock information) read from the RTC. The FSM first writes the time and date into the RTC; then it reads continuously the clock-related data produced by the RTC.

The 16 registers (each eight bits long) of the PCF8593 RTC are detailed in figure 14.13. Register 0 is used for setup information; register 1 stores subseconds, with units of subseconds in bits 3:0, and tens of subseconds in bits 7:4; register 2 stores seconds, with units of seconds (0 to 9 values) in bits 3:0, and tens of seconds (values from 0 to 5) in bits 7:4; and so on.

Register name	Address	Bit7	Bit6	Bit5	Bit4	Bit3	Bit2	Bit1	Bit0
Control	0	RTC setup							
Subseconds	1	10subsec				1subsec			
Seconds	2	10sec				1sec			
Minutes	3	10min				1min			
Hours	4	24h/12h	AM/PM	10hour		1hour			
Year / Date	5	year		10date		1date			
Weekday/Month	6	weekday			10mo	1mo			
Timer	7	(timer register)							
Alarm	8-15	(alarm registers)							

Figure 14.13
PCF8593 registers.

Data is written to registers 0 to 7, which comprise the clock and calendar; then registers 2 to 4 are read (thus the initial address for reading is different from that for writing), which contain clock information concerning seconds, minutes, and hours. The following data are written (assuming that the present time and date are 1:30 pm of Christmas day): Control = "00000000"; Subseconds = "00000000" (0.00 s); Seconds = "00000000" (00 s); Minutes = "00110000" (30 min); Hours = "00010011" (13 h, 24-h option selected); Date = "00100101" (date 25); Month = "00010010" (month 12).

A detailed state transition diagram for this problem is presented in figure 14.14, based directly on figures 14.10 and 14.11. Either a category 2 or a category 3 machine can be used to implement this kind of circuit; the latter option was chosen here, whereas the former option will be employed in the next section, which deals with the SPI interface. This FSM is simple enough to also be implemented using the pointer-based technique described in chapter 15.

Figure 14.14 was divided into three parts. The overall FSM is presented in figure 14.14a, where six common states plus write and read blocks are shown. Because the *wr* and *rd* commands are produced by two switches (long signals) in the experiments, the state called *hold* was included after *stop* to force the machine to wait until *wr* = '0' occurs before returning to *idle* (long *rd* = '1' is accepted because continuous reading is wanted here, although this could also be done by repeating only the data-reading states). It was chosen not to have *hold* wait for t_{BUF} because another immediate write sequence is very unlikely to be needed in this kind of application. The write sequence is shown in figure 14.14b; seven bytes of data (listed under the dashed rectangle) must be transmitted, so the pair of states inside the dashed rectangle must be repeated seven times. Finally, the read sequence is presented in figure 14.14c; three bytes of data (listed under the dashed rectangle) must be received (and stored), so the pair of states inside the dashed rectangle must be repeated three times. We have elected to use a pointer (*j*) to repeat the *wr_data-ack3* pair in the transmitter and to use three separate pairs to repeat the *rd_data-ack* (or *noack*) pair in the receiver.

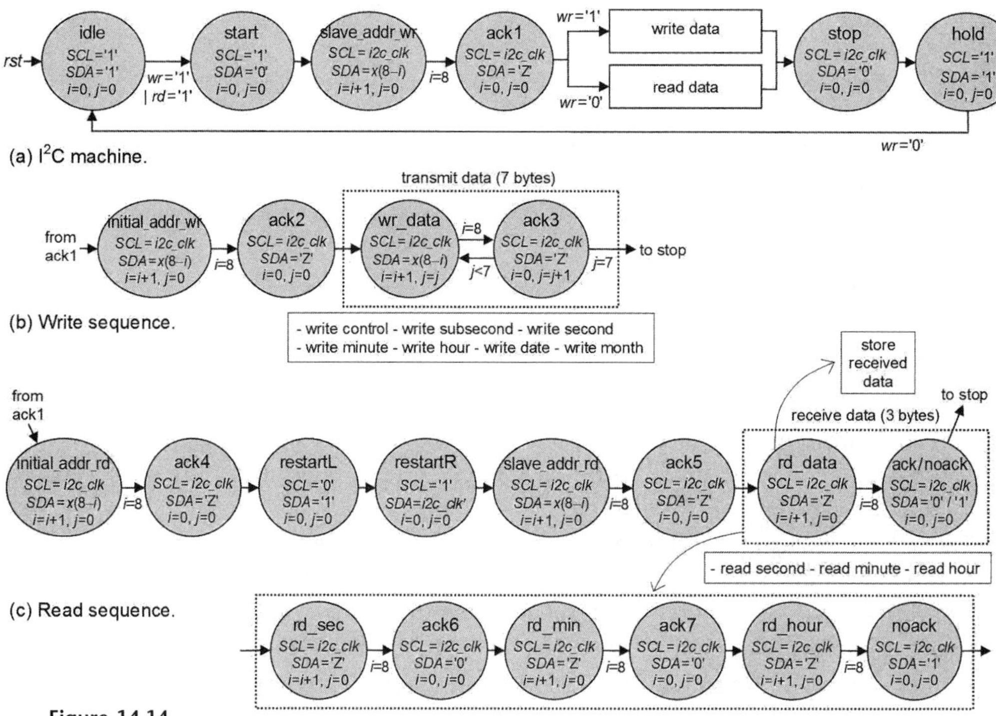
rst
idle
SCL='1'
SDA='1'
i=0, j=0
wr='1' | rd='1'
start
SCL='1'
SDA='0'
i=0, j=0
slave_addr_wr
SCL=i2c_clk
SDA=x(8–i)
i=i+1, j=0
i=8
ack1
SCL=i2c_clk
SDA='Z'
i=0, j=0
wr='1'
write data
read data
wr='0'
stop
SCL=i2c_clk
SDA='0'
i=0, j=0
hold
SCL='1'
SDA='1'
i=0, j=0
wr='0'
(a) I²C machine.
transmit data (7 bytes)
from ack1
initial_addr_wr
SCL=i2c_clk
SDA=x(8–i)
i=i+1, j=0
i=8
ack2
SCL=i2c_clk
SDA='Z'
i=0, j=0
wr_data
SCL=i2c_clk
SDA=x(8–i)
i=i+1, j=j
i=8
ack3
SCL=i2c_clk
SDA='Z'
i=0, j=j+1
j<7
j=7
to stop
(b) Write sequence.
- write control - write subsecond - write second
- write minute - write hour - write date - write month
store received data
from ack1
to stop
receive data (3 bytes)
initial_addr_rd
SCL=i2c_clk
SDA=x(8–i)
i=i+1, j=0
i=8
ack4
SCL=i2c_clk
SDA='Z'
i=0, j=0
restartL
SCL='0'
SDA='1'
i=0, j=0
restartR
SCL='1'
SDA=i2c_clk'
i=0, j=0
slave_addr_rd
SCL=i2c_clk
SDA=x(8–i)
i=i+1, j=0
i=8
ack5
SCL=i2c_clk
SDA='Z'
i=0, j=0
rd_data
SCL=i2c_clk
SDA='Z'
i=i+1, j=0
i=8
ack/noack
SCL=i2c_clk
SDA='0' / '1'
i=0, j=0
- read second - read minute - read hour
(c) Read sequence.
rd_sec
SCL=i2c_clk
SDA='Z'
i=i+1, j=0
i=8
ack6
SCL=i2c_clk
SDA='0'
i=0, j=0
rd_min
SCL=i2c_clk
SDA='Z'
i=i+1, j=0
i=8
ack7
SCL=i2c_clk
SDA='0'
i=0, j=0
rd_hour
SCL=i2c_clk
SDA='Z'
i=i+1, j=0
i=8
noack
SCL=i2c_clk
SDA='1'
i=0, j=0

Figure 14.14
Complete FSM for the I²C RTC interface.

Two pointers are used in figure 14.14. The first (*i*) is employed in all states where more than one bit must be transmitted or received. The second (*j*) can be useful when multiple bytes are involved, which is the case of the dashed rectangles in figures 14.14b,c; as mentioned above, we have chosen to use *j* only in the transmitter. Recall that in a category 2 (timed) machine the pointer would run from 0 to 7 (for one byte of data), whereas here it runs from 1 to 8 because the pointer is immediately incremented when the FSM enters a multibit state (then the $i = 8$ condition in the state diagram would be $i = 7$ if it were a category 2 machine).

Observe that the values for *SDA* in figure 14.14 are those that must be produced by the master. For example, when the slave is supposed to answer with *SDA* = '0' in the ack states, the master must produce *SDA* = 'Z' (FPGA implementation). Note also that *SDA* = $x(8–i)$ was used in the eight-bit writing states, indicating that the MSB will be transmitted first [recall that *i* ranges from 1 to 8, so $x(7)$ will go first and $x(0)$ will go last]. Here, *x* is just a generic name; the actual signal name varies from one state to another. Finally, note that the received data must be stored somewhere, as indicated by an arrow and a box over the dashed rectangle in figure 14.14c.

VHDL Code

Because of space limitations, only the VHDL code is presented. However, with this code and the SystemVerilog codes seen in chapters 7, 10, and 13, writing the corresponding SystemVerilog code is relatively simple.

A complete VHDL code for the FSM of figure 14.14 is presented below. Because it is a category 3 machine, it was based on the template of section 12.2.

Initially, a function called *bcd_to_ssd*, to convert BCD (binary-coded decimal) values into SSD values (display driver), was built in a separate package (called *my_functions*), which is called in the main code (lines 252–257) to make the corresponding conversions.

The entity, called *RTC_with_I2C_bus*, is in lines 6–37. A number of system parameters were entered as generic constants (lines 7–23), including the clock frequency (50 MHz, line 9) and the desired I^2C speed (100 kbps, line 10), so the I^2C clock (*i2c_clk*) is automatically adjusted (in lines 71 and 75) when these parameters change. They also include the RTC addresses of interest (lines 12–15) and the time and date to be stored in the RTC registers (lines 17–23).

The circuit ports, all of type *std_logic* or *std_logic_vector* (industry standard), are in lines 24–36. They are exactly as in figure 14.12b.

The architecture, called *moore_fsm*, is in lines 39–259. As usual, it contains a declarative part and a statements part, both commented on below.

The declarative part of the architecture (lines 41–63) contains FSM-related and auxiliary-register-related declarations plus other system declarations. In the FSM declarations (lines 42–50), the enumerated type *state* is created to represent the machine's present and next states. In the auxiliary-register declarations (lines 53–54), the signals needed to build the pointers *i* and *j* are created. Finally, the other declarations (lines 57–63) include the I^2C clock, the signals that will store the values read from the RTC (test circuit), and also a 1D×1D type called *data_array*, used to build a ROM called *data_out* containing the data to be sent to the slave (to set the clock and calendar).

The statements part (lines 65–259) contains five processes. The first process (lines 70–80) produces *i2c_clk*, with frequency 100 kHz (desired data rate).

The second process (lines 83–94) builds the FSM's state register plus the auxiliary registers for *i* and *j*.

The third process (lines 97–237) implements the entire combinational logic section of the FSM, following the state transition diagram of figure 14.14 exactly. Note that because some of the output values are repeated a number of times, they were entered as default values in lines 100–102, so the actual list of outputs *(SCL, SDA, i, j)* is indeed exactly the same in all states.

The fourth and final process (lines 240–251) plus associated statements (lines 252–257) constitute the test circuit. It stores the data read from the RTC and sends it to the display.

Observe the correct use of registers and the completeness of the code, as described in comment 10 of section 6.3.

The reader is invited to set up this (or an equivalent) experiment and play with it in the FPGA board.

```
----Package with function "bcd_to_ssd":-----------------
library ieee;
use ieee.std_logic_1164.all;
package my_functions is
   function bcd_to_ssd(input:std_logic_vector)
      return std_logic_vector;
end my_functions;
---------------------------------------------------------
package body my_functions is
   function bcd_to_ssd(input: std_logic_vector)
      return std_logic_vector is
   begin
      case input is
         when "0000" => return "0000001";   --"0" on SSD
         when "0001" => return "1001111";   --"1" on SSD
         when "0010" => return "0010010";   --"2" on SSD
         when "0011" => return "0000110";   --"3" on SSD
         when "0100" => return "1001100";   --"4" on SSD
         when "0101" => return "0100100";   --"5" on SSD
         when "0110" => return "0100000";   --"6" on SSD
         when "0111" => return "0001111";   --"7" on SSD
         when "1000" => return "0000000";   --"8" on SSD
         when "1001" => return "0000100";   --"9" on SSD
         when others => return "1111110";   --"-" on SSD
      end case;
   end bcd_to_ssd;
end package body;
---------------------------------------------------------
```

```
----Main code:-------------------------------------------------------
library ieee;
use ieee.std_logic_1164.all;
use work.my_functions.all; --package with "bcd_to_ssd" function
---------------------------------------------------------------------
entity RTC_with_I2C_bus is
   generic (
      --Clock parameters:
      fclk: positive := 50_000_000;  --Clock frequency in Hz
      data_rate: positive := 100_000;  --Desired I2C bus speed in bps
      --RTC addresses:
      slave_addr_for_wr: std_logic_vector(7 downto 0) := "10100010";
      slave_addr_for_rd: std_logic_vector(7 downto 0) := "10100011";
      initial_addr_for_wr:std_logic_vector(7downto0):="00000000";
      initial_addr_for_rd:std_logic_vector(7downto0):="00000010";
      --Values to store in the RTC clock/calendar registers:
      set_control: std_logic_vector(7 downto 0)  := "00000000";
      set_subsec:std_logic_vector(7downto0):="00000000"; --0.00sec
      set_sec: std_logic_vector(7 downto 0) := "00000000";  --00 sec
```

```
      set_min: std_logic_vector(7 downto 0) := "00110000";  --30 min
      set_hour: std_logic_vector(7 downto 0) := "00010011";  --13 h
      set_date: std_logic_vector(7 downto 0) := "00100101";  --date 25
      set_month:std_logic_vector(7downto0):="00010010"); --month12
   port (
      --Clock and control ports:
      clk, rst, wr, rd: in std_logic;
      --I2C ports:
      SCL, CH_RSTn: out std_logic;
      SDA: inout std_logic;
      --Display ports (test circuit):
      ssd_1sec: out std_logic_vector(6 downto 0);  --units of seconds
      ssd_10sec: out std_logic_vector(6 downto 0);  --tens of seconds
      ssd_1min: out std_logic_vector(6 downto 0);  --units of minutes
      ssd_10min: out std_logic_vector(6 downto 0);  --tens of minutes
      ssd_1hour: out std_logic_vector(6 downto 0);  --units of hours
      ssd_10hour: out std_logic_vector(6 downto 0));  --tens of hours
end entity;
---------------------------------------------------------------------
architecture moore_fsm of RTC_with_I2C_bus is

   --FSM-related declarations:
   type state is (
      --common states:
      idle, start, slave_addr_wr, ack1, stop, hold,
      --write-only states:
      initial_addr_wr, ack2, wr_data, ack3,
      --read-only states:
      initial_addr_rd, ack4, restartL, restartR, slave_addr_rd, ack5,
      rd_sec, ack6, rd_min, ack7, rd_hour, no_ack);
   signal pr_state, nx_state: state;

   --Auxiliary-register-related declarations:
   signal i, i_reg: natural range 0 to 8;
   signal j, j_reg: natural range 0 to 7;

   --Other declarations:
   signal i2c_clk: std_logic;
   signal sec: std_logic_vector(7 downto 0);
   signal min: std_logic_vector(7 downto 0);
   signal hour: std_logic_vector(7 downto 0);
   type data_array is array (0 to 6) of std_logic_vector(7 downto 0);
   constant data_out: data_array := (set_control, set_subsec, set_sec,
      set_min, set_hour, set_date, set_month);

begin

   CH_RSTn <= not rst; --chip reset

   --i2c_clk (100kHz):
   process (clk)
      variable count: natural range 0 to fclk/(2*data_rate);
   begin
      if rising_edge(clk) then
         count := count + 1;
         if count=fclk/(2*data_rate) then
```

```
            i2c_clk <= not i2c_clk;
            count := 0;
         end if;
      end if;
   end process;

   --FSM state register + Auxiliary register:
   process (i2c_clk, rst)
   begin
      if rst='1' then
         pr_state <= idle;
         i_reg <= 0;
         j_reg <= 0;
      elsif falling_edge(i2c_clk) then
         pr_state <= nx_state;
         i_reg <= i;
         j_reg <= j;
      end if;
   end process;

   --FSM combinational logic:
   process (all)
   begin
      --Default values:
      SCL <= i2c_clk;
      i <= 0;
      j <= 0;
      case pr_state IS
         --Common states:
         when idle =>
            SCL <= '1';
            SDA <= '1';
            if wr='1' or rd='1' then
               nx_state <= start;
            else
               nx_state <= idle;
            end if;
         when start =>
            SCL <= '1';
            SDA <= '0';
            nx_state <= slave_addr_wr;
         when slave_addr_wr =>
            SDA <= slave_addr_for_wr(8-i);
            i <= i_reg + 1;
            if i=8 then
               nx_state <= ack1;
            else
               nx_state <= slave_addr_wr;
            end if;
         when ack1 =>
            SDA <= 'Z';
            if wr='1' then
               nx_state <= initial_addr_wr;
            else
               nx_state <= initial_addr_rd;
            end if;
         when stop =>
```

```
                SDA <= '0';
                nx_state <= hold;
            when hold =>
                SCL <= '1';
                SDA <= '1';
                if wr='0' then
                    nx_state <= idle;
                else
                    nx_state <= hold;
                end if;
            --Data-write states:
            when initial_addr_wr =>
                SDA <= initial_addr_for_wr(8-i);
                i <= i_reg + 1;
                if i=8 then
                    nx_state <= ack2;
                else
                    nx_state <= initial_addr_wr;
                end if;
            when ack2 =>
                SDA <= 'Z';
                nx_state <= wr_data;
            when wr_data =>
                SDA <= data_out(j)(8-i);
                i <= i_reg + 1;
                j <= j_reg;
                if i=8 then
                    nx_state <= ack3;
                else
                    nx_state <= wr_data;
                end if;
            when ack3 =>
                SDA <= 'Z';
                j <= j_reg + 1;
                if j<7 then
                    nx_state <= wr_data;
                else
                    nx_state <= stop;
                end if;
            --Data-read states:
            when initial_addr_rd =>
                SDA <= initial_addr_for_rd(8-i);
                i <= i_reg + 1;
                if i=8 then
                    nx_state <= ack4;
                else
                    nx_state <= initial_addr_rd;
                end if;
            when ack4 =>
                SDA <= 'Z';
                nx_state <= restartL;
            when restartL =>
                SCL <= '0';
                SDA <= '1';
                nx_state <= restartR;
            when restartR =>
                SCL <= '1';
```

```
            SDA <= not i2c_clk;
            nx_state <= slave_addr_rd;
         when slave_addr_rd =>
            SDA <= slave_addr_for_rd(8-i);
            i <= i_reg + 1;
            if i=8 then
               nx_state <= ack5;
            else
               nx_state <= slave_addr_rd;
            end if;
         when ack5 =>
            SDA <= 'Z';
            nx_state <= rd_sec;
         when rd_sec =>
            SDA <= 'Z';
            i <= i_reg + 1;
            if i=8 then
               nx_state <= ack6;
            else
               nx_state <= rd_sec;
            end if;
         when ack6 =>
            SDA <= '0';
            nx_state <= rd_min;
         when rd_min =>
            SDA <= 'Z';
            i <= i_reg + 1;
            if i=8 then
               nx_state <= ack7;
            else
               nx_state <= rd_min;
            end if;
         when ack7 =>
            SDA <= '0';
            nx_state <= rd_hour;
         when rd_hour =>
            SDA <= 'Z';
            i <= i_reg + 1;
            if i=8 then
               nx_state <= no_ack;
            else
               nx_state <= rd_hour;
            end if;
         when no_ack =>
            SDA <= '1';
            nx_state <= stop;
      end case;
   end process;

   --Store data read from RTC and send it to display:
   process (i2c_clk)
   begin
      if rising_edge(i2c_clk) then
         if (pr_state=rd_sec) then
            sec(8-i) <= SDA;
         elsif (pr_state=rd_min) then
```

```
            min(8-i) <= SDA;
         elsif (pr_state=rd_hour) then
            hour(8-i) <= SDA;
         end if;
      end if;
   end process;
   ssd_1sec <= bcd_to_ssd(sec(3 downto 0));
   ssd_10sec <= bcd_to_ssd(sec(7 downto 4));
   ssd_1min <= bcd_to_ssd(min(3 downto 0));
   ssd_10min <= bcd_to_ssd(min(7 downto 4));
   ssd_1hour <= bcd_to_ssd(hour(3 downto 0));
   ssd_10hour <= bcd_to_ssd("00" & hour(5 DOWNTO 4));

end architecture;
-------------------------------------------------------------------
```

14.3 SPI Interface

Serial peripheral interface (SPI) is another synchronous serial bus for communication between integrated circuits (installed next to each other, normally on the same board). Like I^{2}C, it operates in a master-slave architecture, but it is simpler to implement and can operate at higher speeds (up to around 100 Mbps), although it requires more interconnecting wires. Developed by Motorola for its 68HC family of microcontrollers, it is now in widespread use.

14.3.1 SPI Bus Structure

The SPI bus general structure is depicted in figure 14.15. In figure 14.15a, a single slave is shown (normally, the master is a microcontroller), so four wires are needed, called *SCK* (serial clock, always generated by the master), *MOSI* (Master Out Slave In), *MISO* (Master In Slave Out), and *SSn* (Slave Select, active low). When *SSn* is low, the slave is selected, to/from which the master sends/receives messages through the *MOSI*/*MISO* wires. In figure 14.15b, a multislave system is depicted, so multiple *SSn* wires are needed. Examples of ICs with SPI support are also shown in the figure, which are essentially the same categories as for I^{2}C (e.g., microcontrollers, EEPROM and Flash memories, A/D and D/A converters, RTCs, and accelerometers).

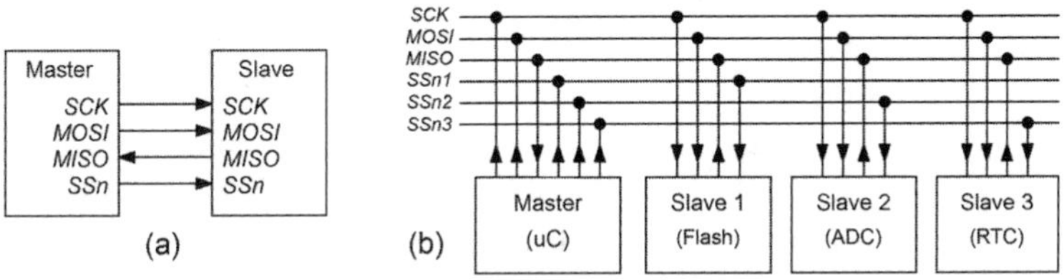
Master
SCK
MOSI
MISO
SSn
Slave
SCK
MOSI
MISO
SSn
(a)
SCK
MOSI
MISO
SSn1
SSn2
SSn3
Master
(uC)
Slave 1
(Flash)
Slave 2
(ADC)
Slave 3
(RTC)
(b)

Figure 14.15
General SPI bus structure with (a) single and (b) multiple slaves.

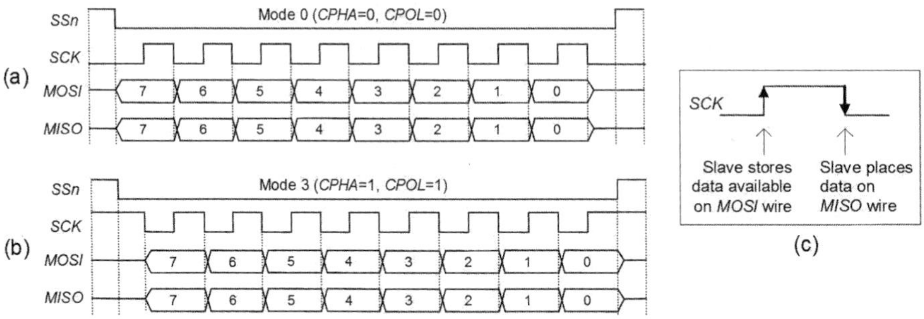
(a)
SSn
SCK
MOSI
MISO
Mode 0 (CPHA=0, CPOL=0)
7 6 5 4 3 2 1 0
7 6 5 4 3 2 1 0
(b)
SSn
SCK
MOSI
MISO
Mode 3 (CPHA=1, CPOL=1)
7 6 5 4 3 2 1 0
7 6 5 4 3 2 1 0
SCK
Slave stores data available on MOSI wire
Slave places data on MISO wire
(c)

Figure 14.16
(a, b) Main SPI operating modes. (c) Slave's registers operation.

SPI is simpler than I^2C because there is no bidirectional line, and the device selection is made with a separate wire for each slave rather than with a transmitted address. On the other hand, SPI demands more I/O pins, can operate with only one master, has no message acknowledgment, and because there is no standard message, format validation would be more difficult. SPI is said to be a four-wire bus, but that is indeed the *least* number of wires, whereas I^2C is truly two wires. In some cases a bidirectional line can be used for *MOSI* and *MISO* together, resulting then a three-wire bus.

14.3.2 SPI Bus Operation

There are four SPI operating modes, determined by the clock phase (*CPHA*) and clock polarity (*CPOL*). They are called *mode 0* (*CPHA* = 0, *CPOL* = 0), *mode 1* (*CPHA* = 0, *CPOL* = 1), *mode 2* (*CPHA* = 1, *CPOL* = 0), and *mode 3* (*CPHA* = 1, *CPOL* = 1). The two most common modes are 0 and 3, illustrated in figures 14.16a,b; note that in mode 0 *SSn* is lowered with *SCK* low, whereas the opposite occurs in mode 3.

Figure 14.16c shows how the slave operates. It stores the data available on the MOSI wire at positive clock edges and places data on the MISO wire at negative clock transitions. Consequently, the FSM used to implement the master side of this interface must operate at the negative clock edge, so the data provided by the machine will be ready for the slave at the positive clock edge. Likewise, a register that records the data issued by the slave must operate at the positive clock edge, so the data (issued at the negative clock edge) will be ready for storage.

Part of the communication between master and slave is ruled by information stored in eight-bit registers at both ends. These registers are not standardized, neither in number nor in content. For example, the SPI in the Motorola MC68HC908GT microcontroller contains three registers (for status, called SPSCR, control, SPCR, and data, SPDR), whereas the SPI in the Maxim DS1306 RTC has two registers (for status and control), and the SPI in the Ramtron FM25L512 FRAM memory contains only one (for status).

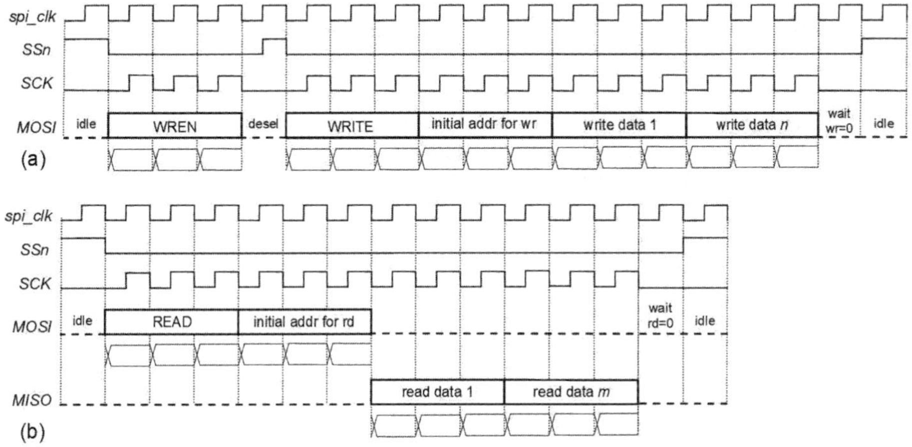
spi_clk
SSn
SCK
MOSI
idle
WREN
desel
WRITE
initial addr for wr
write data 1
write data n
wait
wr=0
idle
(a)
spi_clk
SSn
SCK
MOSI
idle
READ
initial addr for rd
wait
rd=0
idle
MISO
read data 1
read data m
(b)

Figure 14.17
Examples of SPI behavior for (a) writing and (b) reading (FM25L512 FRAM device).

The general communication procedure consists of a number of opcodes transmitted by the master to the slave, followed by a data-write or data-read procedure with any number of data bytes. The only particularity is that each opcode must be preceded by a deselect-reselect sequence.

As an example, the FRAM memory used in the design example presented ahead requires two opcodes, called WREN (sets the write enable latch) and WRITE (enables writing to the memory—at the next positive clock edge), before actual data writing takes place. The same device requires one opcode, called READ (enables reading from the memory—at the next negative clock edge), before actual data reading occurs. Consequently, a typical flow for the SPI interface for this FRAM is that depicted in figure 14.17. Note in figure 14.17a that the device is deselected-reselected between two consecutive opcodes. Dashed lines indicate "don't care" or high-impedance values for the MOSI/MISO wires. Observe the safe distance between the high-to-low transitions of *SSn* and the next positive edge of *SCK*, as well as between the low-to-high transitions of *SSn* and the previous negative edge of *SCK*, both required to be at least 10 ns in this particular device.

14.3.3 Complete Design Example: FRAM (Ferroelectric RAM) Interface

To illustrate the use of SPI, the FM25L512 FRAM mentioned above, from Ramtron, is used as an example. It is a 64k × 8 bits nonvolatile memory with serial access through an SPI bus. Its pinout, list of opcodes, and contents of the status register are shown in figure 14.18. An important feature of this technology (FRAM) is that data can be written into it at high speed (20 MHz in the present example), contrasting with EEPROM, which generally takes a few milliseconds/page.

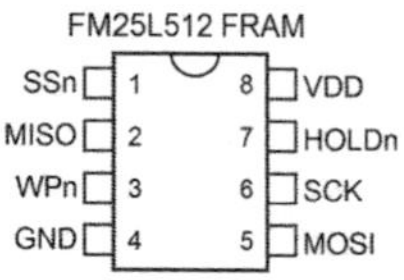
FM25L512 FRAM
SSn 1
MISO 2
WPn 3
GND 4
8 VDD
7 HOLDn
6 SCK
5 MOSI

Opcode name	Value
WREN (sets write-enable latch through *WEL*=1)	0000 0110
WRITE (enables writing to memory)	0000 0010
READ (enables reading from memory)	0000 0011
WRSR (enables writing to status register)	0000 0001
RDSR (enables reading status register)	0000 0101
WRDI (disables writing through *WEL*=0)	0000 0100

Status register							
bit 7	bit 6	bit 5	bit 4	bit 3	bit 2	bit 1	bit 0
WPEN	1	0	0	BP1	BP0	WEL	0

Figure 14.18
FM25L512 FRAM memory: Pinout, opcodes, and status register.

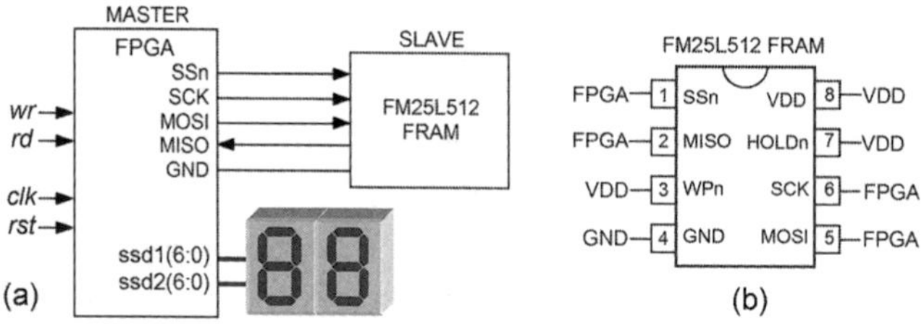
MASTER
FPGA
SSn
SCK
MOSI
MISO
GND
wr
rd
clk
rst
ssd1(6:0)
ssd2(6:0)
(a)
SLAVE
FM25L512
FRAM
FM25L512 FRAM
FPGA 1 SSn
FPGA 2 MISO
VDD 3 WPn
GND 4 GND
VDD 8 VDD
HOLDn 7 VDD
SCK 6 FPGA
MOSI 5 FPGA
(b)

Figure 14.19
(a) Setup for the experiment. (b) FRAM wiring.

Note in figure 14.18 that besides the SPI pins (*SCK, SSn, MOSI, MISO*), the chip contains also two other control pins, called *WPn* (write protect) and *HOLDn*. The purpose of *WPn* is, together with bits 7 (*WPEN*), 3 (*BP1*), and 2 (*BP0*) of the status register, to allow several protection modes against writing to both the memory and the status register. For example, with *WPn* = '1' and *WPEN* = *BP1* = *BP0* = '0', all writings are allowed (see other protection options in the device's datasheets). The role of *HOLDn* is to handle interrupts.

In addition to the bits mentioned above, there is another programmable bit in the status register, called *WEL* (write enable latch), which determines whether writing is allowed (when '1') or not (when '0'). Only when *WEL* = '1' are the protection options mentioned above in place (any writing is forbidden while *WEL* = '0'). Because this bit is automatically zeroed at power up or at the upward transition of *SSn* after a WRITE, WRSR, or WRDI opcode, any write action must start with the WREN opcode because that is the only way of setting *WEL* to '1' (writing to the status register does not affect this bit).

Figure 14.19 shows the setup for the experiment. The inputs are *wr* (write) and *rd* (read) commands plus the traditional clock (assumed to be 50 MHz) and reset (*wr, rd,*

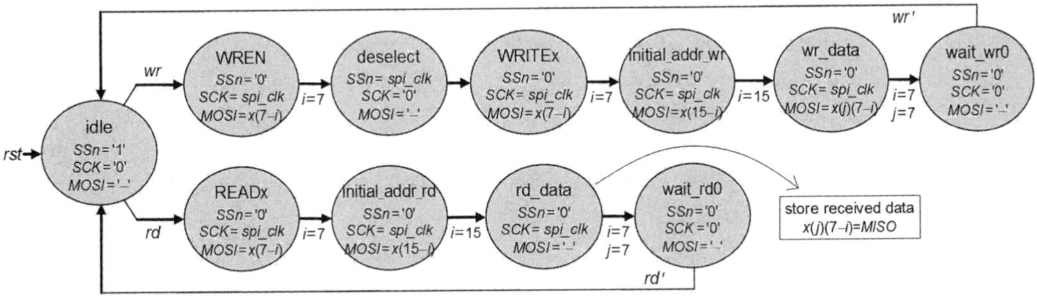
wr'
WREN
SSn='0'
SCK= spi_clk
MOSI=x(7–i)
deselect
SSn= spi_clk
SCK='0'
MOSI='–'
WRITEx
SSn='0'
SCK= spi_clk
MOSI=x(7–i)
initial_addr_wr
SSn='0'
SCK= spi_clk
MOSI=x(15–i)
wr_data
SSn='0'
SCK= spi_clk
MOSI=x(j)(7–i)
wait_wr0
SSn='0'
SCK='0'
MOSI='–'
idle
SSn='1'
SCK='0'
MOSI='–'
rst
wr
rd
i=7
i=15
j=7
READx
SSn='0'
SCK= spi_clk
MOSI=x(7–i)
initial_addr_rd
SSn='0'
SCK= spi_clk
MOSI=x(15–i)
rd_data
SSn='0'
SCK= spi_clk
MOSI='–'
wait_rd0
SSn='0'
SCK='0'
MOSI='–'
store received data
x(j)(7–i)=MISO
rd'

Figure 14.20
FSM for the SPI FRAM interface.

and *rst* are from switches). The outputs are the SPI signals (*SSn*, *SCK*, *MOSI*, and *MISO*—the last one is in fact an input), connected to the slave, plus *ssd1* and *ssd2*, which feed two SSD displays to exhibit the data retrieved from the FRAM. A frequency of 5 MHz is used for *spi_clk* (as mentioned, this device can operate at up to 20 MHz). The figure also shows how the device was wired.

A detailed FSM for this problem is presented in figure 14.20 (employing mode 0). The data-write sequence is in the upper branch, while the data-read sequence is in the lower branch. *MOSI* = *x*(7–*i*) in some of the states is just a symbolic way of saying that vector *x*, with eight bits, starting with the MSB, must be transmitted. Note that this is a category 2 (timed) machine, so the timers (here represented by *i* and *j*) run from 0 to i_{max} and 0 to j_{max}.

In this experiment a total of eight bytes are written into the FRAM, starting at address zero; note that state *wr_data* lasts from $\{i = 0, j = 0\}$ up to $\{i = 7, j = 7\}$, hence transmitting eight bytes, corresponding to *x*(0)(7:0) up to *x*(7)(7:0). A test circuit is also included, which reads all eight bytes from the FRAM and displays them sequentially onto the two SSDs; note the arrow and box associated with state *rd_data*, informing that data must be recorded (from the *MISO* wire) while the FSM is in that state.

The FSM of figure 14.20 can be implemented with a category 2 or category 3 machine (the former is employed in the VHDL code below). It is also simple enough to be implemented using the pointer-based technique described in chapter 15 (see section 15.5).

VHDL Code

A complete VHDL code for the FSM of figure 14.20 is presented below, following the template for timed (category 2) machines introduced in chapter 9.

Initially, a function called *hex_to_ssd*, to convert hexadecimal values into SSD values (display driver), was built in a separate package (called *my_functions*), then called in the main code (lines 217–218) to make the corresponding conversions.

The entity, called *FRAM_with_SPI_bus*, is in lines 6–22. The SPI parameters were declared as generic constants (lines 9–13). The circuit ports (lines 15–21) follow figure 14.19a and are all of type *std_logic* or *std_logic_vector* (industry standard).

The architecture, called *moore_fsm*, is in lines 24–224. As usual, it contains a declarative part and a statements part, both commented on below.

The declarative part of the architecture (lines 26–41) contains FSM-related and timer-related declarations plus other system declarations. In the FSM declarations (lines 27–29), the enumerated type *state* is created to represent the machine's present and next states. In the timer declarations (lines 32–33), the signals needed to build the timers *i* and *j* are created. Finally, the other declarations (lines 36–41) include the SPI clock, plus a 1D×1D type called *data_array* used to build a ROM called *data_out* (lines 38–40) whose contents are sent to the FRAM. A similar 1D×1D signal is declared in line 41, which is used to create a register in which all data read from the FRAM is stored during the tests.

The statements part (lines 43–224) contains five processes. The first process (lines 49–59) creates the SPI clock (5 MHz, assuming that the system clock is 50 MHz; as mentioned, this FRAM can operate at up to 20 MHz). As seen in figure 14.17, this is the only clock needed in the entire SPI circuit. Because the slave stores data at the rising clock edge, the FSM (and therefore its associated timers too) must operate at the negative edge.

The second process (lines 62–80) implements the timers. In this example, the timer-control strategy #1 (section 8.5.2) was adopted.

The third process (lines 83–90) implements the FSM state register. Like the timers, it too operates at the negative clock edge.

The fourth process (lines 93–193) implements the entire combinational logic section. It is just a list of all states, obeying the state transition diagram of figure 14.20 exactly. As usual, in each state the outputs (*SSn, SCK, MOSI*) and the time parameters (i_{max}, j_{max}) are specified, and the next state is defined. Note that because some of the output values are repeated a number of times, default values were entered in lines 96–100, so the actual list of outputs and time parameters is exactly the same in all states.

The fifth and final process (lines 196–223) builds the test circuit. First, the data read from the FRAM (while the machine is in the *rd_data* state) is stored into the *data_in* 1D×1D register (at the positive clock edge, line 201, because the slave places the data on the *MISO* wire at the negative clock transition). Because the machine operates at 5 MHz, an independent slow counter (2 Hz) is produced in lines 205–214, which allows the read data to be sequentially displayed on two SSDs while the machine remains in the *wait_rd0* state (that is, while the *rd* switch remains on). Note in the ROM of lines 38–40 that the first byte contains the values 0 (last half) and 1 (right half), the second contains 2 and 3, the third contains 4 and 5, and so on, so these pairs of values are the values expected to be seen on the display (during 0.5 s each).

Observe the correct use of registers and the completeness of the code, as described in comment 10 of section 6.3.

The reader is invited to set up this (or an equivalent) experiment and play with it in the FPGA board.

```
--Package with function "hex_to_ssd":------------------------------------
library ieee;
use ieee.std_logic_1164.all;
package my_functions is
   function hex_to_ssd(input:std_logic_vector) return std_logic_vector;
end my_functions;
--------------------------------------------------------------------------
package body my_functions is
   function hex_to_ssd(input: std_logic_vector) return std_logic_vector is
   begin
      case input is
         when "0000" => return "0000001";   --"0" on SSD
         when "0001" => return "1001111";   --"1" on SSD
         when "0010" => return "0010010";   --"2" on SSD
         when "0011" => return "0000110";   --"3" on SSD
         when "0100" => return "1001100";   --"4" on SSD
         when "0101" => return "0100100";   --"5" on SSD
         when "0110" => return "0100000";   --"6" on SSD
         when "0111" => return "0001111";   --"7" on SSD
         when "1000" => return "0000000";   --"8" on SSD
         when "1001" => return "0000100";   --"9" on SSD
         when "1010" => return "0001000";   --"A" on SSD
         when "1011" => return "1100000";   --"b" on SSD
         when "1100" => return "0110001";   --"C" on SSD
         when "1101" => return "1000010";   --"d" on SSD
         when "1110" => return "0110000";   --"E" on SSD
         when "1111" => return "0111000";   --"F" on SSD
         when others => return "1111110";   --"-" on SSD
      end case;
   end hex_to_ssd;
end package body;
--------------------------------------------------------------------------
```

```
--Main code:-------------------------------------------------------------
library ieee;
use ieee.std_logic_1164.all;
use work.my_functions.all; --package with "hex_to_ssd" function
-------------------------------------------------------------------------
entity FRAM_with_SPI_bus is
   generic (
      --Device's SPI parameters:
      WREN_opcode: std_logic_vector(7 downto 0) := "00000110";
      WRITE_opcode: std_logic_vector(7 downto 0) := "00000010";
```

```
      READ_opcode: std_logic_vector(7 downto 0) := "00000011";
      initial_addr_for_wr: std_logic_vector(15 downto 0) := (others=>'0');
      initial_addr_for_rd: std_logic_vector(15 downto 0) := (others=>'0'));
      --Assumed: fclk=50MHz, desired SPI speed=5MHz
   port (
      --System ports:
      rd, wr, clk, rst: in std_logic;
      ssd1, ssd2: out std_logic_vector(6 downto 0);
      --SPI ports:
      SCK, SSn, MOSI, WPn, HOLDn: out std_logic;
      MISO: in std_logic);
end entity;
-------------------------------------------------------------
architecture moore_fsm of FRAM_with_SPI_bus is

   --FSM-related declarations:
   type state is (idle, WREN, deselect, WRITEx, initial_addr_wr,
      wr_data, wait_wr0, READx, initial_addr_rd, rd_data, wait_rd0);
   signal pr_state, nx_state: state;

   --Timer-related declarations:
   signal i, imax: natural range 0 to 15;
   signal j, jmax: natural range 0 to 7;

   --SPI clock and test signal declarations:
   signal spi_clk: std_logic;
   type data_array is array (0 to 7) of std_logic_vector(7 downto 0);
   constant data_out: data_array :=
      ("00000001", "00100011", "01000101", "01100111",
       "10001001", "10101011", "11001101", "11101111");
   signal data_in: data_array;

begin

   WPn <= '1';
   HOLDn <= '1';

   --Generate 5MHz clock for SPI circuit:
   process (clk)
      variable counter1: natural range 0 to 5;
   begin
      if rising_edge(clk) then
         counter1 := counter1 + 1;
         if counter1=5 then
            spi_clk <= not spi_clk;
            counter1 := 0;
         end if;
      end if;
   end process;

   --Timers (using strategy #1):
   process (spi_clk, rst)
   begin
      if (rst='1') THEN
```

```
            i <= 0;
            j <= 0;
         elsif falling_edge(spi_clk) then
            if pr_state /= nx_state then
               i <= 0;
               j <= 0;
            elsif not (i=imax and j=jmax) then
               if i/=imax then
                  i <= i + 1;
               elsif j/=jmax then
                  i <= 0;
                  j <= j + 1;
               end if;
            end if;
         end if;
      end process;

      --FSM state register:
      process (spi_clk, rst)
      begin
         if (rst='1') THEN
            pr_state <= idle;
         elsif falling_edge(spi_clk) then
            pr_state <= nx_state;
         end if;
      end process;

      --FSM combinational logic:
      process (all)
      begin
         --Default values:
         SSn <= '0';
         SCK <= spi_clk;
         MOSI <= '-';
         imax <= 0;
         jmax <= 0;
         --Other values:
         case pr_state IS
            when idle =>
               SSn <= '1';
               SCK <= '0';
               if wr='1' then
                  nx_state <= WREN;
               elsif rd='1' then
                  nx_state <= READx;
               else
                  nx_state <= idle;
               end if;
            --Data-write sequence:
            when WREN =>
               MOSI <= WREN_opcode(7-i);
               imax <= 7;
               if i=imax then
                  nx_state <= deselect;
               else
                  nx_state <= WREN;
               end if;
```

```
         when deselect =>
            SSn <= spi_clk;
            SCK <= '0';
            nx_state <= WRITEx;
         when WRITEx =>
            MOSI <= WRITE_opcode(7-i);
            imax <= 7;
            if i=imax then
               nx_state <= initial_addr_wr;
            else
               nx_state <= WRITEx;
            end if;
         when initial_addr_wr =>
            MOSI <= initial_addr_for_wr(15-i);
            imax <= 15;
            if i=imax then
               nx_state <= wr_data;
            else
               nx_state <= initial_addr_wr;
            end if;
         when wr_data =>
            MOSI <= data_out(j)(7-i);
            imax <= 7;
            jmax <= 7;
            if i=imax and j=jmax then
               nx_state <= wait_wr0;
            else
               nx_state <= wr_data;
            end if;
         when wait_wr0 =>
            SSn <= '0';
            SCK <= '0';
            if wr='0' then
               nx_state <= idle;
            else
               nx_state <= wait_wr0;
            end if;
         --Data-read sequence:
         when READx =>
            MOSI <= READ_opcode(7-i);
            imax <= 7;
            if i=imax then
               nx_state <= initial_addr_rd;
            else
               nx_state <= READx;
            end if;
         when initial_addr_rd =>
            MOSI <= initial_addr_for_rd(15-i);
            imax <= 15;
            if i=imax then
               nx_state <= rd_data;
            else
               nx_state <= initial_addr_rd;
            end if;
         when rd_data =>
            imax <= 7;
            jmax <= 7;
```

```
            if i=imax and j=jmax then
               nx_state <= wait_rd0;
            else
               nx_state <= rd_data;
            end if;
         when wait_rd0 =>
            SSn <= '0';
            SCK <= '0';
            if rd='0' then
               nx_state <= idle;
            else
               nx_state <= wait_rd0;
            end if;
      end case;
   end process;

   --Test circuit:
   process (spi_clk, pr_state, data_in)
      variable counter2: natural range 0 to 2_500_000;
      variable counter3: natural range 0 to 8;
   begin
      --Read FRAM and store data:
      if rising_edge(spi_clk) and pr_state=rd_data then
         data_in(j)(7-i) <= MISO;
      end if;
      --Generate slow (2Hz) pointer for displaying data:
      if rising_edge(spi_clk) then
         counter2 := counter2 + 1;
         if counter2=2_500_000 then
            counter2 := 0;
            counter3 := counter3 + 1;
            if counter3=8 then
               counter3 := 0;
            end if;
         end if;
      end if;
      --Send read data to display @2Hz:
      if pr_state=wait_rd0 then
         ssd1 <= hex_to_ssd(data_in(counter3)(3 downto 0));
         ssd2 <= hex_to_ssd(data_in(counter3)(7 downto 4));
      else
         ssd1 <= "1111110";
         ssd2 <= "1111110";
      end if;
   end process;

end architecture;
-------------------------------------------------------------
```

14.4 Exercises

Exercise 14.1: Reference-Value Definer with LCD Display

a) Solve exercise 8.11 if not done yet. The reference values should be {000, 005, 010, 050, 100, 200, 400, 800}.

b) Draw the state transition diagram for a second circuit, which should implement an LCD driver to have the reference value displayed on an alphanumeric LCD.
c) Implement the complete circuit using VHDL or SystemVerilog and test it in the FPGA development board.

Exercise 14.2: I²C Interface for an RTC

Repeat the design of section 14.2.5, this time with a category 2 machine instead of a category 3.

Exercise 14.3: I²C Interface for an EEPROM

Develop an experiment (as in section 14.2.5), including VHDL or SystemVerilog code and physical implementation in the FPGA development board, for a master circuit that interfaces with an EEPROM device through an I²C bus. The device can be, for example, AT24C01 or AT24C02, from Atmel.

Exercise 14.4: I²C Interface for an ADC

Develop an experiment (as in section 14.2.5), including VHDL or SystemVerilog code and physical implementation in the FPGA development board, for a master circuit that interfaces with an analog-to-digital converter through an I²C bus. The device can be, for example, AD7991, from Analog Devices, or PCF8591, from NXP.

Exercise 14.5: I²C Interface for a Temperature Sensor

Develop an experiment (as in Section 14.2.5), including VHDL or SystemVerilog code and physical implementation in the FPGA development board, for a master circuit that interfaces with a temperature sensor through an I²C bus. The device can be, for example, LM75A, from NXP, or AD7416, from Analog Devices.

Exercise 14.6: I²C versus SPI

Make a brief comparison between I²C and SPI interfaces. Include at least the following topics in your analysis: synchronous or asynchronous, number of wires, duplex or simplex, with data acknowledgment or not, which hardware is simpler and why, who generates clock and data, which operates at higher speed.

Exercise 14.7: SPI Interface for a FRAM

Repeat the design of section 14.3.3, this time with a category 3 machine instead of a category 2.

Exercise 14.8: SPI Interface for an ADC

Develop an experiment (as in section 14.3.3), including VHDL or SystemVerilog code and physical implementation in the FPGA development board, for a master circuit that interfaces with an analog-to-digital converter through an SPI bus. The device can be, for example, AD7091R, from Analog Devices, or MAX1242, from Maxim.

Exercise 14.9: SPI Interface for a Flash Memory

Develop an experiment (as in section 14.3.3), including VHDL or SystemVerilog code and physical implementation in the FPGA development board, for a master circuit that interfaces with a flash memory through an SPI bus. The device can be, for example, AT45DB011, from Atmel, or S25FL128, from Spansion.

Exercise 14.10: SPI Interface for an Accelerometer

Develop an experiment (as in section 14.3.3), including VHDL or SystemVerilog code and physical implementation in the FPGA development board, for a master circuit that interfaces with an accelerometer through an SPI bus. The device can be, for example, ADXL345, from Analog Devices.

Exercise 14.11: I^2C Interface for an Accelerometer

The accelerometer mentioned in exercise 14.10 (ADXL345) supports both SPI and I^2C interfaces. Repeat that exercise, this time using the I^2C alternative.

15 Pointer-Based FSM Implementation

15.1 Introduction

In the preceding chapters we have established and used a *standard* and *generic* design approach for *any* FSM. In the particular case of machines with a simple state transition diagram (a single loop, for example), a simpler but possibly equivalent implementation can be adopted, which consists of building a counter that acts as a pointer to a lookup table (LUT) that contains the desired output values. This implementation technique, identified as *pointer-based FSM implementation*, is illustrated by means of a series of examples in the sections that follow.

It is important to mention that although this technique can ease the implementation of FSMs with few loops and repetitive states (like those in chapter 14), it does not eliminate the other design steps, including the development of a precise state transition diagram. Also, it does not mean that a simpler circuit will result. A limitation of this technique is that it is difficult (or awkward, at least) to use an encoding scheme other than regular sequential encoding; moreover, the encoding is set during the design phase, so it cannot be chosen/modified and experimented with at compilation time.

15.2 Single-Loop FSM

The general technique is illustrated with the help of figure 15.1a, which shows an FSM with just one loop. Note that the states' names were replaced with numeric values (to be produced by the counter/pointer). Note also that the machine must run whenever *run*='1' occurs while the machine is in the idle state, and that it must stay in a certain state (pointer in the 3-to-8 range) during six clock cycles (thus this is similar to a timed state in a timed state machine, with *t* running from 0 to 5). The output values are summarized (repeated) in the LUT of figure 15.1b. To implement this FSM, we can build a counter (pointer) ranging from 0 to 9 and use it to retrieve the corresponding values from the LUT.

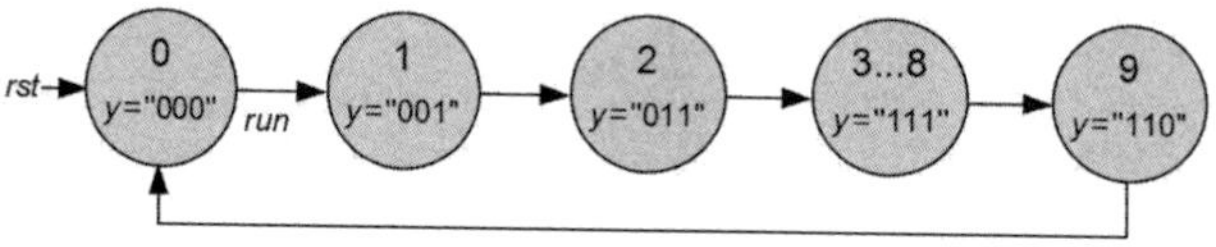
rst
0
y="000"
run
1
y="001"
2
y="011"
3...8
y="111"
9
y="110"

(a)

(b)

state (pointer *i*)	output (*y*)
0	"000"
1	"001"
2	"011"
3-8	"111"
9	"110"

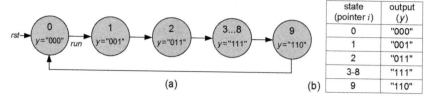

Figure 15.1
(a) Single-loop FSM with a timed state and (b) LUT containing its output values.

A VHDL code for the machine of figure 15.1 is presented below. The code contains just one process, which builds the pointer (called *i*, lines 19–27) and the LUT (for *y*, lines 30–36).

Note: To save space, only VHDL codes are shown in this chapter. However, based on these VHDL codes and on the SystemVerilog codes seen in chapters 7, 10, and 13, writing the SystemVerilog codes for the examples described here is straightforward.

```
1   ---------------------------------------------
2   library ieee;
3   use ieee.std_logic_1164.all;
4   ---------------------------------------------
5   entity simple_machine is
6      port (
7         run, clk, rst: in std_logic;
8         y: out std_logic_vector(2 downto 0));
9   end entity;
10  ---------------------------------------------
11  architecture pointer_based of simple_machine is
12  begin
13
14     process (clk, rst)
15        variable i: natural range 0 to 9;
16     begin
17
18        --Pointer (i):
19        if (rst='1') then
20           i:= 0;
21        elsif rising_edge(clk) then
22           if (i=0 and run='0') or i=9 then
23              i:= 0;
24           else
25              i:= i + 1;
26           end if;
27        end if;
28
29        --LUT (for y):
30        case i is
31           when 0 => y <= "000";
32           when 1 => y <= "001";
33           when 2 => y <= "011";
34           when 3 to 8 => y <= "111";
```

```
            when 9 => y <= "110";
         end case;

      end process;

   end architecture;
   ----------------------------------------
```

15.3 Serial Data Transmitter

Another example is presented in figure 15.2a, which is a kind of serial data transmitter. The output (*y*), which is a single-bit signal, must send out a predefined single-bit value in states 0, 2, and 6 (recall that the states' "names" are determined by the pointer), whereas in states 1 and 3...5 bit-vectors must be serially transmitted (see the data arrays under those states).

In figure 15.2a three pointers are shown, called *i* (main pointer, representing the states), *j* (column index for the arrays of states 1 and 3...5), and *k* (row index for the array of state 3...5). Note, however, that *k* can be replaced with *i*–3, so only two pointers are actually needed, resulting in the LUT of figure 15.2b. This machine too is simple enough to be implemented using the pointer-based technique.

A corresponding VHDL code is shown below. The data to be transmitted was placed in an array of constants (called *x*, lines 12–13). The first word (i.e., *x*(0)="0101") is transmitted in state 1, while the whole array is transmitted in state 3...5. As in the previous example, only one process is used, which builds the pointers (lines 22–38) and the LUT (lines 41–47). Note that in this example the LSB is transmitted first.

```
1   ---------------------------------------------------------------
2   library ieee;
3   use ieee.std_logic_1164.all;
4   ---------------------------------------------------------------
5   entity serial_transmitter is
6      port (
7         run, clk, rst: in std_logic;
8         y: out std_logic);
```

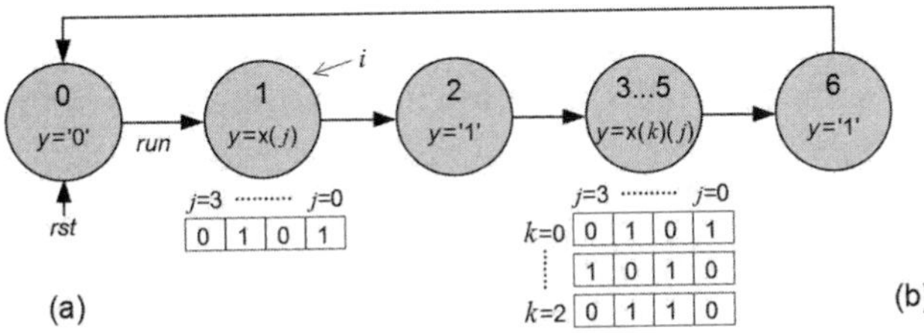

0
y='0'
run
1
y=x(j)
i
2
y='1'
3...5
y=x(k)(j)
6
y='1'
rst
j=3 .... j=0
0 1 0 1
j=3 .... j=0
k=0 0 1 0 1
1 0 1 0
k=2 0 1 1 0
(a)

(b)

state (pointer *i*)	output (*y*)
0	'0'
1	*x*(*j*)
2	'1'
3-5	*x*(*i*–3)(*j*)
6	'1'

Figure 15.2
A serial data transmitter and (b) its output LUT.

```
end entity;
--------------------------------------------------------------
architecture pointer_based of serial_transmitter is
   type data_array is array (0 to 2) of std_logic_vector(3 downto 0);
   constant x: data_array:= ("0101", "1010", "0110");
begin

   process(clk, rst)
      variable i: natural range 0 to 6;
      variable j: natural range 0 to 3;
   begin

   --Pointers (i, j):
   if rst='1' then
      i:= 0;
      j:= 0;
   elsif rising_edge(clk) then
      if (i=0 and run='1') or i=2 then
         i:= i + 1;
      elsif i=1 or (i>=3 and i<=5) then
         if j/=3 then
            j:= j + 1;
         else
            j:= 0;
            i:= i + 1;
         end if;
      elsif i=6 then
         i:= 0;
      end if;
   end if;

   --LUT (for y):
   case i is
      when 0 => y <= '0';
      when 1 => y <= x(0)(j);
      when 2 => y <= '1';
      when 3 to 5 => y <= x(i-3)(j);
      when 6 => y <= '1';
   end case;

   end process;

end architecture;
--------------------------------------------------------------
```

Simulation results obtained after compiling the code above are presented in figure 15.3. The reader is invited to examine the waveforms to check the operation of the inferred circuit.

15.4 Serial Data Receiver

Another example is presented in figure 15.4. The serial data receiver of figure 3.16c is repeated in figure 15.4a. When the data-valid (*dv*) bit is asserted, the circuit must store

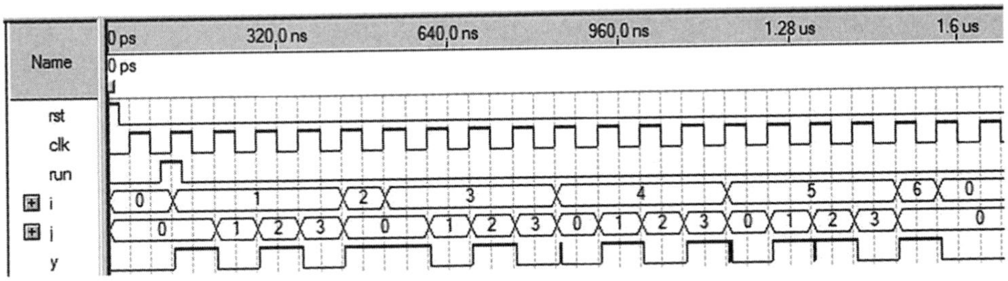
Name
0 ps
320,0 ns
640,0 ns
960,0 ns
1.28 us
1.6 us
0 ps
rst
clk
run
i
j
y

Figure 15.3
Simulation results from the VHDL code for the serial data transmitter of figure 15.2.

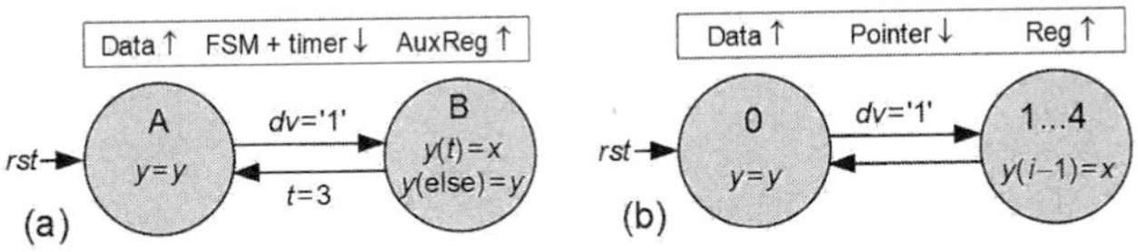
Data ↑ FSM + timer ↓ AuxReg ↑
A
y=y
dv='1'
t=3
B
y(t)=x
y(else)=y
rst
(a)
Data ↑ Pointer ↓ Reg ↑
0
y=y
dv='1'
1...4
y(i−1)=x
rst
(b)

Figure 15.4
(a) Serial data receiver of figure 3.16c. (b) Adapted version for pointer-based implementation.

four consecutive bits received at input *x* (hence, this is a timed machine, with *t* running from 0 to 3—a small range was adopted in order to ease the inspection of the simulation results). Because it is assumed that the first data bit is made available at the same time that *dv* is asserted (both updated at the positive clock edge), which is more difficult to detect, one must be careful with respect to the clock edges (see rectangle in the upper part of the figure and also the discussion in section 3.10). This machine is simple enough to be implemented using the pointer-based approach, for which an adapted state diagram is presented in figure 15.4b. Note that the counter (pointer) must run from 0 to 4. Again, care must be taken with respect to the clock edges.

A VHDL code for the machine of figure 15.4b is presented below. It contains just one process, which builds the pointer (lines 19–27) and the *registered* LUT (lines 30–35). Because *x* must be stored (producing *y*) in a deserializer, the **case** statement (lines 31–34) was placed inside an **if rising_edge(clk)** statement (lines 30 and 35), which is responsible for inferring flip-flops.

```
1    ---------------------------------------------
2    library ieee;
3    use ieee.std_logic_1164.all;
4    ---------------------------------------------
5    entity serial_receiver is
6       port (
7          x, dv, clk, rst: in std_logic;
8          y: buffer std_logic_vector(3 downto 0));
```

```
end entity;
-----------------------------------------------
architecture pointer_based of serial_receiver is
begin

   process (clk, rst)
      variable i: natural range 0 to 4;
   begin

      --Pointer (i):
      if (rst='1') then
         i:= 0;
      elsif falling_edge(clk) then
         if (i=0 and dv='0') or i=4 then
            i:= 0;
         else
            i:= i + 1;
         end if;
      end if;

      --Registered LUT (for y):
      if rising_edge(clk) then
         case i is
            when 0 => y <= y;
            when 1 to 4 => y(i-1) <= x;
         end case;
      end if;

   end process;

end architecture;
-----------------------------------------------
```

Simulation results are shown in figure 15.5. Note that the data (*dv* and *x*) and the register (*y*) are updated at the positive clock edges, whereas the pointer (*i*) changes at the negative clock transitions. Note also that the sequence received in *x* is '1', '0', '1',

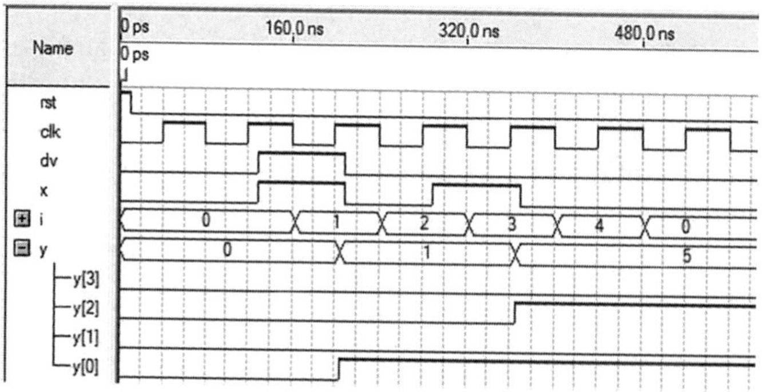
Name
0 ps
160,0 ns
320,0 ns
480,0 ns
0 ps
rst
clk
dv
x
i
y
y[3]
y[2]
y[1]
y[0]
0
1
2
3
4
0
0
1
5

Figure 15.5
Simulation results from the VHDL code for the serial data receiver of figure 15.4b.

and '0', with the first bit considered to be the LSB, hence resulting $y(3:0)$ = "0101" after the pointer's 0-to-4 run is completed.

15.5 SPI Interface for an FRAM

A final example is presented in figure 15.6, which is an equivalent (pointer-based) implementation for the FRAM SPI interface circuit studied in section 14.3.3. Note that the machine of figure 15.6 is exactly the same as that in figure 14.20, just with the adjustments needed for pointer-based implementation (main pointer i ranging from 0 to 75, secondary pointer j ranging from 0 to 7). The values (either 8 or 8×8) under the arrows indicate the number of clock cycles spent in the preceding state (which is the same as the number of bits transmitted or received in that state).

Observe that the enumeration of the states was done differently from that in figure 15.2 (just to illustrate another alternative). In state 1 of figure 15.2, for example, the main pointer (i) stays fixed (i = 1), whereas the secondary pointer (j) sweeps the data. Here, in state WREN, for example, i sweeps the data, while j is not used (so only in states *wr_data* and *rd_data* are both pointers needed). Recall that, as in all FSM designs, the crucial point is to develop a complete and precise state transition diagram (as in figure 15.6), after which any of the implementation techniques can be applied straightforwardly.

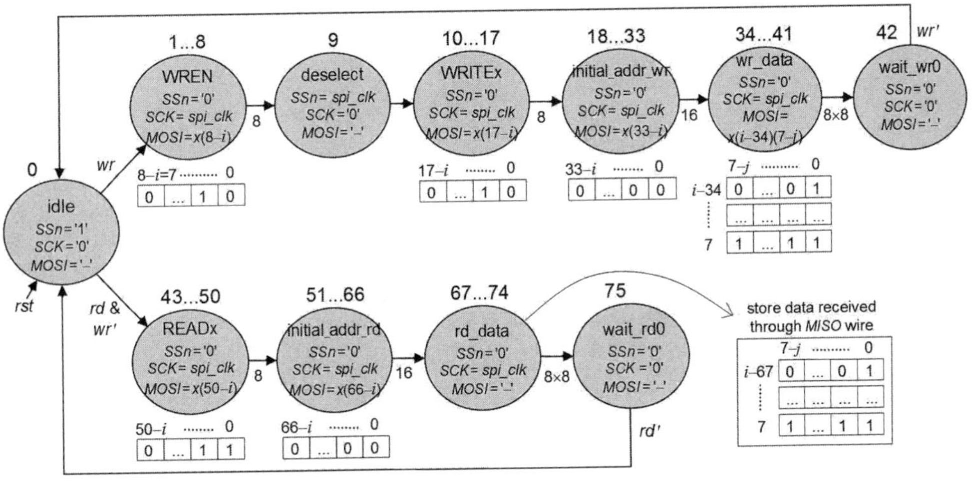

1...8
9
10...17
18...33
34...41
42
wr'
WREN
SSn='0'
SCK= spi_clk
MOSI=x(8–i)
deselect
SSn= spi_clk
SCK='0'
MOSI='–'
WRITEx
SSn='0'
SCK= spi_clk
MOSI=x(17–i)
initial_addr_wr
SSn='0'
SCK= spi_clk
MOSI=x(33–i)
wr_data
SSn='0'
SCK= spi_clk
MOSI=
x(i–34)(7–j)
wait_wr0
SSn='0'
SCK='0'
MOSI='–'
8
16
8×8
0
wr
idle
SSn='1'
SCK='0'
MOSI='–'
rst
rd &
wr'
43...50
51...66
67...74
75
READx
SSn='0'
SCK= spi_clk
MOSI=x(50–i)
initial_addr_rd
SSn='0'
SCK= spi_clk
MOSI=x(66–i)
rd_data
SSn='0'
SCK= spi_clk
MOSI='–'
wait_rd0
SSn='0'
SCK ='0'
MOSI='–'
rd'
store data received
through MISO wire

Figure 15.6
FSM for the FRAM SPI interface circuit seen in section 14.3.3 (figure 14.20), with all adjustments for pointer-based implementation. Values under the arrows indicate the number of clock cycles spent in the preceding state.

A VHDL code for this machine is presented below (function *hex-to-ssd* not shown—bring the package *my_functions*, from section 14.3.3, to your design). This code is equivalent to that in section 14.3.3, except for a small difference in the test circuit (here the stored values are sent to the display continuously). The entity (lines 6–22) is the same as that in section 14.3.3, and so are the list of declarations for the SPI and test signals (lines 27–32) and the *spi_clk* generator (lines 40–50). The FSM, constructed using the pointer-based technique, is in the process of lines 53–115, with the pointers built in lines 59–78 and the LUT in lines 82–113. Note in lines 108–110 that the received data is stored in the *data_in* array while the machine is in the *rd_data* state. The final part of the test circuit (data storage was embedded in the LUT) is in the process of lines 118–140 (see comments in section 14.3.3).

```
1   --Main code:-----------------------------------------------------------
2   library ieee;
3   use ieee.std_logic_1164.all;
4   use work.my_functions.all; --package from sec. 14.3.3
5   -----------------------------------------------------------------------
6   entity FRAM_with_SPI_bus is
7      generic (
8         --Device's SPI parameters:
9         WREN_opcode: std_logic_vector(7 downto 0):= "00000110";
10        WRITE_opcode: std_logic_vector(7 downto 0):= "00000010";
11        READ_opcode: std_logic_vector(7 downto 0):= "00000011";
12        initial_addr_for_wr: std_logic_vector(15 downto 0):= (others=>'0');
13        initial_addr_for_rd: std_logic_vector(15 downto 0):= (others=>'0'));
14        --Assumed: fclk=50MHz, desired SPI speed=5MHz
15     port (
16        --System ports:
17        rd, wr, clk, rst: in std_logic;
18        ssd1, ssd2: out std_logic_vector(6 downto 0);
19        --SPI ports:
20        SCK, SSn, MOSI, WPn, HOLDn: out std_logic;
21        MISO: in std_logic);
22  end entity;
23  -----------------------------------------------------------------------
24  architecture pointer_based of FRAM_with_SPI_bus is
25
26     --Clock for SPI and test signal declarations:
27     signal spi_clk: std_logic;
28     type data_array is array (0 to 7) of std_logic_vector(7 downto 0);
29     constant data_out: data_array:=
30        ("00000001", "00100011", "01000101", "01100111",
31         "10001001", "10101011", "11001101", "11101111");
32     signal data_in: data_array;
33
34  begin
35
36     WPn <= '1';
37     HOLDn <= '1';
38
```

```
--Generate 5MHz clock for SPI circuit:
process (clk)
   variable counter1: natural range 0 to 5;
begin
   if rising_edge(clk) then
      counter1:= counter1 + 1;
      if counter1=5 then
         spi_clk <= not spi_clk;
         counter1:= 0;
      end if;
   end if;
end process;

--FSM (complete SPI circuit):
process(spi_clk, rst)
   variable i: natural range 0 to 75;
   variable j: natural range 0 to 7;
begin

   --Pointers (i, j):
   if rst='1' then
      i:= 0;
      j:= 0;
   elsif falling_edge(spi_clk) then
      if (i=0 and wr='0' and rd='0') or (i=42 and wr='0') or
         (i=75 and rd='0') then
         i:= 0;
      elsif (i=0 and wr='1') or (i>0 and i<34) or (i>=43 and i<67) then
         i:= i + 1;
      elsif i=0 and rd='1' then
         i:= 43;
      elsif (i>=34 and i<=41) or (i>=67 and i<=74) then
         if j/=7 then
            j:= j + 1;
         else
            j:= 0;
            i:= i + 1;
         end if;
         end if;
   end if;

   --LUT (for outputs):
   --Default values:
   SSn <= '0';
   SCK <= spi_clk;
   MOSI <= '-';
   --Other values:
   case i is
      when 0 =>
         SSn <= '1';
         SCK <= '0';
      when 1 to 8 =>
         MOSI <= WREN_opcode(8-i);
      when 9 =>
         SSn <= spi_clk;
```

```
            SCK <= '0';
         when 10 to 17 =>
            MOSI <= WRITE_opcode(17-i);
         when 18 to 33 =>
            MOSI <= initial_addr_for_wr(33-i);
         when 34 to 41 => --transmit data
            MOSI <= data_out(i-34)(7-j);
         when 42 =>
            SCK <= '0';
         when 43 to 50 =>
            MOSI <= READ_opcode(50-i);
         when 51 to 66 =>
            MOSI <= initial_addr_for_rd(66-i);
         when 67 to 74 => --store received data
            if rising_edge(spi_clk) then
               data_in(i-67)(7-j) <= MISO;
            end if;
         when 75 =>
            SCK <= '0';
      end case;

   end process;

   --Test circuit:
   process (spi_clk, rst)
      variable counter2: natural range 0 to 2_500_000;
      variable counter3: natural range 0 to 7;
   begin
      --Generate slow (2Hz) pointer for displaying data:
      if rst='1' then
         counter2:=0;
         counter3:=0;
      elsif rising_edge(spi_clk) then
         counter2:= counter2 + 1;
         if counter2=2_500_000 then
            counter2:= 0;
            if counter3/=7 then
               counter3:= counter3 + 1;
            else
               counter3:= 0;
            end if;
         end if;
      end if;
      --Send data continuously to the display:
      ssd1 <= hex_to_ssd(data_in(counter3)(3 downto 0));
      ssd2 <= hex_to_ssd(data_in(counter3)(7 downto 4));
   end process;

end architecture;
----------------------------------------------------------------------
```

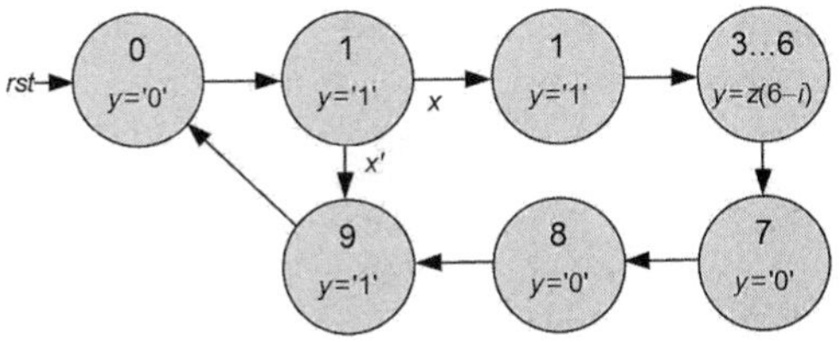

rst
0
y='0'
1
y='1'
x
1
y='1'
3...6
y=z(6–i)
x'
9
y='1'
8
y='0'
7
y='0'

Figure 15.7

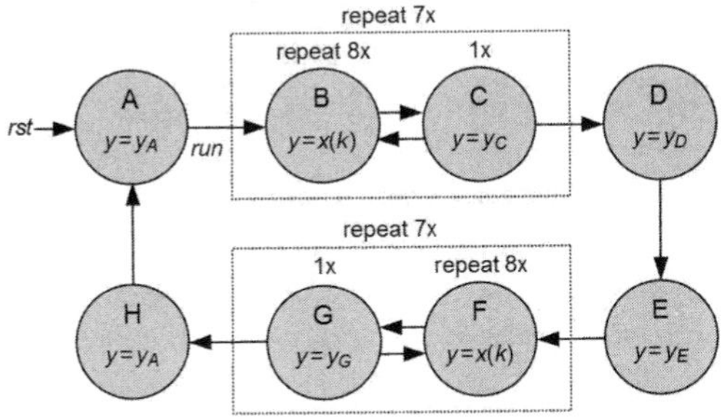

repeat 7x
repeat 8x
1x
rst
A
y=yA
run
B
y=x(k)
C
y=yC
D
y=yD
repeat 7x
1x
repeat 8x
H
y=yA
G
y=yG
F
y=x(k)
E
y=yE

Figure 15.8

15.6 Exercises

Exercise 15.1: Number of Flip-Flops

How many DFFs are needed to build the FSMs of figures 15.1, 15.2, and 15.6?

Exercise 15.2: Two-Loop FSM

Figure 15.7 shows a two-loop FSM, with input x and output y. When in state 3…6 the machine must transmit four bits from a $z(3{:}0)$ array, starting with the MSB.

a) How many flip-flops are needed to construct this FSM? Does your answer depend on the implementation approach (generic, seen in the previous chapters, or pointer-based, seen here)?
b) Implement it using VHDL or SystemVerilog (pointer-based technique). Enter z in your code as a constant. After compilation, check whether the number of flip-flops inferred by the compiler matches your prediction.
c) Show simulation results.

Exercise 15.3: FSM with Repetitive States

Figure 15.8 shows an FSM with an *apparent* single loop. Note that state B must be repeated 8 times, then state C must occur, with this sequence (B–C) repeated 7 times before proceeding to state D. A similar procedure must occur in states F–G.

a) Implement this machine using the pointer-based technique and VHDL or SystemVerilog. Start by making the proper adaptations (using pointer(s)) in the state transition diagram. Create an array of constants to be placed at the output in states B and F, and chose numeric values for y_A, y_C, y_D, etc.

b) Show simulation results. To ease the inspection of the results, use 3 instead of 8 and 2 instead of 7 in the repetitions.

Exercise 15.4: Clock with LCD Display

Redo the design of section 14.1.3 using the pointer-based technique. Start by drawing the adapted (using pointer(s)) state transition diagram. After compilation, compare the resources usage (especially the number of flip-flops) against the results obtained after compiling the code of section 14.1.3.

Exercise 15.5: I²C Interface for an RTC

Redo the design of section 14.2.5 using the pointer-based technique. Start by drawing the adapted (using pointer(s)) state transition diagram. After compilation, compare the resources usage (especially the number of flip-flops) against the results obtained after compiling the code of section 14.2.5.

Exercise 15.6: SPI Interface for an ADC

Solve exercise 14.8 using the pointer-based technique.

Exercise 15.7: SPI Interface for an Accelerometer

Solve exercise 14.10 using the pointer-based technique.

Exercise 15.8: I²C Interface for an ADC

Solve exercise 14.4 using the pointer-based technique.

Exercise 15.9: I²C Interface for a Temperature Sensor

Solve exercise 14.5 using the pointer-based technique.

Exercise 15.10: I²C Interface for an Accelerometer

Solve exercise 14.11 using the pointer-based technique.

Bibliography

On VHDL

Ashenden, P. (2008). *The Designer's Guide to VHDL* (3rd ed.). Burlington, MA: Morgan Kaufmann.

Pedroni, V. A. (2010). *Circuit Design and Simulation with VHDL* (2nd ed.). Cambridge, MA: MIT Press.

IEEE. (2008). *1076-2008 Standard VHDL Language Reference Manual*. Washington, DC: IEEE (www.ieee.org).

On SystemVerilog

Harris, D. M., & Harris, S. (2012). *Digital Design and Computer Architecture* (2nd ed.). Burlington, MA: Morgan Kaufmann.

Sutherland, S., Davidmann, S., Flake, P., & Moorby, P. (2010). *SystemVerilog for Design* (2nd ed.). Berlin: Springer.

IEEE. (2005). *1800-2005 Standard SystemVerilog Language Reference Manual*. Washington, DC: IEEE (www.ieee.org).

On Digital Electronics

Pedroni, V. A. (2008). *Digital Electronics and Design with VHDL*. Burlington, MA: Morgan Kaufmann.

Patterson, D. A., & Hennessy, J. L. (2011). *Computer Organization and Design: The Hardware/Software Interface* (4th ed.). Burlington, MA: Morgan Kaufmann.

Index